GRASS-FED CATTLE

GRASS-FED Cattle

How to Produce and Market Natural Beef

JULIUS RUECHEL

Storey Publishing

The mission of Storey Publishing is to serve our customers
by publishing practical information that encourages
personal independence in harmony with the environment.

Edited by Marie Salter, Elaine Cissi, and Deborah Burns
Art direction by Cindy McFarland
Cover design by Vicky Vaughn
Text design and production by Erin Dawson
Cover photograph by © Peter Dean/Grant Heilman Photography
Illustrations © Elayne Sears
Infographics by Erin Dawson
Indexed by Daniel Brannen
Technical review by Lee Rinehart, Livestock Program Specialist,
    National Center for Appropriate Technology

Printed in the United States by Versa Press
10 9 8 7 6 5 4 3 2 1

To Anne

For a dream, a vision, a whole lot of spice,
and a most exciting journey of discovery

CONTENTS

Foreword

THE HARDEST THING FOR MOST LIVESTOCK producers to realize is that we are not in the cattle business. We are in the grass business.

We are, in effect, grass farmers. Grass is the beginning, the end, and everything in between in natural cattle production. It deserves both our respect and our attention. For example, the annual growth cycle of your grass will determine when you should calve, breed, and wean. How closely you mesh these three activities will largely determine your cost of production.

Pasture subdivision and controlled rotational grazing give you a steering wheel, clutch, and brake for your four-legged grass harvesters. Consider the folly of trying to harvest a grain crop with a motorized combine without the use of a steering wheel, clutch, and brake. Now, consider that most ranchers attempt to do exactly this.

The mineral balance in the soil will largely determine the quality of the grass you grow and its protein and energy content. These two elements will in turn determine the range of potential grass-based enterprises you can realistically consider — and the choices don't end there. For instance, how far you plan to take your cattle on grass will determine the range of cattle genetics from which you can select. If you plan to take them all the way to slaughter on grass, you will need to choose an early-maturing, easy-fattening breed.

The list of considerations for anyone entering a grass-based enterprise goes on and on and is well covered in this book. The primary point to remember is that virtually every decision you will make in cattle production will come back to grass. *Grass-Fed Cattle* serves as an excellent how-to book for grass-based natural cattle production. It is one you will want to read and reread many times.

— ALLAN NATION
Editor, *The Stockman Grass Farmer*

Preface

IN MANY RESPECTS, NATURAL GRASS-BASED cattle farming is an entirely new trade that is emerging as an alternative to conventional beef production and marketing. It requires a whole new set of management skills and an entirely different knowledge base about the unique evolutionary relationship between cattle and grass, all of which are foreign to the technology-dependent, commodity-driven, conventional beef industry. But perhaps it is more accurate to say that the lessons first learned by our prehistoric ancestors about cattle and grass are finally converging with our modern understanding of the natural world and our technological advantages. As such, natural grass-based cattle farming is an exciting journey of discovery into our distant pastoral history while taking us to the forefront of our ecological understanding about plants and animals and relying on our most innovative technological advances. This is the frontier of the modern natural/organic agriculture movement.

Grass-Fed Cattle: How to Produce and Market Natural Beef is the guide I wish I had had when I was searching for advice on how to create a profitable, financially stable, and environmentally sustainable business in the new trade of natural grass-based beef production. It contains all the information vital to the setup, design, management, and marketing of the low-cost, high-profit grass-based cattle enterprise, from calf birth right through to the steak on the customer's barbecue. It is much more, however, than just a comprehensive production and marketing guide. It is also an essential business management tool that will accompany you through your individual process of designing and managing your own natural grass-based cattle enterprise. It is designed to give you independence, to provide you with choices, and to allow you to confidently take charge of your future, a future that you will have the skills, knowledge, and tools to design yourself by the time you reach the end of this book.

The knowledge, techniques, and skills described are universal, regardless of farm size or climate. They are suited equally to farmers new to the trade, established organic farmers looking to integrate cattle into their production systems, experienced family farms seeking to transition from conventional to organic production, and even to the herds of the largest conventional farms looking for new, environmentally sustainable and financially stable production and marketing options.

Within these pages, I hope to help you see the cow's world through new eyes. You will never look at a simple weed the same way again. Raindrops will carry a new meaning. A cow's foot will become the most powerful tool at your disposal. You'll discover the fascinating world of cow psychology. Calving season may actually become your favorite and most relaxing time of year. And I'll help you design your own plan to make a profitable and financially stable living in the most exciting and emotionally rewarding business in the world.

Part 1, The Fundamentals of Grass-Based Beef, takes an in-depth look at grass and cattle and the interdependent relationship that evolved between them during the millions of years before domestication, grain surpluses, and modern technological Band-Aids. It also explores the effect this relationship has on grass, soil, cattle fertility, genetics, and the seasonality of the natural cattle production year. This foundation of knowledge serves as the blueprint for all the production techniques described in the remainder of the book and provides you with the confidence to trust nature as your guide.

Like any practical trade, natural grass-based beef production relies on certain tools that allow us to harness this natural relationship between cattle and grass and turn it into a profitable business capable of sustaining our families, communities, and environment. Part 2, Infrastructure and Management, is about this toolbox. Some of the tools are the low-cost infrastructure of natural grass-based beef productions; others are skills and management principles that allow you to manage your cattle herd and create a salable product from it. These are the unique tools of your new trade.

Part 3, Business Planning and Marketing, describes how to convert the salable products produced by your farm (cattle) into a profitable business enterprise. This section is about the business end of the trade: the grass-finishing, slaughter, marketing, and financial management techniques unique to natural grass-based beef. It is also about building a cohesive plan that allows you to meet your personal, business, and financial goals with the resources available to you, in the environmental and market conditions unique to your specific situation.

By the time you reach the final section of the book, you will have begun developing an image in your mind about how you want your new natural grass-based beef enterprise to look. Part 4, Your Business Plan, allows you to turn theory into reality. This section provides you with a step-by-step planning framework to identify your goals and guide you through the process of designing your own profitable and financially stable natural grass-based cattle enterprise, built on the foundation of knowledge, skills, and tools from parts 1 through 3. By using the framework provided in this section, you can build the blueprint of your production, management, and marketing strategy as it applies to your unique circumstances, your unique market, and your unique personal goals and ambitions.

I wish you an exciting journey of discovery.

ACKNOWLEDGMENTS

An enormous amount of work goes into putting together a book like this. I would like to thank the good people at Storey Publishing for their hard work and for sharing my vision in creating this unique guide. Their mission and my goals for this book are an ideal partnership. In particular I would like to thank my editors, Deborah Burns, Marie Salter, and Elaine Cissi, my illustrator, Elayne Sears, and designer Erin Dawson for transforming my writing, concepts, and ideas into a cohesive package.

I give my appreciation to the Saskatchewan Grazing and Pasture Technology Program and in particular Zoheir Abouguendia and Bob Springer; Dr. Dick Diven and his Low-Cost Cow Calf Program and School for Profitable Beef Cattle; and Prof. Harlan Hughes, Extension livestock economist at North Dakota State University. Their efforts introduced me to low-cost grass-based livestock production. I also thank Lance Brown and Ted W. Van der Gulik at the BC Ministry of Agriculture, Food and Fisheries and Dr. Temple Grandin of the Department of Animal Science at Colorado State University for their generous copyright permissions. Thanks too to Lee Rinehart for his technical input.

Special thanks to my wife, Anne, whose endless encouragement during the long hours of writing and whose talent at editing cannot be overstated. She has been invaluable to me throughout the entire process of developing the vision that went into this book, from cheerfully building electric fences with me at -35°F and patiently sitting through the thousands of miles of Canadian and U.S. highways that formed the research phase of this project to inspiring the planning framework featured in part 4 of this book.

A big thank you to my mother, Charlotte, and my siblings, Charles, Lorna, and Emily, and her husband, Don, for trusting me with the responsibility of managing the family farm at its most difficult moment and for their confidence in me during the growing pains of transitioning the farm from conventional to organic production while I developed my vision. Particularly I thank my dad, Michael, who, before being interrupted by his head injury, opened the door to new agricultural ideas more than fifteen years ago when organic farming, rotational grazing, and grass-finishing were still bad words not to be spoken on Sundays. His smiles of support have meant the world to me, even if his speech and mobility have been stolen by his injury. Thanks, Dad, I hope you enjoy the book — you saw it unfold from the sidelines and here is the completed vision.

I would also like to extend my gratitude to Stan Grimshire, Ron Dalhuisen, and David Jupp for putting their faith in my ideas. It's not easy to "ride for the brand" when everything you've been taught as true about beef ranching is suddenly being challenged and turned on its head. It takes courage to question long-held beliefs and be willing to step into the unknown. I could not have done it without you.

And a special thanks to my mentor, Prof. Dr. Rainer Newberry, University of Alaska, Fairbanks, the most remarkable rock and ore deposit detective, who taught me always to question conventional wisdom. Whether about ore deposits, grass, cattle, or people, the lessons of being a good detective remain the same.

Introduction

CATTLE AND GRASS EVOLVED TOGETHER FOR millions of years, each adapting to the other to create an efficient partnership that shaped the landscape of our planet. Long before humans arrived on the scene, cattle were already among the grazing species that roamed the savannah in vast herds, defying predators, keeping encroaching trees at bay, and creating the fantastically rich soils of the ancient grasslands. The remarkable ability of cattle to convert green grass to meat, fat, and milk; their calving ease; their resistance to disease, predators, and drought; and their predictable nature compelled ancient peoples to choose cattle as an ideal species for domestication.

Our prehistoric ancestors did not have the technological advantages that our modern conventional beef industry is so dependent upon. Consequently, for their own survival, they jealously guarded the small amount of grain they did grow and never used it to feed their cattle, which had to forage for themselves just as they had always done, their lives determined by nature's seasons and the natural world around them. Human intervention was limited to shepherding the cattle and harvesting the excess meat and milk produced by the herd. (Grain did not become part of the cattle diet until after the Industrial Revolution, when machinery created vast grain surpluses, which were recycled through livestock.)

When we developed our modern technological advantages and began interfering with the natural, seasonal, grass-dependent life cycle of cattle, there was actually a dramatic *increase* in problems in the cattle industry, ranging from increasingly severe disease outbreaks to widespread calving difficulties, which have become the norm rather than the exception. This trend continues even today. Sadly, we have become so accustomed to the prevalence of all these problems that many farmers now pride themselves more on their emergency response to crises than on their ability to prevent difficulties through good management practices.

The Problems We Face

Today, diseases — from pink eye and mastitis to pneumonia, coccidiosis, shipping fever, and scours — abound. Less common but extremely frightening illnesses such as foot-and-mouth disease and mad cow disease (bovine spongiform encephalopathy) have become part of our reality as we have adopted increasingly unnatural cattle-production methods. Tractors, fertilizers, feedlots, pesticides, herbicides, antibiotics, diesel fuel, tilled soils, and all of the environmental problems that accompany these modern "advantages" now dominate the same North American landscape that supported vast numbers of wild bison just a few centuries ago.

What's wrong with this picture? With a more advanced scientific understanding of the natural world, agricultural food production should be more

— not less — efficient, problem-free, and profitable. Technology should help us capitalize on livestock's natural advantages rather than imposing upon it and forcing it to adapt to artificial production regimens. Instead of spending our time trying to overcome nature from the seat of a tractor, we should be out in our pastures, cooperating with nature to produce grass-fed cattle in an enjoyable, stress-free way modeled after nature's example. Our greater understanding of cattle and the natural world should translate into grass-based cattle enterprises that are resistant to extreme weather, market turmoil, drought, disease pressure, feed crises, predation, and other common causes of economic downfall. But these enterprises are few and far between. How can we change?

My Experience

I returned to my family's beef ranch in 1998 to take over the farm management. The farm had struggled immensely during the ten years since my father was incapacitated by a severe head injury. Despite valiant efforts by previous managers to turn around the financial situation, the farm had become unprofitable. My quest to save the farm and restore its profitability took me on a fascinating, though desperate, journey that dramatically changed my conventional perspective on beef production and marketing and significantly altered my understanding of the relationship between cattle and grass.

Having an education in economic geology, not agriculture, I felt quite unprepared for the task of fixing the farm: I was trained as a rock and ore detective, not as a conventional farm operator. So I set out to rectify my lack of agricultural education by diving into the piles of conventional agricultural literature and sources of information that I was familiar with from my childhood: agriculture newspapers, agriculture magazines, equipment dealership publications, plant variety trials, farm fairs, equipment demonstrations, agricultural Extension services, and advice from the fertilizer dealer and feed mill.

I thought I could balance the unprofitable accounting books with my enthusiastic efforts: I'd take on more enterprises, find the glitches in those already on my plate, and try to grow "bigger and more efficient." I soon realized that this approach was only taking the farm farther down the same unprofitable path. It was a road of drudgery that often seemed like voluntary slavery.

The new equipment, new infrastructure, bigger workload, constantly changing technologies, extra-long work hours, aborted new ventures, persistent disease and calving problems, and lack of tangible results were wearing everyone thin. I knew farming wasn't supposed to be a get-rich-quick scheme — I had heard that preached often enough at the coffee shops and auction barns. But if farming was supposed to be about a healthy lifestyle and being closer to the land and animals, I sure didn't see where that came in while spending fourteen hours a day sitting in the cab of a tractor or lying on my back underneath one, soaking up oil in my beard and cursing the next mechanical disaster.

If it was such a wonderful lifestyle, then how could I justify the long hours, mud, veterinary medicines, stress, and enormous financial cost of having such an enviable life? If that was not the road to success and enjoying life, what was? There had to be a more natural, healthy way to work.

A Revelation

Then a number of occurrences and changes coincided to completely alter my outlook on farming. It began when my wife and I switched to eating organic food, despite my displeasure about the higher grocery bill. The changes to my health were remarkable. I had been a severe hay fever and allergy sufferer since early childhood, but organic nutrition began to change that.

As I first began exploring organic farming as an alternative solution to the farm's problems, my mind became focused on fulfilling the rules of organic certification and enduring the minimum three-year transition period required to achieve certified organic status. Although the concept of organics theoretically made sense and my improving health convinced me that I was on the right track, the requirements for certification seemed like a merciless checklist of things that I would have to eliminate, that I had come to depend on for the very survival of the farm. This, however, is the crux for every new organic farmer: How do we eliminate the herbicides, pesticides, hormones, synthetic fertilizers, and so much else while still producing crops and livestock that are in good enough shape to make it to market without withering away or dying en route?

As my health improved, I began noticing its delicate balance. Although I was leading a relatively allergy-free life among all the farm pollens and dust, I still had a tremendous sensitivity to chemicals. Exposure to certain chemicals, whether the heady vapors of a new carpet, secondhand smoke, an herbicide application, or the smell of fresh paint, could completely disrupt my equilibrium. Fatigue, stress, poor-quality food, nervousness, and injury seemed to magnify these triggers a hundredfold. As the old allergies reappeared, I'd grow irritable and distracted, and I was no longer able to perform at my peak capacity.

I began to view my immune system as a bucket that I could fill with any variety of stresses without ill effects — but only to a certain point. Once the bucket was full, a single additional drop would make it overflow and cause my ill health to return. I realized, then, that the key to my health and productivity lies in managing my life to avoid overfilling the bucket so there is always room for the normal stress of unexpected daily challenges. The same is true for animals and plants.

If animals and plants are under stress, chemical or otherwise, their resilience is compromised, making them prone to pests, disease, and poor nutrient uptake. They struggle to compete with weeds, pests, predators, and even each other if they are exposed to foreign substances and environmental conditions that they have not had the opportunity to adapt to over thousands of years. What is the true natural potential of our plants and livestock, and how many problems can we alleviate by raising them in natural conditions as similar as possible to those in which they evolved to thrive?

Health and productivity in any ecosystem, whether it is my own body, that of a cow, or the whole farm, is driven not only by the elimination of chemicals and other ingredients restricted by the organic-certifying agencies, but also by eliminating

stress so that natural evolutionary advantages can be fully expressed. Cattle's stress comes in a multitude of forms: nutritional, chemical, social, climatic or weather-related, the stresses of light deprivation or excess, heat and cold, pests — even the simple removal of a key player in the function of the soil, plants, or rumen or other part of the larger ecosystem. In sum, any departure from cattle's natural balance with their optimal evolutionary habitat can induce stress.

Modern medicine teaches us to think symptomatically — we focus on *reacting* to the *symptoms* of disease. Thus, while we have become experts at resolving various ailments and diseases, these are really just the symptoms of a much greater underlying problem. We rarely seek to discover what knocked out of balance our perfectly designed systems in the first place so that disease could find its way in. After all, aren't disease, pests, weeds, and predators designed to remove those individuals that are not able to function optimally in their particular environmental niche? Isn't that how evolution works? Actually, crisis is merely nature's way of pointing out that its delicate balance has been disturbed. When the symptoms of crisis appear, consider them giant arrows directing your attention to the underlying problem. We need to address not only these symptoms, but also whatever caused the weakness that precipitated the crisis, so we can avoid it in the future.

For example, a predator problem with coyotes is a symptom of an out-of-balance livestock management system, not a signal that there are too many coyotes. Something in the balance of the farm ecosystem is making our farm a target for coyotes or is giving them an unnatural advantage. The underlying cause might be that we are calving in late winter or early spring, lean times when these calves are the only easy food source. To recognize this, however, we have to stop blaming the coyote for our problems and look more closely at our management role.

Similarly, we should not consider a pneumonia outbreak as merely an incidence of disease to be "fixed" by a course of antibiotics. It is a clear sign that something in our management style is compromising the immunity of our herd. What is causing this stress? Perhaps we are weaning during the rainy season or perhaps the collection of manure and mud around the feed bunks needs to be addressed by a different wintering system.

In order for a farm to be a thoroughly viable enterprise, the farmer him- or herself and the farm's profitability must also be factored into its ecosystem. In the search for a more natural approach to raising cattle, it is not realistic simply to open the gate so our cattle can regain their natural balance while we turn to hunting and gathering. We have to live in this world too, which means that we have to consider our financial goals, marketing, personal priorities, and involvement on the farm when we analyze the source of symptomatic farm problems. Personal stress, loss of profitability, a sagging fence, and a lineup of machines needing repair all suggest that we have a role in the problem.

A New Way of Thinking

As my outlook on farming evolved, I realized that I had confused profitability with productivity, an easy mistake to make in our volume-oriented, commodity-driven marketplace. How many head of cattle we own, how many pounds each calf weighs when we sell it, how many hours of work we perform each day, how many gallons of milk each cow produces, how many bushels of grain or tons of forage we grow per acre, and how much each individual calf is worth on sale day are often how we measure the success of our farms and our relative success as farmers.

But high productivity does not automatically equal high profit. The more effort we put in, the better results we get, right? Not necessarily. We have to think in terms of high profit, not high productivity. The difference may seem subtle at first, but it requires an enormous shift in focus.

We must begin by measuring our success in terms of net profit per acre, not on the illusion of profitability created by our focus on maximizing yields (such as our obsession with producing the biggest possible calves at weaning). We should value ourselves by the results of our efforts instead of priding ourselves on the effort itself. It isn't enough to eke out a living from our maximum productivity. To maximize our profit, we need to adjust our priorities toward minimizing expenses and addressing logjams in our production systems before we focus on boosting production.

Switching to a natural grass-based beef production and marketing scenario isn't easy. We have to replace time, machinery, effort, and many unnecessary expenses with knowledge. The greatest tools at our disposal are the lessons we can draw from the wild herds that roam the world's grasslands much as our livestock's ancestors did before we domesticated them — before we became so enamored of our modern technological solutions that we let them eclipse nature's solutions to our production challenges.

Although natural grass-based beef production may be new to us, nature's great herds, grasslands, and soils have been succeeding at it for a very long time without our help. Their examples provide us with the reference, insight, and large-scale working model against which we can measure our ideas and will guide us toward sound, positive, simple solutions to our production challenges.

Let's begin.

The FUNDAMENTALS of Grass-Based Beef

PART ONE

The Great Herds
and Their Grasslands

LONG BEFORE HUNTER-GATHERERS BEGAN roaming the earth, the ancestors of our cattle, sheep, goats, pigs, and domestic birds lived much as their wild counterparts do today, flourishing within the balance provided by natural selection, nature's seasons, their specific environmental adaptations, and competition for resources with other species. They thrived remarkably well without us. When our forebears finally arrived on the scene, they interacted with game as predators but barely made an impact on nature's vast abundance. Modern humans, however, have been far more intrusive.

When we decided to domesticate animals, it was for our convenience, not their benefit. Domestication changed their lives in obvious ways (their range was limited with property boundaries, for example), but it is a fallacy to think that we improved them or their lives in any way. Before we interjected ourselves, nature's creatures lived and evolved together to create a vibrant, healthy, self-sustaining balance among soil, grass, microbes, herds, and predators. They were remarkably successful and much can be learned from them.

Despite our dismal record of environmental stewardship, today we persist in inventing expensive technological quick fixes and artificial solutions for the troubling problems we face on our farms and in the environment. We have forgotten how to look to nature — to the great wild herds and their rich grasslands — for guidance and solutions.

The good news is that we don't have to run to fertilizer dealers, seed companies, Extension agents, and equipment dealers every time we want to increase our productivity and efficiency or try to resolve an issue on our farms. Technology has its place, but our first thought should be to look to nature for practical and ecologically sustainable solutions. Animals and plants evolved for millions of years to live in sync with their environment. We certainly have not changed them so much in our short period of influence that they have lost the specialized adaptations, characteristics, and natural traits that made them so successful during their long history.

The Ticking of Geologic Time

Our seemingly arrogant preoccupation with our technological solutions and human-contrived cattle production philosophies and our lack of trust in nature's answers to our production challenges can be traced directly to our biased view of evolutionary time. We mistakenly believe that we are central to history, that we are the glorious end product of a long, linear progression of events. We believe that we have been around for a very long time; we even call the time before the evolution of modern humans *prehistory,* as if it is less important because we weren't part of it. Yet this prehistorical period stretches back through vast spans of time; our human history is but a blink of the eye in comparison.

This bias is hardly surprising if we consider how we experience the passage of time. I have a sense of how long a minute, a day, a week, and a month are. I also have a feeling for how long a year is. But grasping what ten years feels like is a challenge. I doubt that even my grandfather, at age ninety-six, has a true sense of what the passage of twenty or thirty years feels like. I can vaguely imagine the passage of one hundred years, but a thousand years is beyond my comprehension. I know that ten thousand years is a lot less than a hundred thousand years and that a million years is even more, but it's impossible to understand the experience of such vast spans of time. They simply become numbers that are detached from tangible human experience.

In this same vein, I know that after the dinosaurs roamed, mammals evolved, mammoths and saber-toothed tigers flourished, and then humans evolved and started chasing them. We endured a number of ice ages, we domesticated animals and plants at some point along the way, and finally the Egyptians built pyramids. The rest of history follows more or less the way we remember from history class.

Geologic time happens over such vast periods that we simply cannot grasp its implications. Because we cannot relate to the vast passage of this time, we place the greatest emphasis on the brief, most recent interval we know as human history (a few thousand years). No wonder we overlook the significance of the millions of years of evolutionary history that our domestic livestock have under their belts!

To get a true feeling for geologic time and how briefly we have been part of it, we have to put Earth's history in a context we can understand. The chart

A WRINKLE IN TIME

Although it helps to assign some numbers to the major events of evolutionary history, the vast scale of time associated with them still remains beyond our comprehension. Events shown in boldface type highlight the long joint evolutionary history of grass and cattle and emphasize the very brief impact we humans have had on them.

- 245 million years ago: The first dinosaurs evolve.

- 244 million years ago: The first mammals evolve.

- 144 million years ago: The first birds evolve.

- 66 million years ago: Dinosaurs become extinct.

- **24 million years ago: Grass coevolves with grazing animals.**

- 17 million years ago: The first horses evolve (about the size of a dog, with three toes on each foot).

- 5.3 million years ago: Mammoths, saber-toothed tigers, and **the first wild ancestors of cattle, sheep, goats, and bison evolve on the great grasslands of the world.**

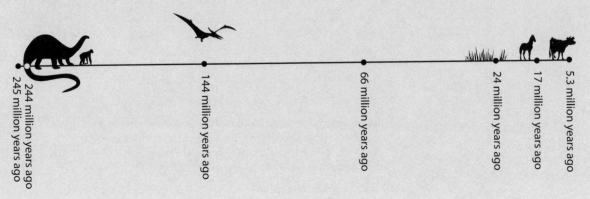

244 million years ago
245 million years ago

144 million years ago

66 million years ago

24 million years ago

17 million years ago

5.3 million years ago

on this page compares the last 5.3 million years of Earth's history (since the ancestors of our domesticated livestock evolved) with a twenty-four-hour day. This analogy makes it clear how recently we domesticated farm animals and how brief the last hundred years of our modern agricultural practices are when compared with the amount of time our domestic livestock and their ancestors spent adapting to specific environmental conditions.

Mind-boggling, isn't it? Our livestock have been domesticated for only 2 minutes and 10 seconds of their 24-hour history. Our modern farm practices have been around for only approximately 1.75 seconds of this 24-hour history. Still, we naively believe that the solutions to our farm and livestock's health, productivity, and production problems lie in technology, biotechnology, petrochemistry, and pharmaceuticals that have yet to stand the test of time. Odds are that the production challenges we face on our farms would be better solved by learning how the wild grazing herds and their grasslands deal with nature's challenges and by exploring the evolutionary history of our domestic livestock.

The Great Herds

Until very recently in our busy but short history, much of the world's landscape was dominated by great grazing herds of one species or another. Today, we can still see remnants of these herds, which retain adaptations to their grassland ecology and characteristics that are shared by domesticated livestock.

Seeing the great herds gathering and moving through the plains, a hypnotizing, awe-inspiring experience, triggers in us a passion that may be a window into our long-forgotten past as hunters and predators. Certainly, the animal kingdom still recognizes humans as predators.

THE HISTORY OF MODERN LIVESTOCK COMPRESSED INTO 24 HOURS

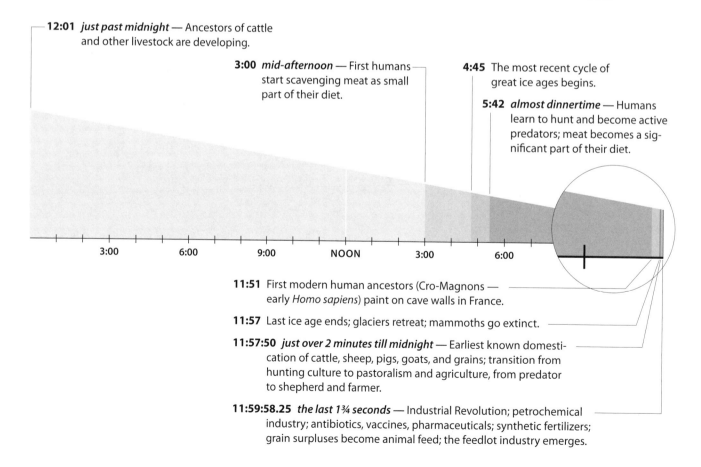

12:01 *just past midnight* — Ancestors of cattle and other livestock are developing.

3:00 *mid-afternoon* — First humans start scavenging meat as small part of their diet.

4:45 The most recent cycle of great ice ages begins.

5:42 *almost dinnertime* — Humans learn to hunt and become active predators; meat becomes a significant part of their diet.

11:51 First modern human ancestors (Cro-Magnons — early *Homo sapiens*) paint on cave walls in France.

11:57 Last ice age ends; glaciers retreat; mammoths go extinct.

11:57:50 *just over 2 minutes till midnight* — Earliest known domestication of cattle, sheep, pigs, goats, and grains; transition from hunting culture to pastoralism and agriculture, from predator to shepherd and farmer.

11:59:58.25 *the last 1¾ seconds* — Industrial Revolution; petrochemical industry; antibiotics, vaccines, pharmaceuticals; synthetic fertilizers; grain surpluses become animal feed; the feedlot industry emerges.

Wild herds have much to teach us. Their environment, food, herd dynamics, calving, breeding, synchrony with the seasons, and even their relationship to predators all help us learn about our cattle. Because domesticated cattle are so similar to the wild herds, we can immediately apply these lessons to improve the profitability of our livestock enterprises.

In North America, the best-known migratory herds were the plains and woodlands bison. Numbering close to sixty million, they shaped the Great Plains prior to their mass extermination. Even as late as the 1870s, *individual* herds occupying fifty square miles or more were sighted in the western Dakotas. Their enormous appetites and hoof power helped maintain the vast expanses of healthy grasslands and kept trees at bay. Their grazing impact caused rapid nutrient cycling, which in turn created the extraordinarily high organic content of the prairie soils.

Capable of storing huge reservoirs of plant-accessible nutrients, these soils are the North American grain belt's secret of success. Without the bison, the plains left behind by the receding ice-age glaciers would have slowly turned to brush and forest, which recycle nutrients and build organic matter at rates far slower than grassland under the influence of a migratory grazing herd. Only small, spread-out remnants of these herds still exist.

Elk also once formed vast migrating herds on the Great Plains, as did pronghorn antelope west of the Rocky Mountains. The Four Corners region (where Utah, Arizona, New Mexico, and Colorado meet) may once have been home to migrating herds of bighorn sheep. In the Arctic, herds of musk oxen and migrating caribou define the high Arctic landscape, providing an invaluable source of meat to creatures along their migration routes and shaping the tundra that is their home.

In Patagonia, at the southern tip of South America, there were once migrating herds of guanacos. Southern Africa had the springbok, which can still be found in small herds today. In sub-Saharan Africa there are remnant herds of topi (also known as tiang) antelope, and in southern Sudan there is still a yearly migration of up to a million white-eared kob antelope, rivaling the great herds of the Serengeti. We can still see the herds of wildebeests, zebras, and Thomson's gazelles migrating across the Serengeti in Kenya and Tanzania. Remnant herds of dorcas gazelles live on the edges of the Sahara Desert. There are even elephant herds roaming now in parts of Africa. Asia still has remnant herds of chiru antelope on the Tibetan Plateau, saiga antelope on the steppes of southern Russian and Kazakhstan, and Mongolian gazelles, also known as zeren, on the steppes and in the sub-deserts of Mongolia, northern China, and southern Russia.

All these herds have been greatly reduced from their former sizes by hunting, habitat encroachment, and competition with domestic stock for resources and space. But in their glory they were truly great. Not so long ago, these herds were accompanied by an even greater variety of species. From woolly mammoths and woolly rhinoceroses to prehistoric horses, woolly camels, and aurochs (wild cattle), a host of fantastic ice-age creatures molded the great plains of their time with their grazing, manure, and pounding feet.

Most of us can relate to at least one of these great herds, and knowing about them will provide a benchmark to which we compare our ideas about cattle husbandry and livestock production. The legacy of these great herds will accompany us throughout this book, from our discussions of genetic selection and grazing practices to electric fences and water sites, as we try to replicate in our domestic livestock what we see in the wild.

Lessons from the Herd

A number of years ago I had an opportunity to watch the Porcupine caribou herd migrating south into the foothills of the Brooks Range in Alaska as part of its fall migration pattern. Over the course of a few days, I saw close to a hundred thousand migrating caribou. Though the herd had split into smaller traveling bunches ranging in size from a few hundred to a few thousand, these groups were all within a few miles of one another, traveling across the tundra toward the same distant destination. From the valley floor I could see many individual dramas, such as cows searching for their calves and individual caribou panting as they strained to keep pace with the herd. I saw hunters harvesting caribou and wolves harrying the weak members along the herd's flank.

But from the hills above, another view unfolded. The individuals merged, their identities lost within the massive herds. Thousands and thousands of caribou were bonded together by a single purpose of mind, linked as if by some invisible glue, the individual dramas blending into the masses like little whirlpools in a giant river. The herd had gained an identity of its own. The individual caribou within it seemed like little more than tiny cells within a much larger body, unaware of their role within this giant living organism that slowly snaked its way across the tundra.

Nor did the herd as a whole seem aware of the individual caribou within its midst. It swelled and flexed in response to the terrain, winding its way over the ridges and valleys, heading south, driven by a higher collective consciousness. From a distance, the herd itself had become an individual, interacting instinctively on a grand scale with entire weather systems, vegetation zones, mosquito plagues, river courses, and wolf packs, just as an individual caribou might react to a gust of cold wind, the grass beneath its feet, the mosquito on its ear, the water in its path, and the lone wolf harrying its flank.

As a collective, the Porcupine caribou herd is capable of shaping the landscape and vegetation of the Arctic and sustaining entire populations of wolves and other predators. Through its calving grounds, which lie in the middle of the proposed controversial oil-drilling programs in the Arctic National Wildlife Reserve, it even influences the politics of global oil economics. To fully understand the wide-ranging impact of the caribou herd, we cannot limit our focus to the individual members of the herd; we must recognize the herd's identity as a whole.

The relationship of the herd to grass, soil, water, nutrient cycles, climate, vegetation, microbes, and predators can teach us much more than an individual cow can.

If we watch a flock of birds, we can observe the same phenomenon. With a rush of furiously beating wings, the birds lift into the sky and suddenly individuals disappear into the flock, now a cohesive whole. Instead of crashing into each other, individual birds fly in perfect harmony as the flock twists and turns; they move as if driven by a single mind, working in unison for the benefit of the group.

If we focus on an individual, we do not see its connection to the larger group. Watching the caribou mother calling her calf, we see an animal looking for food, struggling for survival, and seeking the

The Porcupine Caribou herd travels south into the foothills of Alaska's Brooks Range.

companionship of her young. We see wolves feeding on the weak and vegetation being trampled into the ground. From this vantage point we can study individuals within the herd, become experts at caribou calls and the hunting strategies of wolves, and learn about vegetation growth, but we will not gain an understanding of the instincts driving the herd. Only after we have stepped back and looked at the herd as a whole can we understand how the individuals are shaped by the dynamics of the group: The whole is greater than the sum of its parts. The relationship of the herd to grass, soil, water, nutrient cycles, climate, vegetation, microbes, and predators can teach us much more than an individual cow can.

Of Microbes, Humidity, and Feet

Have you ever looked at the ground with your nose inches from the soil and poked around to see what is happening beneath the surface? Have you ever sat in your pastures and tried to figure out how vegetation is recycled? Have you considered what it takes to recycle nutrients back into your soil? What does grass have to go through on your land to grow, flourish, reproduce, die, and be reincorporated into the soil so its nutrients become available to the next generation of plants? Not surprisingly, the decomposition and recycling process varies greatly throughout the world because of climate. Temperature and

humidity are at the heart of the great herd's existence and therefore play pivotal roles in the herd's ability to create and maintain the earth's grassland environments.

Where I grew up in British Columbia's Okanagan Valley, uneaten grass quickly browns and crisps in the relentless summer heat. By early fall, though, the rains start, and by early winter, they've washed many of the nutrients from the plants. The wet snow crushes the plants to the ground, and within a year the stalks are indistinguishable from the rest of the organic layer in the soil — gone, reincorporated, from dust to dust. Untreated fence posts rot from the top down almost as quickly as from the bottom up, and a tree that falls in the forest turns into a mushy, rotten mass full of centipedes and beetles in just a few short years. In the rain forest on Canada's west coast, the nutrient cycling occurs even more quickly. The microbes that break down dead organic materials are extremely active year-round, above and below ground, in this warm, humid environment.

In arid regions, however, the story is very different. Although fence posts still rot where they come in contact with the soil, material above the soil's surface seems to last forever. The dry, dead grass oxidizes, turns gray, crumbles, and blows away in the wind, never returning to the soil. Years later, dry grass still stands, almost as if it died just the day before. Dead

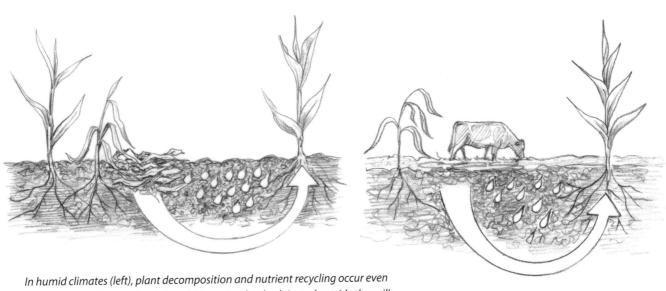

In humid climates (left), plant decomposition and nutrient recycling occur even without animal impact due to flourishing microbes both in and outside the soil's protective environment. In arid climates (right), animal impact must break down and return plant debris to the soil because soil microbes are not as broadly active.

trees in the forest seem to be permanent fixtures. Gray on the outside, the dry wood hardly seems to change, only growing less dense with time until it finally disintegrates. Things don't seem to rot in such a climate. Why?

In order for nutrients to be recycled back to the earth quickly, the microbes that decompose dead plant and animal remains must be active. Like us, these microbes need water to function. In humid areas such as rain forests, microbes can do their work in the open air, but in dry regions the microbes work efficiently only in the moisture zone below the soil surface. Until dead material can make contact with the soil, it remains untouched by these microbes. Nutrients from these dead organic remains are not recycled back to the soil for future plants; instead, they disappear into the atmosphere through oxidization or are broken down by the wind, ultraviolet sunlight, and physical weathering until they blow away as dust.

Herds are nature's steamrollers and plant crushers.

Rainfall and warm temperatures alone do not drive this process. More important is the humidity in the air between rainfalls. The decomposition microbes need the right balance of moisture and temperature to survive and work efficiently. If the air is very dry, the microbes will be confined to the moist soil, becoming less and less active as the soil dries out. Some areas may get high rainfall amounts over a relatively short period and have tremendous plant growth, but if the air is dry for the rest of the year, the dead material won't decompose and be recycled unless it is physically pushed down into the moisture zone in the soil. Other areas may get less rainfall, but if they are more humid, the microbes can continue to work aboveground, breaking down and recycling dead plant material even before it contacts the soil.

As an area becomes drier, microbial nutrient recycling becomes less efficient and we have to look to some other process to help break down and recycle dead plant material. We can certainly turn on sprinklers or use heavy equipment to mash the material

and bring it into contact with the soil, but at what economic cost? Mother Nature has a much simpler solution: hungry animals, sharp feet, and manure.

If we look at the distribution of animals around the world, we recognize an interesting trend. In humid areas such as the rain forests around the globe, where microbes can be active outside the soil year-round, we see more and more solitary animals or small groups of animals spread uniformly throughout the area. Yet in areas where microbial activity is limited by decreasing rainfall and humidity or by the onset of a dormant winter season, when temperatures drop below the microbes' comfort zone, we see larger herds of animals clumped together. There is a great advantage to this massing of feet and mouths: Herds are nature's steamrollers and plant crushers.

Animal grazing plays an important role in ecosystems: Animals eat grass, before it can become old, dry, and unpalatable. Periodic grazing maintains grass in its growth (or *vegetative*) stage, during which the plant roots spread out, much as they do in a lawn. As the grass extends across the soil, it becomes an insulating layer that shields the earth from direct heat, which in turn helps to retain moisture. When the rains finally come, the carpet of live grass and dead grass litter slows the water runoff, giving it more time to be absorbed by the soil. The more water is absorbed, the more water is stored and the longer it will take the soil to dry out after the rains.

It sounds like an ideal arrangement because it is. Grazing animals and grass are a perfect match; they coevolved twenty-four million years ago to take advantage of each other's best traits.

Animal feet knock over the dead plant material, driving it into the ground so it contacts the microbes in the moist soil. Looking closely at the feet of the majority of animals that make up the great herds (including cattle), we can see that most have two toes on each foot. As they step, and especially as they step violently, these flexible toes twist and flex, particularly at the front edges of the hooves, where they are the sharpest. The sharp hooves slice up the dead plant material as they push it down into the moist soil, where the active microbes are waiting for lunch, and also fracture the ground, allowing rainfall to penetrate easily through the hard crust on the soil surface. Plant material that has been trampled fur-

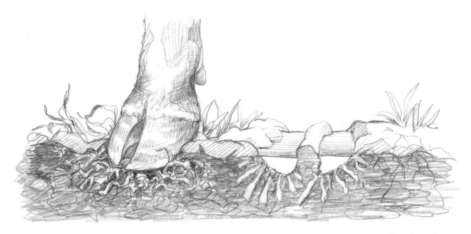

A grazing animal's foot in action: The toes twist and flex, slicing up dead plant material and pushing it into contact with the moist soil layer. The footprint behind shows water being absorbed by the fractured soil.

ther slows rainfall runoff, and the depressions left by the animals' hooves create little pools to hold water.

But that's not all. The animals also leave behind their manure and urine, pure gold to the microbes. After dead plant material has passed through the digestive tract of a grazing animal, the finely chewed and partially digested material becomes much easier for soil microbes to digest. Manure piles also provide a perfect moist environment in which soil microbes can flourish. This is why cattle, bison, and so many other grazing animals have such seemingly inefficient digestive processes that appear almost wasteful (*poor conversion rate,* in scientific jargon): Partially digested plant remains play a vital role in feeding the soil microbes, which keep the soil fertile.

Plant digestion in the soil has a great deal in common with plant digestion in the rumen of the cow; in fact, they are intimately linked in a healthy grazing environment. In the first step of plant digestion in the cow, grass must pass through the rumen, the first chamber of the cow's stomach, which contains a host of microorganisms responsible for fermenting and breaking down grass's tough cellulose structure so the cow can extract nutrients from the cellulose as it passes through the rest of the digestive tract. Many of these rumen microorganisms are the same as those that live in the soil, where they are responsible for the decomposition of plant materials, likewise breaking down or "digesting" the tough cellulose structure of dead plant remains to extract nutrients from them for the use of future plants and other soil microorganisms.

The soil can support only as many decomposition microbes as it can feed from dead plant remains in the earth. Very little dead plant material is accessible to the soil microorganisms until grazing animals trample the plants and return partially digested plant remains to the soil via their manure. This trampling creates a surge of dead plant material in the soil, but unfortunately it takes time to rebuild the populations of soil microorganisms responsible for digesting these dead remains. Complicating matters is that there is only a limited time period during which the soil microorganisms can unlock nutrients contained within the dead plants' tough cellulose structure to make them available to new plants in time for the next year's growth season. This time period is shortened even further by the arrival of the drought season, when soil moisture levels drop substantially and cause a marked decrease in microbial activity within the soil, or the arrival of the winter season, when cold soil temperatures also decrease microbial activity (which drops off sharply when soil temperatures fall below 9°C/48°F).

Nature offers a solution, however. Every time manure falls onto the ground, an enormous number of rumen microorganisms are carried with it and are incorporated into the soil by the animals' trampling hooves. Because these microorganisms are the same as the soil microbes responsible for plant decomposition, this flood of rumen microorganisms boosts soil microorganism populations at the very instant when passing herds' migration provides a surge of dead plant remains. Thus, plant decomposition can take place quickly within the limited time frame available. The flood of microorganisms in the manure provides an enormous kick-start to the decomposition process in the soil, allowing it to occur in the available window.

To add yet another reason to marvel at this perfect match among animal, plant, and microbe, we need only look at the role of urea (nitrogen) in both the soil decomposition process and the ruminant digestive process. Urea contains an easily digestible source of nitrogen-rich protein, the primary fuel that feeds the rumen and soil microorganisms responsible for fermenting and breaking down grass's cellulose structure. The cow's body produces urea naturally in the liver. Some is sent to the rumen directly though the rumen wall and some is introduced via the cow's saliva into the plant materials it consumes. Excess urea produced by the liver is excreted via the kidneys in the form of urine, which in turn feeds the microorganisms responsible for digesting dead plant remains in the soil. Thus, urea in the cattle's urine not only provides nitrogen fertilizer for the next growth phase of plants, but it also fuels the microbes working in the soil. Without enough nitrogen in the soil, the microbes become inactive, as they do when the soil is too dry. Clearly, large grass-eating herbivores, grassland, and the digestive microbes in both stomach and soil evolved simultaneously.

How These Lessons Translate into Practice

Nature does not have one hard-and-fast rule for everything. The process of plant decomposition changes dramatically depending primarily on the humidity and temperature levels that keep the digestive microbes active. As a result, we farmers have to adjust our grazing impact and management practices to compensate for conditions that become increasingly challenging to the microbes, such as humidity and temperature changes from one climate to the next or from one season to the next.

On one end of the scale are extremely humid rain forests, where nutrient recycling and organic decomposition occur very quickly without animal impact. From this we can see that a humid environment allows great flexibility in how we manage the grazing habits of our livestock. On the other end of the scale

are landscapes that are utterly dependent on grazing herds to maintain the nutrient-recycling process. Without sufficient animals in large migrating herds, such landscapes become less and less fertile over time. In arid environments like these, without a well-managed grazing strategy, the heat (or cold) and drought take a toll on plants and soil. Spaces between plants become bigger, and lush grasses are slowly replaced by scrubby plants and bushes that are able to cling to life without the help of periodic grazing. As more soil is exposed, the cycle spirals downward until a new balance is struck with the conditions that remain. Extreme outcomes of this downward spiral are seen in places like the Sahara Desert, where little more than a vast wasteland of sand remains in an area where, only six thousand years ago, vast grasslands flourished as in the Serengeti in East Africa or on our North American prairie, kept healthy by the great herds of elephants, wildebeests, gazelles, zebras, and giraffes that roamed them.

Grazing intervals, herd sizes, plant species, and herd impact are not the same throughout Europe or Africa or in the American Midwest, Southeast, dry desert rangelands, or intermountain regions. Yet, management practices that are developed in one climate and socioeconomic environment are often precisely exported to other regions with little or no consideration of the long-term negative effects they will have in a different environment. This cookbook approach to farming may be convenient in the short term, but in the long term it does significantly more harm than good. For example, many European practices have been exported worldwide, including those of summer fallowing; a preference for small, evenly distributed cattle herds; making and storing winter feed; and using barns to house animals in winter. These European practices are as much cultural as they are products of a wet, relatively warm climate where microbes are active above the soil surface almost year-round.

It could be argued that these practices evolved to address the climatic challenges in Europe, where winter grazing reserves are easily leached of their nutrients and wet fields are susceptible to trampling damage during the muddy winter. Yet the cultural reality of medieval Europe played an even greater role: In the damp, cold climate, people were exceed-

THE DOWNWARD SPIRAL OF POOR GRAZING MANAGEMENT

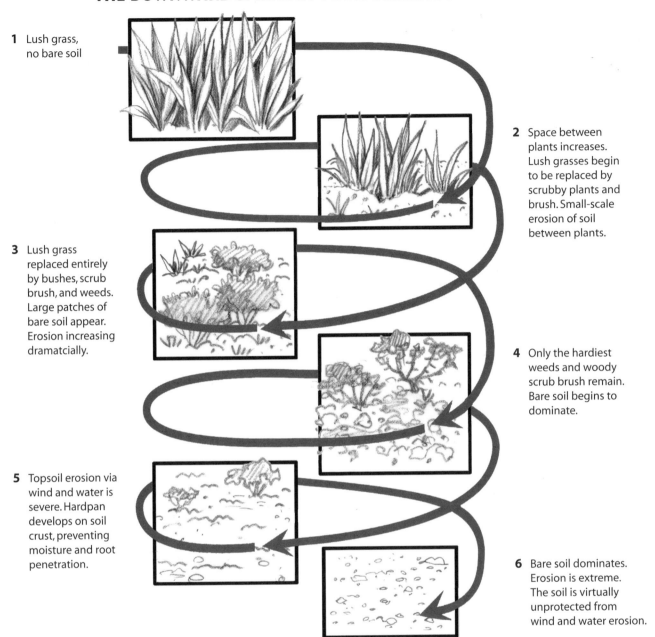

1 Lush grass, no bare soil

2 Space between plants increases. Lush grasses begin to be replaced by scrubby plants and brush. Small-scale erosion of soil between plants.

3 Lush grass replaced entirely by bushes, scrub brush, and weeds. Large patches of bare soil appear. Erosion increasing dramatcially.

4 Only the hardiest weeds and woody scrub brush remain. Bare soil begins to dominate.

5 Topsoil erosion via wind and water is severe. Hardpan develops on soil crust, preventing moisture and root penetration.

6 Bare soil dominates. Erosion is extreme. The soil is virtually unprotected from wind and water erosion.

A poorly managed grazing strategy causes a downward spiral of decreasing soil productivity, weed infestation, loss of vegetation, soil crusting, and erosion.

ingly poor. They built their homes alongside and on top of animal barns so they could keep warm; homes were heated in part by the rising warmth thrown off by the barn animals and the heat emitted by the composting manure pile under the house.

Hay production and feeding were developed to warm meager homes during winter and to allow farmers to oversee their animals at night and avoid theft in a starving society. European winters are less than ideal for shepherding cattle not because the cattle cannot graze through the snow, but because the cold, wet climate would kill any human shepherd trying to spend a day or, worse yet, a night outdoors in those conditions. Stored feed in barns kept people out of the weather during winter. It also allowed cattle to maintain a high enough level of nutrition during winter so that people could continue milking them as a source of food for themselves and to produce a salable product year-round. Milk and cheese were often two of the medieval farmer's few sources of income; no matter how inefficient and labor-intensive this stored-feed production system was, farmers could not afford to stop milking their cattle. Ironically, this approach to cattle husbandry has persisted and been exported to almost every corner of the world, regardless of differences in cultural and climatic realities.

Herd Migration

Livestock are tools to create and maintain healthy grasslands, but they must be integrated properly in order to have the desired effect. There is no single correct form of integration. In rain forests, countless species of animals are scattered throughout the whole ecosystem; though they do not congregate in large migrating herds, their presence works. On the other hand, in dry regions, we see enormous herds of grazing animals that are constantly on the move as giant grazing units, intermingling with other grazing species and never spreading out evenly across the plains. These dry areas need animals — and lots of them — to keep the land healthy, but for their presence to work they have to be concentrated to provide maximum impact on the land over the shortest time possible. But how do we make sure that the violent impact of the herd is in the right proportion — not so much that it tramples the living daylights out of the grass and not so little that its impact makes no difference?

The herd's impact must be measured or rationed. The key is migration; animals have to move on. Migration in the wild is driven by the search for fresh grass as the grass is quickly depleted around the giant herd. If individuals in a herd spread out across the grassland to graze, each animal picks and chooses which plants it eats and where it steps, causing the herd's total impact to be very irregular. When a large herd grazes together, however, the impact of the animals is evenly distributed across the land.

Competition and the need to keep up with the group before all the good stuff is eaten compel the animals to trample and eat almost everything in their path: "I can't be too picky or everything will be gone when I get there, and I can't take too long looking for the best place to put my feet or the herd will push me aside and leave me behind." The herd benefits from this arrangement just as much as the grass does: Grazing as a compact, migrating group guarantees that it will continue to find abundant, high-quality grass all along its migratory route. (Selective grazing would cause desirable plant species to be overgrazed and would give undesirable plant species the competitive advantage, causing a slow deterioration of the quality and quantity of the herd's grazing supply.) Grazing as a group also decreases the vulnerability to predators of each individual animal in the group. This reduction in stress makes them much more nutritionally efficient than when they are spread out and graze individually, which easily compensates for the reduced selectiveness in their nutritional intake. So like a giant beast, the herd moves on, always searching for fresh food, leaving behind a grazed and trampled swath.

Livestock are tools to create and maintain healthy grasslands, but they must be integrated properly in order to have the desired effect.

If the herd stays too long in one place or returns to an area too soon, it grazes the tender regrowth while it is still small and weak, making the desirable plant species less competitive than undesirable species by grazing it before it is strong enough to recover quickly. If the herd doesn't return soon enough, though, the dry, dead material will choke out the live regrowth underneath it, suffocating it and blocking necessary sunlight. Staying too long or returning not soon enough, then, leads to takeover by less desirable plants. Weeds dominate and the good grass becomes patchy. Likewise, too much, too little, or inconsistent trampling by the herd means that the ground does not efficiently absorb rainwater. The herd's impact must be applied in the right intervals and in the right amount for the grassland to remain healthy.

Predators: Nature's Shepherds

We see the value of the herd, but try telling your cows to stay together when you let them loose on your farm. They spread out over every square inch of available land, targeting the choice grass species and chewing them completely to the ground before they will even consider touching the less desirable plants. On their own, they don't have enough common sense to stick together and migrate as a herd.

So are wild animals smarter? If the great herds migrate as a unit to give the land the right measure of impact and rest, why doesn't the same thing happen when humans get involved?

Remember the wolves I described harrying the caribou herd? Therein lies the answer. Not only do predators fulfill the well-known role of purging weak members from the herd, thereby keeping the bloodlines strong, but they also play the vital role of keeping the herd bunched together out of fear. But they are commonly missing from our farms, and wild herds naturally disperse whenever predators are eliminated by guns, poisons, and traffic. And so while the predator's fangs replace the shepherd's protection, in the predator's effect on the herd as a whole, it plays the role of nature's shepherd. Both the shepherd and the predator keep the herd bunched together and migrating as a group, albeit for different reasons.

The individual is most secure in a large herd. The likelihood of being eaten by predators is significantly smaller when there are many others in the herd from which predators can choose and when weak animals lag behind or are jostled outward to the exposed edges of the group. The wolf, lion, hyena, and bear all have roles to play along the flanks of the great herds. These predators incite the fight-or-flight fear response that glues together the herd and causes the animals to migrate as one. Fear of predators keeps the herd milling around, breaking up the soil and dead grass underfoot and uniformly grazing the food supply, and also helps drive the herd forward in search of greener pastures. By contrast, a calm herd without predators drifts apart, each animal seeking the easiest route and avoiding brush and prickly vines.

Through their rigorous selection process, predators also keep the genetics of the herd strong. The herd, in turn, places within its reach along the herd's periphery a vast supply of meat in the form of old, weak, injured, and newborn animals.

There is a perfect symbiotic relationship among the soil, grass, microbes, herds, and predators. It is this relationship that we must mimic in our domestic farm environment, replacing the predator with shepherds (or the bite of electric fences) to drive migration (or pasture rotation) and to focus the herd's grazing impact by keeping the herd bunched together as a group.

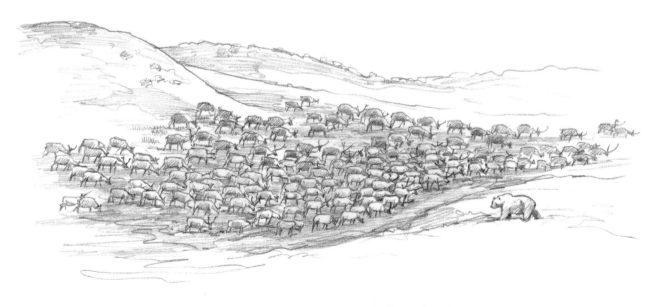

Fear of predators keeps the herd bunched up and moving.

Genetics and Breeding:
Selecting the Right Animals for Your Herd

IF WE MODEL OUR DOMESTIC CATTLE AFTER their distant ancestors and nature's wild herds and provide them with living conditions and a management strategy that mimic their historical natural environment, they should thrive. Therefore, our objective as an industry should be to produce cattle capable of grazing year-round on the grass grown on our land without requiring costly off-farm inputs such as synthetic fertilizers, feed supplements, expensive farming equipment, diesel fuel, antibiotics, and growth hormones, while producing healthy calves that bring us a high return per acre. To accomplish this, our genetic selection process must strengthen our breeding herds.

No matter how pretty they are or how expensive they were at auction, high-maintenance and low-fertility cattle should be removed from the herd.

In the wild, predators, drought, starvation, disease, and other trials eliminate weak individuals from the herd. The least fertile and the least nutritionally efficient are those most vulnerable to nature's brutal selection process. Nature purges their genes, strengthening the gene pool. Because we can no longer rely on the wolf or lion to eliminate the weak, we must train ourselves to identify characteristics that

indicate problems with fertility and maintenance efficiency. (*Maintenance efficiency* is an assessment of nutritional efficiency, particularly as it affects production costs. A low-maintenance animal requires far less quantity and/or quality food than does a high-maintenance one to maintain the same body condition and produce the same *net* income for the farmer.) No matter how pretty they are or how expensive they were at auction, high-maintenance and low-fertility cattle should be removed from the herd.

Among wild animals, including wildebeests, bison, caribou, musk oxen, and moose, individual members of the same species and gender are virtually indistinguishable physically. Yet when these same wild animals are placed in our domestic care, sheltered from natural selection, their physical characteristics begin to change as less fertile, higher-maintenance animals are allowed to survive. A single, tightly constrained set of physical characteristics (see below) defines the most efficient, most fertile physical form of a species in any given environment; deviations from this ideal are less able to cope with the environmental challenges experienced by the herd. Because animals adapt to the unique environmental challenges of an area, the physical characteristics of individuals from herds of the same species vary somewhat from one climate to the next.

In domestic cattle herds, even in purebred herds that are all the same color, there is no uniformity. Our selection criteria have been based as much on the personal biases, moods, and prejudices of breeders as on the fertility and maintenance efficiency of the cattle. Although certain breed-specific characteristics may influence heat and cold tolerance to some degree, the most fertile and efficient cattle all have the same physical characteristics. In domestic cattle, there is as much variation in fertility and maintenance efficiency within a breed as there is among breeds.

The Pitfalls of Single-Trait Selection

When selecting and breeding cattle, we tend to be biased in favor of traits that are of particular concern to us — such as high milk production, wide hips for easy calving, long back, height, high weaning weight, and narrow shoulders — and against traits that have been problematic in the past. This is called *single-*

trait selection, and it is a dangerous trap. The ideal selection characteristics of breeding animals must mesh in a complete, balanced package of low-maintenance, high-fertility characteristics.

Just as males and females are markedly different among wild animal species, our domestic cows should not look like bulls, and vice versa. The physical expression of high-fertility, low-maintenance characteristics is a *gender-specific package.* Although some individual characteristics may be shared, the complete package varies considerably from male to female because they have very different roles to fulfill within the herd. For example, broad hips are a feminine trait (as long as they are balanced within the complete package) so the cow can fulfill her gender-specific role of bearing offspring, but broad hips on a bull are not a masculine trait because they are not required for him to fulfill his gender-specific

MY EXPERIENCE WITH SINGLE-TRAIT SELECTION

When I took over the management of my family's farm, I tried to solve a high incidence of calving problems through what I now realize was single-trait selection among my replacement-bull purchases. I followed the advice of several breeders who recommended selecting bulls with narrow, smooth shoulders, thinking that the bulls' offspring would calve more easily. In fact, quite the opposite occurred. Calving problems increased and so did the bulls' aggressive behaviors. Narrow shoulders in a bull are not a masculine trait; they are a sign of low testosterone.

I also wanted to increase the quality of the calves. My intentions were good, but once again I made poor choices based on single-trait selection. I chose top-dollar bulls that showed exceptional daily rates of weight gain during feed trials at the bull-testing stations, thinking that

this would automatically translate to their calves. Instead, it caused a host of problems.

Young bulls should not be fed a high-grain diet to produce high gains during the growth stage in which they mature sexually. Doing so causes fat to be deposited in the testicles, which enhances scrotal measurements but prevents proper cooling of the sexual organs later in life, resulting in infertility. Moreover, because the less masculine, subfertile bull does not divert as much of his energy resources toward sexual maturation, he produces higher rates of daily gain than do masculine fertile bulls entering this developmental stage. And because feed trials evaluating the weight-gain potential of the young bulls typically coincide with the growth stage responsible for a bull's sexual maturation, my top-dollar young bulls did not mature

to become the most fertile bulls of the herd. Thus, selecting young bulls based on their weight gains during feed trials is a flawed approach to selecting a replacement herd sire.

The most fertile bulls do not have the largest body sizes and highest daily gains; consider how much bigger an ox is than a bull of the same age, for example. In retrospect, it is no surprise that my bulls' fertility deteriorated as they got older, even though they passed their semen evaluations at the bull sale.

Last, high daily gains on a grain diet do not reflect how efficient bulls and their offspring will be on grass. My bulls' performance on grass was terrible, costing a fortune in off-farm feed supplements (mostly grain) just to keep them in good condition — a highly undesirable trait to pass on to the remaining herd.

role of protecting the herd and competing with other males.

When we use single-trait selection, we do not address the various aspects of fertility and maintenance efficiency as a balanced package, nor do we take into consideration gender-specific differences. Such an approach eventually fails because some traits are magnified at the expense of others, while other traits are overlooked until they become a concern. The resulting physiological stress on the animal eventually leads to problems such as calving difficulties, nervous behaviors, lower conception rates, poor health, and lower weight gains, as well as a drop in profit.

In the dairy industry, for example, the focus of most brood-stock selection is high milk production. Consequently, the selection criterion used for most dairy bulls is high milk production among the bull's female ancestors, even though milk production is a feminine trait. Selecting bulls based only on a *single* feminine trait on the bull's maternal side while ignoring the masculine traits of the bull itself has led to a loss of masculinity — a feminization — among dairy bulls. This in turn has led to the bulls passing on high-maintenance characteristics to their offspring (the inevitable result of weakening genetics), which we can see expressed as the extraordinarily high nutritional requirements of dairy cows in much of the mainstream dairy industry. Additionally, it is common to see highly aggressive behavior among the bulls due to hormonal imbalances.

Instead of focusing on single traits, we must focus on characteristics that indicate low maintenance and high fertility, which are gender-specific, as we will see in Characteristics of the Ideal Breeding Animal, later in this chapter. By doing so, we select for the same criteria that are crucial to the survival and genetic continuity of animals living in the wild — two traits that are the only ones that truly affect our profitability.

Environment and Climate

Although fertility and maintenance-efficiency traits are the same among cattle all over the world, all cattle show some breed-specific variations that have resulted from adaptations to various environmental and climatic challenges, including temperature and humidity, disease and parasite loads, wind, soil and nutrient deficiencies, and light intensity and ultraviolet radiation.

Animals in hot areas must release a lot of heat from their extremities to maintain a normal core body temperature. That is why the ears of elephants are so large and full of blood vessels. Animals in hot regions are also less hairy than their cold-climate cousins; any hair they have tends not to have high insulating value, and they usually lack a woolly undercoat. Cattle living in a hot climate, especially if the climate is also humid, need to rid their bodies of excess heat. If body temperature is too high, normal body functions will be depressed, as if the cattle are in a state of persistent fever. The immune system will be compromised, weight gains will be reduced, health will deteriorate, and fertility will be questionable. Unless the cattle can adapt to the heat and regulate their body temperature, they will become poor producers. This heat-induced loss of productivity is called *tropical degeneration*.

Tropical cattle breeds such as Brahman and Zebu have evolved to be heat tolerant. They have big ears with significant surface area to expel heat and sleek hides with a very low insulating value. They are also more resistant to the parasites and diseases endemic to tropical areas, having been exposed to them longer, which has built up a natural immunity. Some of the British and continental breeds do well in hot regions, but it takes considerably more attention to select highly efficient, fertile animals that do not suffer from tropical degeneration, which typically becomes an issue among British breeds when the average yearly temperature exceeds 18°C/65°F. For this reason, cattle in tropical areas usually derive at least a portion of their genetic makeup from tropical breeds.

Cattle in colder climates must insulate themselves against cold temperatures, wind, and heat-robbing humidity. Breeds such as Galloway and Scottish Highland developed extremely long hair and a thick undercoat to protect against the cold wind and bone-chilling humidity typical of northern coastal regions. Other breeds, such as Simmental, evolved the ability to store thick layers of fat over their backs as insulation against the cold instead of relying on their hair coat. The long hair and woolly undercoat

Brahman

Zebu

Galloway

Simmental

seem to have evolved in the humid cold (such as the British Isles and coastal regions of northern Europe), whereas the ability to store thick layers of fat seems to have originated in dry cold areas (such as the inland, high-elevation areas of continental Europe, in particular the Alps). Both adaptations present severe challenges to these breeds if they are moved to a hot climate, but the hairy breeds seem to adapt better because they carry less fat on their backs and are able to shed their undercoats in hot weather.

Soil and nutrient deficiencies also play a big role in traits. It follows that more-fertile soils will grow more-nutrient-rich plants that result in fatter, healthier animals. Cattle moved from poor soils to more fertile ones will improve, but cattle moved from highly fertile to less fertile land will have trouble adjusting. Thus, when selecting breeding stock,

it is best to choose animals that come from conditions that are similar to or poorer than what they will experience on your farm.

Although typically only health and growth rates vary in response to different soil conditions, there is one notable exception: The frame size (bone structure) of Scottish Highland cattle varies dramatically based on the soil quality they grow up on. This difference is most clearly seen when comparing the tiny Kyloe Island variety and the much larger Scottish Highland cattle found on the more fertile mainland. Scottish Highland cattle are one of the few breeds that can maintain overall health by limiting frame size to compensate for lower nutrients. In poor soil conditions, other breeds may also have smaller frames, but unlike smaller Scottish Highland cattle, their health is compromised and they are poor producers.

Scottish Highland breed (left) and the Kyloe Island variety

Some cattle are better able than others to cope with high ultraviolet radiation and intense sunlight, which cause cattle's eyes to water, attracting dust that gets into the eye and causes severe eye irritation. Intense glare and irritation can ultimately lead to cancerous growths in the eye. Though cows can't put on sunglasses, nature evolved dark pigment around their eyes to minimize glare in regions where light intensity is high. This pigment functions much like the charcoal that the traditional Inuit peoples of the high Arctic rubbed under their eyes to reduce the snow's intense glare and to avoid snow blindness. (Susceptibility to cancerous growths in the eye, also known as cancer eye, is determined by pigmentation around the eyes, not by a genetic predisposition for the disease.) If cattle that don't have this pigmentation are raised in areas of intense glare or intense ultraviolet light, they will have an increased likelihood of developing cancer eye over cattle that have protective pigmentation. I have seen plenty of cancer eye among cattle whose eyes are surrounded by white hair but rarely among cattle with full pigmentation around the eyes. Eye irritation from intense sunlight and ultraviolet radiation also increases susceptibility to pinkeye (or infectious bovine keratoconjunctivitis, a highly contagious infectious bacterial disease of the eye); the added stress is often just enough to weaken the immune system and precipitate the disease.

Hair color plays a significant role in how much heat is absorbed or repelled, affecting the heat tolerance of the animal. In addition, dark skin pigmentation is more resistant than light skin pigmentation to damage (sunburn) from intense ultraviolet radiation. I remember frying eggs on black asphalt one hot summer day at my childhood home. In hot, sunny weather, can you imagine wearing a coat that's the same color as asphalt? Cattle with light hair colors — white to red hair (with dark skin pigmentation to prevent sunburn) — absorb less heat and flourish in areas of intense ultraviolet radiation and high temperatures. Cattle with darker hair colors fare better in shaded, forested, or overcast regions, where they are exposed to less-direct sunlight.

When you select animals for your herd, keep in mind the climatic challenges and conditions they will face while in your care. Will they need to keep warm through a long northern winter or survive the muggy heat or scorching desert of the south? What is the quality of the soil they will live on? Will they be exposed to strong, direct sunshine or bright, snowy weather?

Characteristics of the Ideal Breeding Animal

Among most wild species, there are marked physical differences between males and females. These physical characteristics are a direct result of the interaction between sex and growth hormones. As a result, physical appearance of genders is directly related to their gender-specific roles in the herd.

The Cow's Role

The female is responsible for producing offspring to maintain the continuity of the herd. To accomplish this, she must give birth to a live, vigorous calf that gets up quickly, suckles, and follows the herd almost instantly so it will not fall into the clutches of the wolf. The cow must maintain her physical health, fertility, and reproductive functions over the course of the year, which she accomplishes by storing fat on her back during seasons when food is plentiful and rationing it out during seasons when it is scarce and of lower quality. She must resist parasites, regulate her body temperature throughout the seasons, and graze efficiently in competition with the other cows in the herd.

She must be highly fertile and come into estrus during the brief breeding season. Delivery must be timed to coincide with the birthing season of nature's many wild prey animals. During this time, the overwhelming number of babies born reduces predation pressure on her young (safety in numbers). She must produce proportionately sized calves so she can give birth unassisted, and she must have good mothering instincts and produce quality milk for her calf.

The Bull's Role

The bull's function is to breed once a year and to pass on quality genes that produce maintenance-efficient, fertile daughters and sons. To achieve this, he must be large and strong enough to compete successfully with other males for females, but not so large that he risks collapsing young heifers under his weight during breeding.

He must be adept at displaying his physical power and at interpreting the other bulls' displays of aggression in order to minimize physical fights when the dominance hierarchy is being established. This helps the bull avoid being killed or injured in battle (which makes him a target for predators), and allows him to continue fulfilling his function as a breeder. In other words, the bull has to be a well-adjusted male, capable of walking the walk and talking the talk without fighting. Fighting excessively to prove strength is the sign of a subfertile bull.

A bull must keep his sexual organs safe from injury and regulate the temperature of his scrotum to ensure his fertility during the breeding season. He must resist parasites, regulate his body temperature, store fat on his back to survive the drought season, and compete for his share of food.

When we take this information into the pasture to look for breeding stock, the characteristics outlined on pages 26–27 are the ones that indicate high fertility and maintenance efficiency.

Replacement Heifers and Cows

A replacement heifer must be feminine in appearance, develop on schedule, and be balanced in build and proportion for efficiency in size, birthing, and ability to consume large volumes of grass. A feminine, low-maintenance, highly fertile cow or heifer calves unassisted, has good mothering instincts, and is an excellent milk producer. She produces a calf every year; her gestation period is a consistent

length each year; and she produces consistent, quality calves that yield quality carcasses. She is relaxed and resistant to stress, diseases, and parasites. She and her daughters should cycle for the first time at ten months and conceive at fourteen months. She will produce top-notch offspring each year from your grass *without* expensive feed supplements by supplementing herself during the drought season with the fat stored on her back. Finally, she will be long lived, thus minimizing the cost of expensive replacement heifers.

A high-maintenance, subfertile cow lacking femininity

A feminine, low-maintenance, highly fertile cow

The Bull, Your Herd Sire

Your herd sire should be a masculine bull with wide shoulders; a thick, short neck; a coarse head; and curly dark hair on the front and lower half of his body. He should be well muscled, compact, and balanced in his frame and build, and have large, well-formed sexual organs. Do not compromise. If you pick a bull that resembles a steer to be your herd sire, he'll be about as effective as a steer when it comes

Remember, all these points must come together to form a *balanced* package that allows the heifer and/or cow to fulfill her gender-specific role within the herd. Isolating any one of the following points would simply result in single-trait selection.

- *Feminine appearance.* A feminine appearance is a direct result of the interaction between her sex and growth hormones. When her hormonal balance is correct, she will grow up to be well proportioned according to her gender-specific role in the herd. She will have a slender build, and her front quarters and neck will be sleek and light, whereas her hind quarters will have depth (broad rump, lengthy between the hip bones and the pin bones) to fulfill her role of bearing young. Likewise, she will have a large stomach capacity, with the largest diameter of her body being the mid-rib area, both for feed efficiency and for bearing young.

- *Wide rump.* Think "child-bearing hips." A wide rump is an expression of femininity: It indicates high fertility, nutritional efficiency (the ability to store meat and fat easily), and adequate space for a calf to squeeze through during birth.

- *Well-formed udder.* The cow should have well-formed teats and a nicely shaped, hairless udder. The openings in the teats should be well formed — not too big to invite dirt and disease and not too small to restrict milk flow. Coke-bottle teats, pendulous udders, and other deformities are not acceptable: They are signs of disease and weakness. A heifer should have a tight, proportional, hairy udder with tiny teats. After she has cycled a few times, her udder will drop somewhat and lose its hair.

- *Deep heart girth.* This is the sign of a low-maintenance, calm animal. A deep heart girth adds weight to the carcass.

- *Deep flank.* A pinched flank is a sign of a stressed, high-maintenance cow. A pinched flank on a bull is less significant; it does not necessarily indicate a high-maintenance bull.

- *Feminine shoulders.* Shoulders should be lean, loose, and relatively lightly built, and should move freely. Yet they must be proportionate: Narrow, pinched shoulders are the sign of a high-maintenance cow. Wide, heavy shoulders are a masculine trait that indicates a deficiency in female hormones.

- *Well-proportioned neck.* A long or pinched neck creates a disproportionate animal. Think of balance: When balance is disturbed, the result is a stressed, nervous, high-maintenance animal.

- *Proportionate length.* Extra-long animals may seem appealing to the

time to breed your herd. Feminine traits in a bull will bring you grief.

A well-balanced, masculine bull is low maintenance, highly fertile with a high libido, and able to perform his job admirably. He will be able to breed on pasture without losing weight, will not be excessively aggressive, and will be able to assert his place in the social hierarchy without sustaining debilitating battle injuries. After the struggles for dominance are over, he will fulfill his role as a breeder. The dominant herd bull breeds 90 percent or more of his harem. The younger, less dominant bulls have to share the remainder.

The low-maintenance, highly fertile bull produces a uniform crop of calves having uniform birth weight similar to what his was. A masculine bull's daughters will mature early and his sons will yield top-notch carcasses. He will be ready to breed at 14 months — the same time yearling heifers are ready to breed for the first time. He will be disease- and parasite-resistant and will live efficiently off your grass without expensive feed supplements as long as your management system supports this. In contrast, a high-maintenance, subfertile bull has an inconsistent calf crop with widely ranging birth weights, infertile daughters that are prone to calving difficulties, and sons that are slow to mature, slow to finish (fatten for market), and costly to raise.

butcher (he can cut a longer loin), but in truth an extra-long animal is high maintenance and gains weight slowly, cutting into your profit margin.

- *Identical rump and shoulder length.* Balanced proportions are key.

- *Large gut capacity.* A large gut enables her to be an efficient grazer, gathering as much food as possible during a grazing session.

- *Wide mouth.* The larger her mouth, the more food she can stuff into it with each bite. It's all about efficiency on pasture: The grass won't be green forever, so she has to make the most of it while it's there.

- *Big nostrils* make for ease of breathing.

- *Naturally shaped back legs; not post-legged.* Legs are free of structural defects and show a solid posture. Tall, straight-hocked hind legs on a heifer will be your doom at calving time.

- *Compact build.* Tall, big-framed animals are out-of-balance, high-maintenance animals; contrary to popular coffee-shop wisdom, a large-framed animal does *not* calve more easily.

- *Dark hooves* seem to have a lower incidence of lameness.

- *Pigmented skin.* This is particularly important in areas of intense sunlight and/or intense ultraviolet radiation.

- *Short, thick, shiny hair.* Shiny hair is caused by the natural secretion of body oils in a healthy animal and is more parasite-repellent. Long, dull-looking hair is a sign of compromised health and a target for parasites.

- *Hair sheds out completely.* If hair is retained by an animal after the normal shedding season, it is a sign that the animal is either out of balance or has a mineral deficiency.

- *Dark hair around eyes.* This is particularly important in areas of intense sunlight and/or intense ultraviolet radiation.

- *Thick, loose hide.* This indicates a relaxed animal; a loose hide is more parasite-resistant.

- *Unassisted calving.* This is an absolute must; no excuses!

- *Cycling at 10 months and conception at 14 months.* A fertile heifer does not have to be 2 years old before she is bred.

- *Weight of 1,000 to 1,200 pounds.* This is ideal for a cow. A large-framed cow has a much higher maintenance cost but still produces only one calf per year. You want the highest return per acre, not per cow; it's all about efficiency (see A New Way of Thinking in the introduction).

A well-balanced, masculine, low-maintenance, highly fertile bull (left) and a high-maintenance, subfertile bull lacking in masculinity (right)

As with a replacement heifer, all these points must come together to form a balanced package that allows the bull to fulfill his gender-specific role within the herd. Isolating any one of the following amounts to single-trait selection.

- **Masculine appearance.** Think of the manly man on the front cover of a Harlequin romance novel. Like him, the bull should have wide shoulders, a deep chest, thick curly dark hair on the face and chest, thick skin, a wide jaw, and a big gut, and be of ideal size (not huge, not tiny). Yes, it's a stereotype, but I guarantee you will remember it when you go into the bull pasture.

- **Wide shoulders.** To compete with the other bulls in the herd and provide the herd protection, the bull must present a strong front and have the strength to back up that appearance. Thus, bull calves will develop broad, masculine shoulders as they mature as a direct result of a healthy interaction between their sex and growth hormones.

- **Deep heart girth.** This is the sign of a vigorous, low-maintenance, calm animal. Deep heart girth adds weight to the carcass.

- **Deep, wide chest.** Again, this is the sign of a low-maintenance, calm animal. The chest cavity must be large to contain a large (strong) heart and large (strong) lungs, and provide him with a powerful stance with which to face predators and other competing bulls.

- **Large gut capacity.** This allows him to maximize his grazing. A larger gut allows more grass to fit into his rumen in a single day.

- **Short, heavy neck.** Unlike the cow (whose neck should be proportionate), the bull's neck should be short. A short, muscular neck is more powerful than a longer one in a fight and is less prone to injury. It is the product of a healthy interaction between sex and growth hormones during the bull's development, and as such is a good indication the bull will have a healthy libido as an adult.

- **Coarse head.** Because a bull's role is to be powerful in battle and display that power to avoid fights, a powerful, coarse head is part of his complete package. A deficiency in male sex hormones during his growth stage will result in a finer-boned, sleeker (more feminine-looking) animal.

- **Well-fleshed, muscular rump.** The rump indicates the bull's ability to store red meat. An excessively large rump is out of balance, as is a scrawny one. The rump is proportionate to his shoulders, but not equal to the shoulders in size. Remember, the powerhouse of the bull is in his front end, whereas the cow's rump is larger than her shoulders for her to fulfill her role in nature; think of the male ste-reotype of wide shoulders, narrow hips, muscular legs, and a muscular behind.

- **Proportionate length.** Extra-long animals seem attractive because of the longer loin, but if the bull is not proportionate, he will be out of balance: a high-maintenance animal that is slow to finish.

- **Naturally shaped back legs; not post-legged.** Legs are free of structural defects and show a solid posture. Tall, straight-hocked hind legs will be passed on to daughters, which in turn will have difficulty calving unassisted.

- **Compact muscular build.** Tall, long-legged, big-framed animals are out of balance and require too much food.

- **Good muscle definition.** A bull should not have the shapeless bulk typical of a steer's muscling. A bull with excess fat typically has a male hormone deficiency.

- **Pigmented skin.** This is particularly important in areas of intense sunlight and/or intense ultraviolet radiation.

- **Dark hooves** seem to have a lower incidence of lameness.

- **Thick, loose hide.** This indicates a relaxed animal; a loose hide is more parasite-resistant.

- *Wide mouth.* The larger his mouth, the more food he can stuff into it with each bite. It's all about efficiency on pasture.

- *Big nostrils* make for ease of breathing

- *Heavy lower jaw.* A strong jaw is less likely to be injured in battle. (See the description for a coarse head.)

- *Dark hair around eyes.* This is particularly important in areas of intense sunlight and/or intense ultraviolet radiation.

- *Short, thick, shiny hair.* It's parasite-repellent and an indicator of good health. Long, dull-looking hair is a sign of compromised health and a target for parasites.

- *Hair sheds out completely.* Hair retained after normal shedding is a sign of an unbalanced physical form or a mineral deficiency.

- *Coarse, curly hair on the neck, face, and lower half of the body.* The tighter the curls, the higher the fertility. A subfertile bull has wavy hair on his neck, face, and underside; an infertile bull's hair is straight.

- *Darker hair on the lower half of the body, neck, and face.* This is caused by an oil secreted due to a high presence of male hormones.

- *Minimum 36-centimeter scrotal circumference at 12 months of age.* Be sure that fat from a high-grain diet doesn't contribute to scrotal size, which does matter for a bull, and in this case, bigger is better. The bull's teats should never be developed. Enlarged teats are a sign of a hormonal imbalance and a lack of libido.

- *Scrotum hangs straight down without a twist.* If the scrotum is tucked too close to the body, the sperm will overheat; if it's dangling like a bell, it will be injured. A twisted scrotum is out of balance.

- *Both testicles the same size.* This is an absolute must for high fertility in a bull. *Hypoplasia* (underdevelopment or incompleteness) of one testis indicates a deficiency in male sex hormones. The underdeveloped testicle will eventually become sterile.

- *Scrotum covered with very fine sparse hair and having a buckskin look.* Buckskin skin pigmentation on the scrotum makes it more resistant to sunburn, warts, and infections. The fine, sparse hair has no insulating value, which helps keep the scrotum cool.

- *Tight sheath.* An enlarged or floppy sheath is either the sign of a past injury or an invitation to disease. An enlarged sheath typically goes hand-in-hand with weak sphincter muscles, which operate the sheath opening, thus allowing the bull's prepuce to protrude — and dirt and bacteria to enter. An excessively small sheath or sheath opening will impair the bull's reproductive functions. Remember, the healthy development of the reproductive organs is a direct result of the bull's hormonal balance and fertility.

- *Masculine appearance, with a fully developed scrotum by 12 months of age and able to breed comfortably at 14 months.* This is the same age at which young heifers are ready to be bred for the first time.

- *Weight of 1,200 to 1,500 pounds.* This is the ideal size for a mature bull. As is true of bison, the male is bigger than the female, but is not enormous. The male hormones of a masculine bull will halt his growth once he reaches his ideal weight. Enormous oxen grow to that size because they don't reach sexual maturity and lack adequate hormones to regulate their growth.

Culling Animals: Training Your Predatory Eye

In the wild, high-maintenance and low-fertility animals all suffer a similar fate. Certainly, a weak animal is targeted first by predator, disease, and parasite, but the signs of high maintenance and poor fertility are much subtler. Predators — and we — must train our eye to determine which animals we should target and which we should cull.

High-maintenance cows are not as well nourished and have less strength to push their way to the front or center of the herd, which puts them last in line for access to food and water. They are the first to have calving difficulties or to abandon their young, and their calves are weaker, more sickly, or runty. They are always the last animals through the gate and the last to cross a river after the main herd has churned up treacherous mud along the riverbank. In the wild, these weaker cows, and especially their calves, are the stragglers that become the unfortunate objects of prey. Barren females are also pushed to the herd's periphery by the other cows because they don't fit in. They are essentially treated as sacrificial bait by the other cows, who use them as decoys to keep predators away from the young in the midst of the herd.

The subfertile or high-maintenance bull has a similar fate. The other bulls will not allow a weak bull to breed. He will be beaten and chased out, left to face predators alone without the protection of the herd. Because he does not have the stamina or the dominance to take his place as herd sire, he will not succeed in bringing his genetics to the group.

Individuals in domestic cattle herds display the same subtle signs of weakness that tell the predator which animal to select as its prey. We must train ourselves to recognize these subtle signs of weakness, infertility, and maintenance inefficiency. One of the best ways to do this is to make a game of it every time we are around cattle and other wildlife. Doing this will help you note the weak members of the herd during the stressful times of the year: drought, winter, parasite season, breeding, calving, herd moves, and times of competition for resources, among others. By recording ear-tag numbers of animals that show signs of weakness, you can identify them again at a later culling date, even if good weather and lush grass have masked the obvious signs of a high-maintenance or subfertile animal.

Infertile, high-maintenance cows are pushed to the periphery of the herd.

As you train yourself to see like a predator, ask yourself the following questions.

- What do the mothers of your best, most vibrant, healthiest calves look like? You probably already know which mothers are your best cows, producing top-notch calves year after year. Do they fit your mental image of a highly fertile, maintenance-efficient animal? What about the mothers of the runty calves? Do they have similarities?

- What do the cows that have calving difficulties look like? Carefully survey their physical characteristics before removing them from the herd. Knowing how to recognize what you don't want is an extremely powerful skill.

- What do the cows and heifers that failed to conceive look like? What do the late calvers look like?

- What do the cows and bulls look like that are performing poorly during drought, winter, or harsh weather?

- Which animals are always covered by flies and other parasites?

- Which animals are slow to shed their winter coat or fail to shed completely?

- What do the nervous animals look like?

- Which animals are submissive? Which are too aggressive?

- Which animals are slow to fatten up on the lush spring grass?

- Which are the bulls whose calves experienced difficulties at birth? What do they look like?

- Which are the bulls that have poor semen tests and are overly aggressive, meek, in poor condition, or under the weather?

Weak animals are the first to get sick, and a sick animal is the best teacher in identifying the physical traits of a low-fertility, high-maintenance animal.

Breeding for Your Target Market

It is important to know what market your livestock is destined for when selecting animals and designing a breeding program. Is it a feedlot that simply buys calves to fit into various weight classes in a commodity market, or is it a direct market for finished beef animals? How many months of the year do you need to supply meat? Is the meat sold by the side, the cut, or the pound? What kind of meats do your customers want? The end product determines what frame size you will look for when selecting animals, which breeds are easiest to market, how long your breeding season will be, and what kind of breeding program you will choose. (For detailed information on determining your target market, see chapters 15–19.)

Knowing how to recognize what you don't want is an extremely powerful skill.

Crossbreeding has become popular due to the hybrid vigor of the offspring. (*Hybrid vigor* is the increased performance associated with a combination of two dissimilar parents, most noticeable in traits such as fertility, growth, and survivability.) If your calves are destined for a commodity market, the quick growth and extra pounds will be welcomed by your buyers. To get true hybrid vigor, however, the crossbred animal must be a *terminal cross* — that is, both parents are purebred and the crossbred offspring must be used only as beef and never for breeding. If crossbred animals are rebred, the physical appearance of their offspring is highly unpredictable, ranging from purebred (resembling any one of their original purebred grandparents) to somewhere between the purebred ends of the spectrum. The physical size and hybrid vigor seen during the initial crossbreeding are not passed on to subsequent generations of crossbred calves.

A terminal crossbreeding program can get complicated, especially if you intend to produce your own replacement heifers. It necessitates maintaining several separate herds, each requiring individual management and infrastructure. The impact of several small herds will be significantly less beneficial to the land than a single, combined herd. The additional complications, stress, and costs of such a program are serious drawbacks to a profitable and ecologically sustainable beef production enterprise, so I do not recommend this approach. In the long run, focusing on maintenance and fertility in your replacement animals will have the greatest impact on your profitability, no matter what breeds your replacement heifers and bulls come from.

If you grass-finish your beef animals for a specific direct market, your focus will differ from other farmers selling calves or yearlings by the pound at auction. You will want predictable tenderness, flavor, carcass weight, and cut size; predictable finishing weights and grass-finishing time periods for ease of management and direct marketing; and predictable income based on a predictable volume, time frame, and meat quality. Your customers will want a product whose quality and delivery date are reliable. Unless you use true terminal crossbreeding, the variable flavor, tenderness, carcass size, and finishing dates of crossbred cattle will make your breeding program more challenging, with higher costs for direct marketing. Selecting breeding stock from the same breed or from breeds with similar frame size, carcass weight, climate adaptations, growth rates, fat distribution, and muscle fibers — for example, Red Angus, Black Angus, and Galloway — produces the most predictable calf crop for your direct-marketing program.

Purebred cattle have much less variation in frame size, growth rate, flavor, and tenderness from one calf to the next than do crossbred offspring. All the calves finish at roughly the same carcass weight and in roughly the same time frame. The uniformity of the cuts simplifies marketing and results in a more predictable product. It is important to note that by *purebred* I mean that the herd is composed of a single breed. A registered-purebred breeding program, which consists of purebred cattle with known pedigrees (bred according to those pedigrees) and whose records are strictly logged and registered with a purebred breed association, is something entirely different. Selecting breeding stock by pedigree instead of by their individual fertility and maintenance char-

acteristics is analogous to single-trait selection, or even worse. It is also extremely expensive and time consuming to deal with the paperwork, herd segregation, record keeping, and other hoopla required in the registered-livestock business. Registered-purebred breeding is an entirely different business from the low-cost, high-profit, grass-based beef business that I describe in this book.

Registered-Purebred Breeding

Registered-purebred breeding evolved in the eighteenth century as a by-product of the Industrial Revolution, which created a European upper-class aristocracy that had attained wealth but did not share the "pure" bloodlines of nobility. Instead, they created their own noble pedigrees in show-ring live-stock competitions (using their dogs, cats, horses, and cattle) that focused on physical perfection and parental lineages. This resulted in a disturbing amount of intentional and unintentional inbreeding and a willingness to overlook the maintenance and fertility characteristics of individual animals in the interest of creating a particular bloodline pedigree or to achieve a fashionable show-ring look. This selection process had little or nothing to do with an animal's performance on grass in a serious meat-producing business.

Think of your cattle as employees that you hire to turn grass into beef.

Though much has changed since then, many of the selection criteria in the registered-breeding business continue to be based on single-trait selection to emphasize trendy physical traits for the show ring. The bias toward parental lineages also continues to overshadow the importance of maintenance and fertility characteristics in individual animals. The countless health problems of the purebred-dog-breeding industry, ranging from the more common cases of hip dysplasia to rare nervous system disorders such as narcolepsy, should be a warning to us. Many registered purebred cattle are on the same dangerous path, peppered with countless health issues, calving problems, caesareans, and unnecessary calving-related deaths, all because we have bred cattle that never should have been allowed to breed.

These flawed selection criteria are not restricted to the purebred industry. Sadly, contributing to the problem is the fact that selection criteria established by the purebred industry for the show ring are frequently adopted by the mainstream beef industry when selecting replacement animals for the beef market. The purebred industry's role as a supplier of breeding stock (especially bulls) to commercial producers, along with their "registered" status, has given their show-ring-oriented selection criteria an unquestioned, and often undeserved, legitimacy in the commercial beef industry.

I remember one reputable registered-purebred breeder who proudly showed me the "best" cow in the herd: "We have to assist her calving every year and she needed an operation when she was young because of a deformity in her reproductive tract, but she's the most expensive heifer we have ever bought. She's a direct descendant of the famous . . . lineage; she's so pretty and she produces our biggest calves!" I truly felt sorry for this breeder's cattle. This is not to say that all breeders of registered-purebred cattle breed ticking time bombs — many of them have excellent cattle — but it is important to be wary of breeders who emphasize single-trait selection or focus on glowing descriptions of parental lineages and show-ring values.

Think of your cattle as employees that you hire to turn grass into beef. Would you hire an employee to work on your farm based on his or her parents' aristocratic lineage or his or her pretty face without ever asking to see a résumé of qualifications and experiences?

Linebreeding versus Inbreeding

Linebreeding and inbreeding both involve the crossing of two closely related individuals, but the outcome is different with each process. When two unrelated animals are bred, their offspring "choose" the best genes from their parents' gene pools. When two closely related cattle are bred, however, their gene pools carry many of the same traits, resulting in a much smaller variation in genetics from which the offspring can choose. As a result, traits common to both tend to be magnified, and recessive genes common to both may be expressed in the offspring, even if neither parent shows any physical signs of them. If both parents are genetically sound, their offspring will be even better off. Good traits are easily magnified. When the outcome in offspring is positive, we call the process *linebreeding,* a common practice used by breeders of purebred cattle, horses, and dogs. If the parents have poor genetics or carry negative recessive traits, these characteristics can also be expressed and magnified. This is called *inbreeding.*

Producing Your Own Bulls

Grass-based natural beef producers are able to produce their own heifers and bulls. No bull is better adapted to your environment and to your grass-based management program than one that was raised on your land and exhibits low-maintenance and high-fertility characteristics. With a rigorous genetic-selection program to cull less fertile, high-maintenance animals from the herd, some purebred linebreeders have managed to successfully breed their bulls for many generations without introducing new off-farm genetics. Thus, they have produced extremely consistent, exceedingly fertile, efficient bulls and replacement heifers.

In nature, large herds have so many herd sires that inbreeding doesn't happen very often. Because only the strongest genetics are capable of surviving nature's many dangers, the herd is kept sound enough genetically so that most closely related pairings result in linebreeding that improves the herd. A very large cattle herd with many herd sires could be managed relatively safely in a similar manner, but *only* if the breeding-stock selection and culling are extremely rigorous to remove genetically poor animals. If you keep only the very best bulls and heifers, any closely related pairings should contribute positively to the herd's genetics.

Much smaller wild herds can also survive virtually isolated from outside genetics. For example, musk-ox herds in the high Arctic are typically composed of closed family groups of twenty-five to fifty animals per herd. These little herds are very closely related, but the rigors of natural selection (through cold, wolf attacks, and hunger) make their genetics extremely resistant to the effects of inbreeding. In effect, they show a very high incidence of linebreeding — it is only the outcome that separates the two. Periodically, a lone bull from another herd joins the herd and introduces new genetics, preventing the gene pool from becoming stagnant.

With a rigorous culling program to keep your herd genetics clean and a small amount of extra work and record keeping, you should be able to produce bulls in small domestic cattle herds without the ill effects of inbreeding. Here's how.

1. Identify your very best older cows — the ones that have been consistently producing the best calves without any calving difficulties or health problems.

2. Breed these few best cows separately to your best bull after he has completed his full breeding term in the main herd (usually a maximum of three years to avoid inbreeding), and choose your potential herd sires from among their male offspring. To breed them separately, separate them from the main herd for the brief duration of the breeding season or an even shorter period of time (ideally no more than twenty-one days if they are very fertile; see the section on breeding in chapter 3). As soon as you see a cow being successfully bred, she can rejoin the main herd.

3. Breed your top-producing cows separately from the main herd year after year. If you keep a bull calf from one of these top-producing cows, sell any heifer calves by the same mother and father to avoid ending up with full sisters in the main herd. Likewise, if you keep a heifer calf from one of your top-producing cows, don't keep any of that cow's bull calves with the same father. If you frequently change sires among your top-producing cows, these losses will be minimal. (If you change yearly, there will be no possible full sisters; if you change every two years, there will be one possible full sister; etc.)

Breeding top cows separately prevents mother–son breeding in the main herd. If your top-quality bull is related to any of the top-producing cows that you breed separately, his mother can be moved back into the main herd for that breeding season. An off-farm bull should be introduced into the herd periodically to bring genetic variety to the herd.

The Cattle Year on Grass

Every aspect of a wild herd's yearly life cycle is synchronized to nature's seasonal rhythm. Maintaining this careful balance allows these animals to thrive without feed mills, fertilizer plants, and calving barns. Nature, therefore, is the perfect model for a cattle-management program. Working in concert with the seasons allows cattle to live profitably off the land year-round without great off-farm input and supplemental feed expenditures.

To achieve a natural system of management, we must study a year in the life of a grassland cow to understand her nutritional needs over time and the various demands of a growing calf's developmental stages. In this chapter, we will examine the cow's digestive process and learn how cattle metabolism responds to the seasons. We will also learn how to match the cattle's needs with the changing supply and quality of grass over the course of the year. Studying the once thriving wild herds of bison, wildebeests, deer, moose, and caribou, whose yearly life cycles were in sync with nature's rigorous seasons, is key in this process.

The Calving Season and Predator Exposure

Calving season in the wild is a giant all-you-can-eat smorgasbord for predators. Females are vulnerable when pregnant or with newborn offspring underfoot and young prey animals are easy to catch. To reduce their individual vulnerability, all prey animals give birth at the same time. This enormous flood of newborns directly follows the beginning of the spring grass flush or the return of rains that revive the grasses in tropical regions. Because the calving seasons of the various prey animals overlap, newborns are so abundant that predators simply cannot eat all of them. Therefore, an individual calf's risk of predation is significantly reduced. Any animal that calves before or after this brief window of opportunity becomes a target for the hungry predator.

If cows calve before the green grass returns, when food is still scarce, calves will be targeted heavily by predators. But when the calving season is in sync with the birthing season of mice, deer, moose, ducks, and rabbits, the calves — guarded as they are by their protective 1,200-pound mothers in close proximity to humans — no longer look very attractive to the hungry wolf or coyote. There is a vast abundance of prey that are easier to catch. By the time the predator's food supply becomes scarce, the calves are already quite big, strong, and fast and are far less attractive than many other wild prey species.

Incidentally, this smorgasbord of newborns in late spring also dictates when predators give birth: Their young, which are usually quite uncoordinated and incapable of hunting and traveling with their parents until they are several months old, are born several months *before* the green grass arrives so they are strong enough to begin their hunting lessons when the greatest abundance of newborn prey animals is available.

The Calf: Birth to Sexual Development

Calves born on lush green grass during warm weather in late spring and early summer are not as shocked by the change in temperature when they emerge from the warmth of the cow's uterus as they would be in winter. (If nature intended winter births, calves would be born with much more and thicker hair rather than a light hair coat appropriate to the summer weather.) Likewise, when the weather and grass conditions are most favorable, they have the best chance to get up, become strong, and quickly join the herd without the hindrance of dust, mud, drought, cold, or snow. The clean grass of this time does not expose the calf's delicate navel to risk of infection, and bacteria and diseases have the least opportunity to attack the animal's immune system. Conditions ensure that the calf's footing is secure if it has to outrun predators and provide plenty of tall grass to hide behind during the first vulnerable days of the calf's life.

A calf must respond quickly to its mother's promptings to stand, so it can bond with her, drink its first colostrum, and quickly gain sufficient strength to follow the herd. Remaining within the safe confines of the herd is imperative to a calf's survival, but the herd is constantly moving. The mother cannot afford to protect a weak calf for any great length of time; and to be left behind means certain death.

By calving at the time of year when grass is lush and nutrient-rich, the cow ensures that her colostrum is the best possible quality. *Colostrum* provides a concentrated dose of essential nutrients and energy required by the calf to recover from the ordeals of birth, to activate the calf's digestive processes, and to gain strength quickly. It also transfers a cow's disease immunity to the calf so the calf will be protected until its own immune system is strong enough to take over in the next few months.

Colostrum production lasts only a brief time; within twenty-four hours, the udder stops making it and starts producing milk. Warm weather helps ensure that the calf can take full advantage of the short colostrum-production window. Without it, the calf will grow up to be a runt, if disease, starvation, or a wolf doesn't kill it first.

A calf responds to its mother's promptings to stand.

Rumination and the Immune System

The newborn calf lives exclusively on its mother's milk during the first weeks of life. Within a month after birth, the calf's rumen (first stomach chamber) becomes active and the calf begins to supplement its mother's milk with its first mouthfuls of grass. As luck would have it, the second flush of grass growth, which typically follows the most intense summer heat, has just started by this time, so there is plenty of tender grass available to stimulate the calf's small delicate mouth and rumen. By early fall, the calf will be strong enough to deal with the dry, coarse left-over grass that it will have to survive on through the upcoming winter or drought season. The precise timing of these developmental milestones works to the calf's advantage. As opposed to the early-spring or late-winter calving customary in the conventional beef industry, it is best to match calving to this natural schedule.

When the calf is three months old, the immunity offered by colostrum wears off and the calf's own immune system kicks in. Luckily, the weather is still warm and dry and there is plenty of good grass; the winter or drought season has not yet begun so the calf easily makes this transition.

The calf is growing vigorously at this age and its cells are multiplying rapidly. Although the calf is not yet gaining fat, it is forming empty fat cells *within* the fibers of its muscles, which will fill in later during its final growth phase at the end of adolescence (known as the *finishing* stage because it is the last stage of bringing an animal to its desired fat level and slaughter weight for market). The calf reaches 45 percent of its mature body weight just as the grass becomes dormant at the onset of winter (or at the beginning of the drought season). This marks the beginning of the calf's sexual development; it is becoming a teenager. If the calf grows too fast during this stage, fat cells will form in areas of the body where they will harm sexual development. In the wild, the onset of winter and a scarce food supply prevent the calf from gaining weight too rapidly during this growth stage.

The calf is still suckling its mother's milk at this time, and it will not be weaned until winter is almost over, though the exact time when a cow kicks away her calf depends on how much of her fat reserves has been depleted during winter.

Bypass Protein: The Advantages of Suckling

When a calf sucks milk from its mother's teats with its head held in that funny upward kink that allows it to reach her udder, a wonderful thing happens: The combination of sucking, the position of the calf's head, and the stimulation of the warm milk causes the esophageal groove to form in the calf's digestive system, which allows the milk to bypass

the rumen and go directly to the second stomach chamber in a process called *esophageal groove closure*. The rumen contains bacteria that ferment and break down the tough cellulose structure of grass. But these bacteria feed on proteins — including milk protein — that enter the rumen. Only protein excesses and the proteins released by the rumen bacteria after they die are passed on to the second stomach chamber, where they become available to the cow. Suckling allows milk to bypass these bacteria so that all of the milk's protein is available to the calf. (When milk is fed to a calf from a bucket, because the calf does not receive the full combination of stimuli, full esophageal closure often does not occur. Consequently, a large percentage of the milk's protein is lost to the rumen bacteria.

By the time winter arrives, the cow's milk production has dropped off considerably. The little milk still available to the suckling calf allows just enough milk protein to bypass the rumen so the calf can continue growing and developing sexually during the winter season. Without this extra burst of protein, a calf's sexual development will be compromised. The male and female calves' developmental paths diverge during this time period.

When a calf suckles, milk bypasses the rumen.

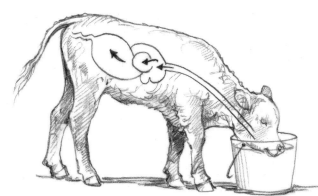

When a calf drinks milk from a bucket, milk protein is lost to rumen bacteria.

The Heifer Calf: Sexual Development to Calving

During the sexual development stage, the heifer's future fertility and reproductive functions are determined. The hormones responsible for the development of her ovaries and mammary tissue are soluble in fat. If the heifer gains weight too quickly (such as on a high-grain diet in a feedlot), she will continue to form fat cells throughout her sexual development, which will absorb some of these important hormones, preventing them from reaching their intended destination in the ovaries and mammary tissue. As such, fewer DNA-containing cells with the genes responsible for fertility and milk production will be formed. Although the heifer must continue to gain enough weight to guarantee healthy sexual development, the more fat cells she develops during this stage, the less fertile she will be and the less milk she will produce for her calves. A delicate balance must be struck between healthy and unhealthy weight gain.

This stage of a heifer's sexual development ends when she reaches 65 percent of her mature body weight, ideally with the arrival of spring. This is also when she has her first estrus cycle, at approximately 10 months of age. Her mammary tissues are completely formed by now, so further fat development will not affect her future milk production. As grass becomes plentiful again, the heifer starts filling all the fat cells within her muscle fibers, which she formed earlier in her life. She accumulates as much fat as possible during the summer months to help her through the next winter. At the end of summer, she becomes fertile in time for her first breeding season, in the fourteenth month of her life. The fertile, low-maintenance heifer discussed in chapter 2 conceives at this time and begins her first pregnancy. The typical term of a cow or heifer's gestation is 280 days (slightly longer if you are calving out of sync with nature's seasons, as is discussed in Calving to Breeding on page 40).

The calf growing inside her will not require a significant portion of her nutrients until it reaches its last trimester. She survives winter by reabsorbing the previous summer's fat reserves to supplement the season's scarce food supply. In the final three months of her pregnancy, her nutritional requirements increase significantly. Because calving season doesn't start until after the new grass begins to grow, she has enough time to replenish her fat reserves and finish her pregnancy on a nutrient-rich diet.

Fat Reserves for the Drought Season

A grazing animal's fat cells exist not just to keep the animal warm during winter or to make a predator's meal tastier, but also to serve as cold cellars and grain bins serve humans: They are storage facilities to help ration last year's harvest until next season's harvest begins. All animals in the wild survive the winter or drought season by reabsorbing the fat inside their fat cells to supplement their lean diet. Although a cow can still find some grass baking under the sun or lying brown and frozen under the snow, without her fat reserves she would not be able to meet all of her nutritional requirements and would abort her calf to ensure her survival.

When we calve out of season, we have to supplement our cattle's nutrition during winter to prepare the cow for calving. Without expensive feed supplementation, the calf's development and the cow's fertility would be compromised. These extra food supplements require money, fuel, time, and labor to grow, harvest, and feed, thereby seriously compromising our profit margin. When our calving season corresponds to summer grass growth, the cow can comfortably supplement her winter grass intake with her fat reserves, allowing her body weight to fluctuate by as much as 150 pounds or even more over the course of the year, because her nutrient requirements will not peak until after the end of winter. In this way, she still has enough time on lush spring grass to refill her fat cells and regain the body condition required to maintain her fertility *before* her calf is born. Certainly, it takes a lot of grass to fatten an animal every year, but because grass is usually in excess during the spring flush and the cow does all the harvesting for you, the extra grass is essentially free.

During the scarcity of the winter season, the cow's body slows its metabolic rate in an effort to conserve energy and reduce nutritional demands to ensure survival. When green grass finally arrives in spring, her body still has a very slow metabolic rate, and it takes some time for her metabolism to speed up again. Until then, her body is able to gain weight with tremendous efficiency. This *compensatory gain* allows animals to quickly make up for food shortages and even catch up to those that didn't experience such a shortage.

Compensatory gain helps to ensure a cow's fertility for next season by fattening her up quickly before lactation begins. Once her calf is born, the high demands of milk production prevent her from gaining any more weight, which is why the calving season has to be timed to occur *after* the grass greens up, not with the beginning of the spring grass flush.

Compensatory gain also allows growing cattle to use fewer calories to reach the same finished carcass weight in the same time frame as growing cattle whose metabolism remained unchanged during their entire growth due to access to a constant feed supply. Thus, you can winter calves very cheaply and still make up for it the following summer. The cows' fat reserves ensure that calves, with the help of their mothers' milk, can continue gaining sufficiently for their sexual development.

Incidentally, compensatory gain is the reason why so many human diet programs don't work. When the food supply is rationed, our bodies compensate by slowing our metabolic rate, following a natural survival mechanism. Once we have lost the desired weight and stop rationing, we experience compensatory gain until our metabolism catches up; in the meantime, we regain all the weight.

Good Body Condition and Fertility

Fertility is strongly influenced by the cow's body condition (called the *body condition score*, or BCS) at the time of calving, *not* at the time of breeding. The following list illustrates how a cow's body condition score at the time of calving influences her probability of conception at the time of breeding, provided that all other variables in this example (that is, health, genetics, access to a balanced mineral supplement, etc.) remain consistent.

- A BCS of 7 at calving = a 90 percent or greater probability of conception at breeding.

- A BCS of 6 at calving = a 75 percent probability of conception at breeding.

- A BCS of 5 at calving = a 50 percent probability of conception at breeding.

- A BCS of 4 at calving = a 25 to 30 percent probability of conception at breeding.

- A BCS of 7 or greater = a decline in the cow's fertility, as her obesity impedes her reproductive functions.

A cow's reproductive system can recover quickly after calving only if her fat reserves are replenished *before* calving. If she has to replenish them after the high demands of lactation begin, she will not have enough nutritional stores left to rebuild her reproductive system in time for her next pregnancy. The nutritional demands of the current calf take priority over a future pregnancy if the cow's body is not fat enough for both at calving.

If a long winter or late rains prevent a cow's body condition from recovering fully before calving, good nutrition after calving will not speed the recovery of her reproductive system in time for the upcoming breeding season, even if she is able to recover all of her body condition in time for breeding. She simply cannot absorb enough nutrients, even on the best possible diet, to simultaneously replenish her fat cells, produce milk, and prepare her reproductive tract for breeding. In such a case, fertility is always priority number three. This is why the time interval between the season's first grass growth and the onset of calving is so important; it gives a cow a chance to replenish her fat cells prior to calving to ensure her fertility in the upcoming breeding season.

If a food shortage prevents a cow from regaining her full body condition before calving, the high demands of lactation will jeopardize her survival in the coming winter if she becomes pregnant again.

Nature compensates for this risk by reducing her fertility until after she has fulfilled her responsibilities to her current calf. If her survival continues to be threatened during the next winter, her milk supply will dry up sooner to conserve nutrition and energy for herself at the expense of her current calf.

Body Condition Scoring

The BCS is a standard by which we can monitor our cows' body conditions as they fluctuate over the course of the year, allowing us to maximize our cattle's compensatory gains, minimize their feed supplements during winter, and still know when to intervene if fluctuating body condition threatens to affect fertility levels in the next breeding season. The nine-point BCS is summarized in the following chart.

A cow has ideal fat cover when her BCS fluctuates between 5 and 7 during the year. Cows should regain a BCS of 7 just prior to calving to ensure maximum fertility at breeding, a measure she can safely regain if she does not drop below a BCS of 5 during winter and if we assume there is enough time between the beginning of the spring grass flush and calving. A BCS of 4 is the borderline between good health and malnutrition. If she drops below that, she will begin to damage body tissues, which heal much more slowly than she can replenish her fat cells. She will be unlikely to return to a BCS of 7 prior to calving, but even if she does, her fertility will be compromised during the next breeding season because it takes time to heal malnourished body tissues after extreme weight loss.

By monitoring BCS throughout the year, we are not surprised by the condition of our cattle at breeding time. If the cattle slim down too rapidly and we can predict that they will reach a borderline BCS before spring grass returns, we can intervene by weaning early, increasing their access to pasture, administering feeding supplements to compensate for protein or energy deficiencies in the grass supply, and avoiding pastures that have a low nutrient value. We can also separately supplement higher-maintenance animals (such as first-calf heifers) and cull high-maintenance cows and heifers that lose body condition more quickly than the rest of the herd or record their numbers for culling at a later date.

BODY CONDITION SCORING

	BCS	Description	
Severely emaciated to thin	1	The bones, ribs, and spine of the cow are visible, and there is no fat anywhere on the body. The muscles are atrophied, and the cow is physically weak and has difficulty walking and standing.	
	2	The cow has all of the above characteristics but she can still stand and walk.	
	3	Similar to a BCS of 1 or 2, but muscles are visible and not atrophied to the same degree. Ribs and spine protrude; there is no fat on the body (see illustration). A BCS of 3 represents approximately 13% body fat. (An average-sized cow with a BCS of 3 is approximately 350 pounds lighter than the same cow with a BCS of 7.)

Body Condition Score of 3 |

	BCS	Description	
Borderline	4	The bones and three to five ribs are visible, but the spine is less pronounced. Individual muscles in the hindquarters are visible; there is no visible fat. A BCS of 4 represents approximately 18% body fat. An average-sized cow with a BCS of 4 is about 270 pounds lighter than a cow with a BCS of 7.	
Optimal: Moderate to very good	5	Slightly fatter than a cow with a BCS of 4. Only one or two ribs are visible; hindquarter muscles aren't distinguishable. A small amount of fat cover can be felt over her pin bones, but there is no fat in her brisket (see illustration). A BCS of 5 represents approximately 22% body fat. (An average-sized cow with a BCS of 5 is approximately 180 pounds lighter than the same cow with a BCS of 7.)	*Body Condition Score of 5*
	6	The cow begins to show some fat in her brisket and flanks, and her ribs are no longer visible. Her pin bones and hips are still visible, but her body has a smooth appearance. A BCS of 6 represents approximately 26.5% body fat, which is approximately 90 pounds lighter than the same cow with a BCS of 7.	
	7	The cow's brisket and flanks are full of fat, and the indent of the spine is visible on the cow's back, which is square with fat. The outline of the pin bones and hips is less pronounced, though they are still visible (see illustration). A BCS of 7 represents approximately 31% body fat.	*Body Condition Score of 7*
Obese	8	The cow is so fat that her pin bones and hips aren't visible, and the brisket is distended. The udder begins to show fat and there are patchy fat deposits around the tail head. Her neck is extremely thick and the indent of the spinal cord is quite pronounced. A BCS of 8 represents approximately 35.5% body fat.	
	9	The cow's fat cover is even heavier than in a BCS of 8. Her udder is fat, and fat protrudes from the tail head and over her pin bones.	

From Calving to Breeding

Calving on grass after the beginning of the growing season (late spring or early summer) is highly advantageous for the cow. It is the most favorable time of year for her to face the stress of labor, meet the high nutritional needs of her calf, and prepare for rebreeding. She is well fed, the grass is full of nutrients and vitamins, she does not have to cope with the adverse conditions of winter weather, predator pressure is at its lowest, and water is abundant.

On such a calving schedule, the cow's gestation period is a few days shorter than that of a cow calving during the winter months (280 days versus 285 days). Perhaps the warm weather and the cow's good health help her relax and come into labor sooner. During this time of year, calves' birth weights are lower, and the cows have significantly fewer calving difficulties. Plenty of space and a relatively stress-free environment allow the cows to concentrate on their new mothering duties. As a result, their maternal instincts are stronger than if they calve out of season. The favorable conditions and absence of stress also speed the recovery of the cow's reproductive tract after calving, allowing her to begin her fertile estrus cycles sooner and have more cleansing cycles prior to breeding, and thus increase the likelihood of conception at breeding time. This inevitably increases the percentage of cows bred during the first cycle of being exposed to the bull.

In addition, the low-stress combination of calving on grass and allowing a cow's body condition to fluctuate with the seasons extends her life so you will

A 21-DAY BREEDING CYCLE?

Nature's breeding seasons put our 42-day breeding seasons to shame. Wild breeding seasons are extremely short to accommodate the very brief window of opportunity that maximizes ideal calving conditions and minimizes the nutritional cost of surviving the winter in good health and high fertility. Calving any earlier or any later significantly increases the risk of predation, disease pressure, parasite loads, starvation, and decreased fertility. This brief window of opportunity becomes even more tightly constrained in more northern climes, resulting in very short breeding seasons that may be shorter than the duration of one full estrus cycle.

The estrus cycles of breeding females actually become synchronized to the brief rut, the annual time period when bulls, rams, and so forth are excited and are physiologically and behaviorally capable of reproduction (sperm are produced). It occurs during the breeding season, in the same time frame as the estrus cycle, its female equivalent.

The moose rut in Alaska's Denali National Park, which I've witnessed, lasts less than 2 weeks. During that time, the moose world breaks into complete pandemonium as hundreds of moose congregate in a single giant valley to participate in a massive free-for-all mating frenzy. Bull moose cruise up and down the valley, bugling and challenging one another, blinded by their sexual frenzy and oblivious to everything except each other and the scent of their cows. Trees, cars, and tourists become invisible to them as they crash around to meet their tight, 2-week deadline. It is a spectacular commotion, but at the end of the 2 weeks the moose have left the valley and peace returns.

Mimicking Nature's Cycles

Couldn't we mimic the wild herds' single-estrus, 21-day breeding cycle in our domestic cattle herds? In fact, this begins to happen naturally in our herds after we adopt a calving season that is in sync with nature's seasons (see chapter 23 for a detailed guide on how to time your calving seasons to your farm's climate and season), and adopt a rigorous selection *and culling* process based on fertility and maintenance characteristics (see chapter 2 for a detailed discussion about selection characteristics and culling). Remember, stress (from nutrition, climate, environment, management, etc.) and poor genetics cause decreased fertility, increased recovery periods following calving, and poor conception rates. The most fertile, low-maintenance cows always recover their estrus cycles most quickly after calving and therefore rebreed first.

After several years of managing your herd according to nature's rhythms and selection characteristics, you will notice that an ever-increasing majority of cows will calve during the need fewer replacement heifers each year. Calving on grass and good genetic-selection practices virtually eliminate calving difficulties, calf mortality, calf disease, and calving-related deaths. Instead of playing night watchman, nanny, and veterinarian for the entire calving season, you can essentially turn your back on the herd and focus on grazing management, marketing, and your profit margin, because now you have time to do so.

After calving, the cow's physiology changes dramatically as the bulk of her nutritional intake is redirected from replenishing her fat reserves to meeting the high demands of lactation. In addition, her body prepares for rebreeding. Following a brief period of infertility, her estrus cycles return, each successive cycle before breeding allowing her body to further cleanse and rebalance itself, making her more fertile and better prepared for her next pregnancy.

Late-calving cows have less time to cleanse and rebalance before being exposed to the bull, so they are less fertile. In a 60-day breeding season (three estrus cycles), the last cow to calve has only one cleansing cycle after calving before the breeding season begins, whereas the early-calving cows will have at least three cleansing cycles.

In a 42-day breeding season (two estrus cycles), the last cow to calve will have an additional estrus cycle during which to recover before being exposed to the bull, allowing her two cleansing cycles prior to breeding. By shortening the breeding season to 42 days, every cow in the herd has a similar opportunity to regain fertility after calving. In the long term

first half of your 42-day calving season, corresponding to the cows' first estrus cycle of the previous breeding season. This happens as a result of removing the added stress of calving out of sync with nature's seasons and increasing the number of fertile, low-maintenance animals in the herd through your culling program and replacement-heifer selection process. The decreasing numbers of cows that continue to calve during the second half of the calving season are physiologically the least fertile and highest-maintenance cows in the herd; you ultimately want to remove them from the herd anyway.

Over time, this remaining number of less fertile, higher-maintenance cows will become small enough that you can shorten your breeding season to 21 days, eliminating these animals from the herd without significantly affecting your conception rates and financial income. This can be done either gradually or within a single breeding season. By recording the number of cows calving on each day of the calving seasons in the years *prior* to shortening the calving season from 42 to 21 days, you can see exactly how many cows will fail to conceive in the first year of a shortened breeding program simply by looking at how many cows calved during the second half of the calving seasons prior to the switch. In this way, you will not be surprised by the sudden drop in conception rates in the first year of your shortened calving season. You can plan your finances accordingly if you plan a switch while a significant portion of your herd is still calving during the second half of the calving season. These late calvers are the less-fertile, high-maintenance cows in the herd, which take longer to conceive each year. Once they are removed from the herd by this switch, conception rates will return to normal or even improve slightly.

In the long term, this dramatically shortens the calving season, makes your grazing management easier, and produces a more uniform calf crop. It also prevents subfertile and high-maintenance animals from rebreeding, so their genetics do not enter the herd. A 21-day breeding cycle works particularly well on replacement heifers, helping you to select only the most fertile heifers to contribute to your herd's genetics (see Rebreeding the First-Calf Heifer, on page 44). A word of warning, however: Switching the main cow herd to a 21-day breeding cycle is advisable only after *all* other parts of the cattle production year have been fine-tuned to mesh with the most efficient seasonal rhythm *and* you have recorded all of your cows' calving dates for *several* calving seasons prior to making the switch to ensure that both your herd and business planning are prepared for the first year of a 21-day breeding season. A 21-day breeding cycle needs to be laid on a very stable foundation or it will topple your production tower.

(after subfertile animals are culled), a short breeding season improves overall conception rates, conception rates during the first cycle of the breeding season, and overall herd health.

Photoperiod: Sunlight and Fertility

The photoperiod, the interval of the twenty-four-hour day during which an organism is exposed to light, is one of nature's most powerful triggers in the wild, influencing migrations and breeding activity. The farther an animal lives from the equator, the more pronounced the effect, because animals have evolved to use the variations in day length to time their life cycles to the dramatic seasonal variations they experience at these latitudes. (In lower latitudes, the effect is less pronounced: There is less daylight variation and less dramatic variation between seasonal extremes, and seasonal variations in climate are influenced little by day length.)

Sexual activity and breeding tend to peak during the equinoxes, so calving occurs during the grass flush around the summer solstice. Even the *acyclic period*, a cow's infertile period after calving, is significantly shorter when calving occurs around the summer solstice.

Day length is such a powerful trigger that many wild animals (and even domestic cattle) go through a "false rut" at the spring equinox, when the number of daylight hours equals those of the fall equinox, although the bulls' dominance challenges never quite reach the same intensity as during the fall rut — hence the term *spring fever*. The powerful combination of increased sexual hormones at the equinoxes, a shorter acyclic period when a cow calves close to the summer solstice, and body conditions that peak on the spring grass flush just before calving helps to guarantee that nature's prey animals are born when the grass is greenest and they have the highest chance of survival. The effect of photoperiod is less pronounced in tropical latitudes, resulting in longer calving seasons that revolve around the cycles of the rainy/drought seasons rather than the photoperiod.

Humans are not exempt from the effects of photoperiod. Anyone who has lived in the Far North can testify to the dramatic fluctuations in energy and cheerfulness as day length changes between summer and winter. In fact, winter carnivals were invented in northern latitudes to counteract depression caused by the limited daylight. The powerful effect of photoperiod is well known in agricultural sectors. The chicken industry manipulates photoperiod by keeping on the lights in egg-laying barns to increase egg production. Dairy producers use the lights in their dairy barns to simulate longer days, thereby maintaining a higher level of milk production.

Although less well known, cattle fertility, estrus cycles, and libido are also profoundly affected by photoperiod. The closer a cow calves to the summer solstice, the shorter is her acyclic period before her estrus cycles recommence. For example, at 40 degrees latitude, a cow with a BCS of 6 will have an acyclic period of thirty-four days if she calves in July. If the same cow calves in January at the same latitude, however, her acyclic period could last as long as seventy-five days.

In more northern latitudes, this photoperiod effect becomes even more pronounced. Calves are influenced by the photoperiod of their birth date. Heifers born around the summer solstice will begin their first estrus cycles at a much younger age than heifers born earlier or later in the year; this in turn influences their fertility (and conception rates) during their first and second breeding seasons, because they will have had more cleansing estrus cycles prior to their first pregnancies. If we consider how dramatically photoperiod affects acyclic periods after calving, it is not surprising that the calving season for cattle left on their own will naturally shift to coincide with the summer solstice and the calving seasons of animals in the wild.

When combined, photoperiod and BCS at calving have an even more dramatic effect on breeding-season fertility and conception rates. The chart on page 43 illustrates the combined effect of photoperiod and BCS *at* calving on conception rates eighty-five days *after* calving. As you move farther from the equator, the combined effect increases, while closer to the equator, the effect of photoperiod weakens until it is overridden by the effect of body condition, which fluctuates in response to rainy seasons, grass growth, and drought cycles.

PROBABILITY OF CONCEPTION BY 85 DAYS POSTCALVING*

Calving Period	BCS at Calving				
	5.0	5.5	6.0	6.5	7.0
Jan 1–Jan 22	0.000	0.000	0.063	0.194	0.321
Jan 23–Feb 13	0.000	0.004	0.123	0.258	0.385
Feb 14–Mar 7	0.085	0.207	0.353	0.503	0.631
Mar 8–Mar 29	0.315	0.482	0.664	0.822**	0.835
Mar 30–Apr 20	0.567	0.690	0.792	0.859	0.872
Apr 21–May 12	0.592	0.720	0.827	0.896	1.000
May 13–Jun 3	0.616	0.890	1.000	1.000	1.000
Jun 4–Jun 25	0.775	0.960	1.000	1.000	1.000
Jun 26–Jul 17	0.806	0.997	1.000	1.000	1.000
Jul 18–Aug 8	0.799	0.989	1.000	1.000	1.000
Aug 9–Aug 30	0.756	0.938	1.000	1.000	1.000
Aug 31–Sep 21	0.608	0.739	1.000	1.000	1.000
Sep 22–Oct 13	0.583	0.709	0.815	0.883	0.896
Oct 14–Nov 4	0.475	0.674	0.780	0.846	0.859
Nov 5–Nov 26	0.227	0.377	0.545	0.709	0.823
Nov 27–Dec 18	0.014	0.121	0.256	0.400	0.528
Dec 19–Jan 9	0.000	0.000	0.072	0.204	0.331

*Assumes 280-day gestation period.

**I've highlighted all the probabilities greater than 80% to emphasize that the probability of conception increases around the summer solstice.

Note: This chart was prepared for my parents' farm, located at approximately 49 degrees latitude in southern British Columbia, by Dr. Dick Diven, Agri-Concepts, Inc., Tucson, for his Low Cost Cow/Calf Production School. It demonstrates just how dramatically conception rates are affected by the calving date and BCS at calving.

Synchronizing Estrus Cycles

When cows are exposed to a bull, his presence synchronizes the cows by triggering their estrus cycles and also shortens the infertile period between calving and their first cycle. This is nature's way to ensure that no cow misses the opportunity presented by the bull's arrival.

Synchronizing a cow herd can be very useful in emergency situations in which you suspect fertility may be compromised by a less-than-ideal BCS prior to calving. It is also a powerful tool to help shift your breeding season to an earlier date (for example, from fall calving to early-summer calving) or if you want to gear your cows to a shorter breeding season.

It is quite difficult to put a dollar value on the extra benefits of raising your own replacement heifers.

There are several natural alternatives to veterinary methods of synchronization. Flushing is a method of stimulating estrus in which the cows receive a sudden and dramatic increase in their energy and protein intake in the week before breeding either by being turned into a particularly lush fresh pasture or by receiving a feed supplement. (A word of caution: When using flushing pastures to stimulate estrus, note that the high estrogen content of some clovers and alfalfas can actually inhibit the estrus cycle and decrease conception rates.)

A more common method is to send a vasectomized (sterile) bull out into the cow herd. The presence of a bull on the opposite side of a very secure fence will have a similar effect as long as the cows can see, smell, and interact with him without being bred.

Short-duration weaning (removing the young calves from their mothers for 24 hours just prior to breeding season) will also synchronize the cow herd. Though it is tremendously stressful for all involved, intense stress tends to trigger the estrus cycle. This adaptation allows a cow whose calf dies or is physically separated from her to become fertile quickly so she can increase her chances of another

successful pregnancy. A lion will kill cubs that are not his own so he can stimulate the lioness to breed sooner, increasing the likelihood that his genetics will survive.

Rebreeding the First-Calf Heifer

When heifers are rebred after their first calving season, they have the lowest conception rates of the entire herd: Because they are still growing when nursing, they are the most nutritionally challenged age group on the farm. Some people try to compensate for this by breeding their replacement heifers a bit earlier than the main cow herd to give them more time to recover before being rebred as first-calf heifers. This is not advantageous to the replacements in the long term because they have less time to develop their reproductive tracts through multiple cleansing cycles prior to being bred the first time. Their conception rates as first-calf heifers will not improve significantly, and their fertility will remain compromised throughout their life because they were bred prematurely the first time.

Another common practice is to wait until replacement heifers are two years old before breeding them the first time. Although this does give a heifer more time to cycle, it is not economical for the farmer and allows the genetics of slow-maturing, high-maintenance, subfertile females to enter the breeding herd. If a heifer requires two years to become fertile, she is not suited to your management conditions and will not be a low-cost, high-profit investment.

To improve the fertility of first-calf heifers, breed replacement heifers at the same time as the main cow herd, but allow them to breed for only the first twenty-one-day cycle. Conception rates will be lower among the replacement heifers due to the short breeding cycle, but this ensures that only the most fertile, most efficient replacement heifers are bred and approximates the natural culling and genetic-selection process. After calving, each heifer will be guaranteed a minimum of three estrus cycles to cleanse and rebalance her reproductive tract before being rebred, giving her the best possible advantage during breeding as a first-calf heifer.

During their first winter nursing a calf, first-calf heifers are the most vulnerable to malnutrition. Monitor BCS closely and provide supplemental protein if needed to protect them from emaciation and to prevent a drop in their fertility during the next breeding season.

Buying versus Raising Replacement Heifers

The economic cost of raising your own replacement heifers can be quite deceiving, particularly if you add the opportunity cost (the missed financial opportunity of not pursuing another, more valuable alternative) represented by grazing another salable livestock in their place. During a market low, when herds are typically downsized to cut losses, it is not uncommon for replacement heifers to cost much less to buy than to raise from your own herd when you add up all the hidden expenses.

Yet it is quite difficult to put a dollar value on the extra benefits of raising your own replacement heifers. Their quality is less of a gamble because they are already accustomed to your environment and management conditions. You know their medical histories. They are familiar with your handling practices and the location of your gates. If you practice year-round grazing, they will not need to be taught how to forage for grass under the snow; they will have already mastered the technique as calves in the company of your herd. And they do not need to go through the stress of adjusting to an unfamiliar social hierarchy.

A multigenerational herd develops very complex social dynamics rarely if ever seen in a herd composed of an agglomerate of off-farm replacement animals. Introduced animals can take years to fully integrate into the tightly knit social fabric of the herd and often will continue forming little subgroups along the fringes of the herd for years after they arrive on the farm. In a multigenerational herd, it is not uncommon to find granddaughters, daughters, mothers, and grandmothers consistently grazing together, babysitting one another's calves, or simply enjoying each other's company. Like the migrating caribou herd or the flock of birds that have learned to fly together, the social glue that binds a multigenerational herd gives the group a much greater sense of security and a tighter connection. Less stress ultimately means a healthier herd, better gains, and a higher profit margin.

The Bull Calf: Sexual Development to Breeding

The bull calf enters sexual development when he reaches 45 percent of his mature body weight, at the beginning of his first winter. Much like the heifer calf, whose reproductive tract and mammary glands develop during this period, the bull calf's testicles develop at this stage. His mother's milk ensures that he gains just enough weight to guarantee healthy sexual development. If he gains too much weight (on a heavy grain diet or feed trial, for example), fat cells will deposit in his scrotum, insulating it and preventing it from regulating the temperature of his sperm, thereby causing infertility. Temperature regulation is the key to a bull's fertility. No matter how impressive his weight gains and scrotal circumference are or how well proportioned and masculine he may be, if fat deposits in his testicles, he will be subfertile or infertile. In addition, by preventing the accumulation of fat cells in the scrotums of young bulls, we guarantee that as adults they will be able to withstand summer heat without negative effects on their fertility during the fall breeding season.

This critical sexual development stage ends when a bull reaches 65 percent of his mature body weight, the following spring. He can then gain weight rapidly without depositing fat cells in his scrotum and the seasonal filling and depleting of his fat reserves no longer affects his fertility.

From Apprentice to Herd Sire

In the wild, young bulls follow the older herd sires, play fighting with them to learn how to challenge dominance, read displays of aggression, and settle fights without serious injury and learning many breeding behaviors. The older bulls continue to serve as role models for the young bulls until the youngsters mature enough to become serious threats to the herd sire's leadership. When that happens, the play fighting becomes serious and the younger bulls are chased out to the herd's periphery.

The dominant herd sire breeds up to 90 percent of the cows in his herd. To accomplish this, he must learn to court his harem efficiently and to assert his dominance over the other bulls without risking serious injury to himself. In the wild, only a tiny

Creep feeding gives calves access to supplemental feed (usually grain) during winter by allowing them to crawl under a divider fence designed to prevent cows from reaching the creep-feeding bunks. This grain is typically fed free-choice to the calves in the hope that the extra feed will increase their weight gains, but because the grain is available, the calves typically stop competing with the adults for hay at the haystack or stop the effort to dig for grass under the snow. Instead, they tend to gorge themselves on the grain, risking depositing fat in their developing mammary glands and scrotums at the expense of future milk production and fertility. Calves that are not creep-fed will experience compensatory gain on the lush grass the following spring and will catch up to their creep-fed counterparts, but at a fraction of the cost per pound of weight gain.

Bull calves are often fed at bull test stations after they are weaned (or are simply placed on high-grain diets at home) in order to test their ability to gain on feed. These high-grain rations typically coincide with their critical sexual development phase. Consequently, they risk forming fat cells in their scrotums, which negatively affects scrotal temperature and compromises their fertility for the remainder of their lives. With this feed practice, an extremely masculine bull whose growth is naturally slowed by his hormone production may be discounted by his feed-trial results in favor of a less fertile bull that does not have an adequate hormone balance to limit his growth. There is more than one show-champion bull that landed in a hamburger grinder without producing any calves because of infertility. High-grain-diet, feed-trial gains are not an indication of a bull's ability to produce low-maintenance, highly profitable fertile offspring and profitable pasture gains on grass.

Calves at a creep feeder gorging themselves

fraction of dominance challenges between bulls results in physical fights; they are the exception, not the rule. Most disputes are avoided by carefully reading aggressive behaviors. A physical battle occurs only if the strength of two bulls is so evenly matched that they cannot settle their dominance disputes by simply gesturing.

Young domestic bulls rarely grow up around older herd sires, so they have no opportunity to play fight with the older, stronger bulls or to observe the herd sires as their role models. Instead, they are grouped together in feed pens and pastures, where they can wrestle only with their playful age-mates. Because they are all so equally matched, they do not develop an ability to gauge each other's strength by behavioral displays alone. In fact, a good shoving match in a feed pen becomes the preferred method of settling a score just to counteract boredom. As a result, these young bulls learn that all dominance disputes are settled by physical conflict, resulting in far more serious injuries (and expensive replacement bulls) than is common in the wild. Letting our young bulls grow up on pasture among older bulls makes them better adjusted and gives them the skills necessary to resolve most dominance struggles through displays, not battles.

How Many Cows Per Bull?

This is the million-dollar question, because good pasture bulls are expensive and hard to find. There is no single right answer. The number of bulls you need for your herd depends on the size of the herd, the age of the bulls, the terrain they will be breeding on, and the size of the breeding pasture.

When I was young, our cattle herd spent its summers on range, dispersed over thousands of acres. At the time, we were calving in spring, so our bulls had to breed on summer range. With the cow herd so widely dispersed among dense brush and mountainous terrain, we required a bull for every twenty-five cows to get the job done successfully.

This bull-to-cow ratio worked well until we switched our cows from extensive rangeland management, in which our herd was widely dispersed over hundreds of square miles of government range, to an intensive pasture-grazing program. The problems started after the herd was bunched into groups of 100 to 150 cows in flat, treeless pastures no larger than forty or fifty acres. The bulls kept getting hurt. At twenty-five cows per bull, the bulls were tripping over each other to reach the cows. A bull could not breed in peace without a rival coming and knocking him off his job. When the cow-to-bull ratio was reduced to one bull for every forty or fifty cows and bulls of a wider age range were included to reduce the number of dominance challenges, the injuries disappeared almost entirely and conception rates were not affected.

So the ideal bull-to-cow ratio depends on the situation. How dispersed is the cow herd? How easy is it for bulls to travel between cows to check on their breeding readiness? For example, do they have to traverse bog, mud, thick brush, steep terrain, or flat meadow to reach them? How easy is it for bulls to see each other wooing cows in the pasture? Is their view obstructed by trees or does low grass allow them a full view? As the bull's job becomes easier and the energy he must invest to find and breed each cow decreases, the competition between bulls increases and fewer bulls are necessary to accomplish the task.

But at what point are there simply not enough bulls to do the job? What is the greatest number of cows a bull can feasibly breed? I have heard a story of a bull that serviced 100 cows with a 90 percent conception rate; in another trial, seventy cows per bull worked just as well as twenty cows per bull. I would not be comfortable with so few bulls, especially if each one is alone in a pasture. If the fertility of one of the bulls drops unexpectedly or if one of the bulls is injured, it's prudent to have a backup. When the breeding season lasts only one or two cycles, even a single missed day of breeding has repercussions, so plan ahead. Match the number of bulls to your herd situation, then closely watch the bulls to determine whether they are getting the job done without too much fighting.

Cattle Processing: The Human Role in the Cattle Year

Before planning the ideal cattle year on our calendars, we have to consider our involvement with and impact on the cattle herd through branding, castrating, dehorning, tagging, vaccinating, weaning, and processing on market days. The extreme stress of a hot iron pressed against the skin, needles stuck in the body, sorting, dust, mud, ice, crowded conditions, rough handling, and noise can wreak havoc with our livestock's demeanor and immune systems. Not only is it crucial to minimize the frequency and stress of these events, but we must also time them to coincide with the rhythm of nature's seasons in order to avoid negatively affecting the health, productivity, and profitability of the herd.

Tagging

When we don't calve in sync with nature's ideal calving season, we have to tag calves to overcome the confusion created by crowded conditions, increased handling, and health problems. But in a scenario of summer calving in sync with nature's window of opportunity, the rules change. When calves are born on warm, dry, spacious pastures, they do not face the same challenges that require calving barns, treatment protocols, and constant handling. On the contrary, disturbing the new cow/calf pair while they are bonding in order to punch a tag through the calf's ear may cause the calf to bolt or to be abandoned before the bond is fully formed. This interruption in bonding results in orphans and runty calves that don't have their fill of colostrum. You could also be seriously injured by a protective cow — and it is unwise to cull cows for displaying strong mothering instincts, because these very instincts are crucial to raising strong, healthy calves that are safe from predators. A cow cannot be expected to differentiate between a predator and you, particularly if you sneak up on her with a vicious tagging tool that makes her calf scream bloody murder. Adjust your

TAG JUDICIOUSLY

Why are we tagging all these animals, anyway? We need a record only for cattle that have been ill, are high maintenance, have had calving problems, or struggle with their fertility — in other words, cows (and their daughters) that have shown signs of weakness — to remove their genetics from the herd. Tag sick calves when you treat them so you can identify them later to prevent them from being sold as antibiotic-free beef and from being kept as breeding animals. All other cattle records are unnecessary. The cow will be more likely to remember which calf is hers if you don't interfere with the calf at calving time. And why create a record you'll never use?

This is where a lot of people tend to panic: "No records and no way to know which calf belongs to which cow?" If you need to sort a cow from the herd and can't identify her calf immediately, the calf will become easy to identify by its behavior as soon as it gets hungry. Within a day, it will be desperately cruising around the paddock, calling to its mother and looking rather unsettled while the rest of the herd is quiet. If you remove a calf from the herd, you will be able to find its mother as soon as her udder fills with milk and she starts bawling for her calf to come and relieve the pressure.

management protocols and leave the rodeo to the arena.

Every mature cow should receive an ear tag when you select her to become part of your breeding herd. This allows you to record those that perform poorly during the winter or drought months so you can find them again after they have fattened on the lush spring grass and remove them from the herd. It also allows you to record cows that show poor mothering instincts, lose a calf to abortion or death, or have health or behavior issues so you can cull them at a later date. But you need to keep records only for problem animals that should be culled. These identify the animal, why it should be culled, and what veterinary treatment, if any, it received, for shipping and marketing purposes. You should also keep a record of all the ear-tag numbers used in your cow herd so you don't hand out doubles by mistake.

For the cow/calf producer who still feels it is necessary to tag calves, there is an alternative that does not require the momentous effort of finding and tagging calves on pasture and does not disturb the crucial bonding process between a cow and her newborn: Simply tag every calf during calf-processing day, when branding, castrating, and all the other niceties happen. One additional procedure doesn't add significantly to a calf's stress level and requires little additional effort on your part. Afterward, you can pair the calf numbers to the cow numbers by spending a morning walking calmly among the herd in a small pasture to observe calves suckling.

The process is quite efficient. With the help of a friend who double-checked the numbers, I paired four-hundred cow/calf pairs on two consecutive mornings (two to three hours each day) by waiting for each calf to suckle until we accounted for all the numbers. If the herd got restless, calf-feeding time ended, or the herd settled down for a nap, we would simply stop and come back later: no stress, no near-death encounters with overprotective cows, and only a tiny fraction of the time investment of most tagging programs.

Castrating, Dehorning, Branding, Vaccinating, and Parasite Control

Processing days are tremendously stressful to cattle. Their weight gains are seriously disrupted

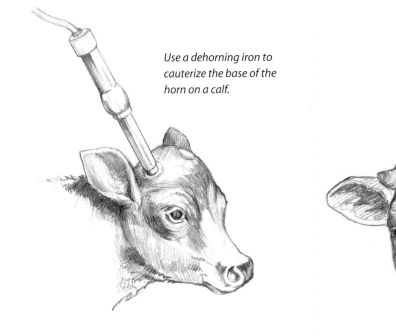

Use a dehorning iron to cauterize the base of the horn on a calf.

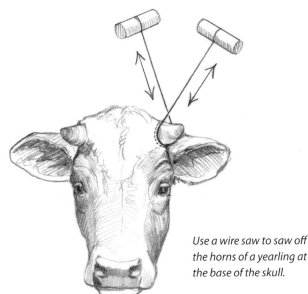

Use a wire saw to saw off the horns of a yearling at the base of the skull.

for several days afterward, the stress depresses their immune systems briefly, and if this type of traumatic handling occurs too frequently, your cattle will become nervous and flighty. If possible, try to time processing activities so you can combine them into a single event. If you are already branding, for instance, a few more unpleasantries won't make the cow's day much worse.

You can reduce recovery periods, stress, and processing-related health consequences by carefully planning the processing day. Wait until the bond between mother and calf is well established and the calf is strong enough to handle the stress and old enough to receive all its vaccinations. Time the event so that the weather will be working in the calf's favor during the recovery period. Heat, cold, food shortage, flies, high humidity, and mud are all extra stresses that will compromise the calf's recovery and invite disease.

Once you decide on a vaccination program (with the help of your veterinarian) to suit your management conditions and specific environment, the vaccination schedule will determine the earliest date you can process your calves. If you are using parasite-control measures to deal with warble fly, worms, or ticks, the life cycle of the parasites must be considered when selecting the processing date. For example, if you apply a warble-fly parasiticide, you should process your calves *after* the first fly-killing frost and *before* the warble grubs start burrowing in their migration through their host's body.

By waiting for a convenient processing day to handle the newborn calves for the first time, producers lose the option of dehorning with dehorning paste because the horns have already begun to grow. On processing day, stop the horns from growing further by cauterizing the base of the horn with a hot dehorning iron, which prevents blood from reaching the growing horn. The skin around the horn should turn a copper color and must form a complete circle around the horn, tight to where the horn emerges from the skull. This is far more painful than even the hot branding iron, but it does allow you to remove the horns at a young age instead of waiting until they have become large enough to cut off. If you wait until the calves are yearlings to dehorn them, the horns can be sawed off at the base of the skull using a wire saw. In most cases, if used correctly, the saw will cauterize the blood vessels to prevent bleeding and the process will be relatively painless compared to dehorning with a hot iron. Nevertheless, because using a wire saw is a tedious and time-consuming task compared to dehorning with an iron and the wire saw poses a much greater risk of serious injury to the animal if the horn is not removed correctly, I still prefer using the dehorning iron on calves at a young age, particularly if I must process a large number of cattle.

Why dehorn? The feedlot industry (buyers of cattle that are fed feed commodities in confined yard areas to supply the slaughterhouse industry with slaughter-ready cattle) has set the standard for dehorning cattle to maximize feed-bunk space, avoid goring, prevent meat losses to bruising, and safeguard livestock handlers. Although grass-finished animals may never see a feedlot, even direct-marketed animals will still be transported and handled in sorting facilities, where horned animals can inflict serious bruising on one another. Even the

regular social jostling of cattle on pasture will cause costly bruising discounts at the slaughterhouse if the cattle are not dehorned. When people unaccustomed to horns have to handle horned animals during transport or at the slaughterhouse, their nervousness is transmitted to the animals, which increases livestock stress and affects meat tenderness. Dehorn for the sake of stress-free handling; for a comfortable rapport with your livestock transporters, auctioneers, and butchers; and to minimize bruised meat — and do it even if your grass-finished natural-beef animals will never see a feedlot.

By not handling the calves at birth, you miss the opportunity to use a castration ring on your bull calves, so it is either back to the old-fashioned method of castrating with a scalpel or using the banding method, which is similar to using a castra-

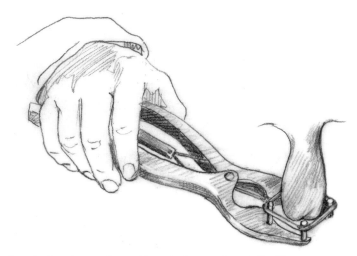

A castration ring can be used to castrate a newborn bull calf.

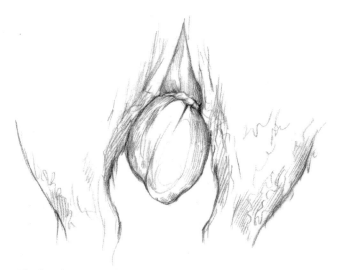

The banding method can be used to castrate an older bull calf.

tion ring except that it requires a larger tool and a sturdy rubber band designed for older animals.

Banding and ringing (even done at birth) are somewhat more of a health risk to the calves than castrating with a scalpel. The rubber band cuts off the circulation to the scrotum until the scrotum dies and falls off on its own. Before the sac falls off, however, it becomes an oxygen-deprived environment that is perfect for anaerobic bacteria such as tetanus and other clostridial diseases. The dying, festering, bloodless sac also becomes a fly magnet, which brings additional bacteria to the wound. A surprising number of calves that are ringed at birth have a low-level infection around their scrotum for weeks after the sac has fallen off. This puts an unnecessary strain on a calf's immune system and reduces its vigor and gains.

Most important, castrating allows you to manage your heifers and steers as a single herd.

Castrating with a scalpel does not leave a festering wound, and because there is no chance of a testicle slipping out, you avoid the risk of winding up with half a bull calf capable of breeding out of season. Although it seems more invasive to castrate with a scalpel instead of a band or a ring, the open wound heals immediately, blood flows to the wound to clean the area and prevent infection, no anaerobic environment is created to attract bacteria, and the wound does not become a festering fly magnet. *If you are unaccustomed to castrating calves with a scalpel, it is extremely important to have your veterinarian show you the correct procedure. A mistake here could easily cause a calf to bleed to death.*

The trick is timing the castration so the calves can immediately access plenty of fresh pasture free of mud, dust, and manure after the procedure so there is less chance for dirt and bacteria to get into their wounds. If they go back to a previously grazed pasture, the leftover manure piles will begin hatching fly larvae that will irritate a healing wound. It also helps if the weather is warm and dry, but in northern climates it is advantageous to wait until the first fall frosts have killed the flies. A lush pasture provides the newly castrated calves with high-quality nutrition that assists their recovery, and the grass will not irritate the calves' wounds as overly mature or seedy grass would.

Castrating makes it easier to produce a consistent meat product because the calf will be less prone to stressful dominance interactions and will not ride (try to breed) the other calves in a feedlot situation, where space is cramped. Although space is not a factor in a grass-finishing scenario, the grass steer still gives you more marketing flexibility because his meat does not toughen as quickly with age as that of a bull. Most important, castrating allows you to manage your heifers and steers as a single herd to maximize the grazing impact on the land and facilitate your grazing management. There is minimal riding among them because the constant supply of fresh grass and the open space keep their minds occupied with grazing.

Until a bull calf is castrated, he experiences an extra growth boost from his testosterone. Thus, the longer you wait to castrate, the faster your bull calves will grow. But castrating becomes more troublesome and traumatic as a calf gets older. The best compromise is to castrate the bull calf at several months of age, when you perform all your other processing activities. The brief time between calving and processing nevertheless provides a bit of a testosterone boost, but you avoid creating an additional stressful and time-consuming processing day. You will disrupt the herd's weight gains only once and your management will be kept as simple as possible. When calving in sync with nature's seasons, calf processing inevitably occurs sometime during fall, an ideal time for castrating; the weather has not yet turned ugly and the grass is still lush so the calves will recover quickly with minimal complications.

Weaning

In the conventional marketing system, we are accustomed to calving in spring and weaning in fall so we can take our calves to market and pay back our bank loans before the end of the fiscal year. A grass-based summer-calving scenario requires a different marketing approach because the calves are still too young to wean in fall and are not big enough to be profitable. Instead, you can either sell the calves as stockers (calves in the stage between weaning and the final pre-slaughter finishing phase that are between 350 and 850 pounds and are being grown on pasture) the following spring, when stocker prices are typically very strong, or you can take advantage of their compensatory gains yourself, graze them through the following summer, and sell them as 800-pound yearlings before the weanlings start to flood the fall market. You can also continue grazing the yearlings throughout the fall to grass-finish them yourself if you develop your direct market for natural or organic beef. See chapter 17 for an in-depth discussion about grass-finishing cattle.

No matter which market scenario you choose, if you are summer calving, the longer you wait to wean, the longer the calves can supplement their diet with their mother's milk and the lower your cost per pound of gain. Milk is free as long as the cows have excess fat reserves to burn. Cows that forage through winter on last summer's pasture stockpile slowly reduce their milk production to compensate for the deteriorating grass quality and their compromised body condition. Often, they will wean their calves to reduce the nutritional drain of lactation, so by the time spring arrives, many calves will already be weaned with none of the costly and dramatic separation stresses that normally accompany the process.

The only reason to wean at all is to guarantee the cows enough time to regain their body condition prior to calving. If a cow's body condition allows it, waiting until as late as a month and half before calving will still give the cow enough time to recover. Weaning is also a powerful tool to manage BCS during winter. If the cows lose weight too quickly, you

Calves weaning across an electric fence

can wean their calves before you have to resort to feeding expensive supplements.

As you observe cows beginning to lose condition, you can also wean individual cows that are nutritionally stressed. This is best done by removing the cow from the herd. The calf remains in the relative security of the herd, where it can continue to steal milk from other cows; its environment is unchanged except for the absence of its mother. This is far less stressful for the calf than removal and it makes sorting much easier. The weaned cow, now separated, can receive nutritional supplements (or be culled) without having to supplement the entire cow herd.

Weaning should be timed to provide the calf with the most favorable weather and nutrition during the transition stage so that weaning stress does not result in disease, reduced weight gain, and so forth. If the cow's BCS allows you to wait until spring, the best time to wean is just as the fresh grass starts to make its appearance at the end of the long winter. The cow still has plenty of time to recover her BCS and the calf will be weaned onto fresh lush grass shoots during mild weather.

Weaning across an electric fence is the most stress-free method. It virtually eliminates all the bawling, broken fences, sleepless nights, and disease pressure that typically follows the process. Calf gains will hardly be interrupted and may not be affected at all. The cows are separated from the calves by a two- or three-strand electric fence that divides the pasture. As long as the two groups can see, smell, and hear one another, and if they have lots of fence line along which to socialize without actually being able to suckle, separation anxiety is kept to a minimum. The first day, both will spend a great deal of time pacing the dividing fence; by the second day, both sides will begin to graze farther and farther from the fence; and as early as four days later, the calves can be quietly moved out of sight of the cows. It really can be this easy!

Marketing Grass-Fed Cattle

Although the specifics of marketing grass-fed cattle are discussed in much greater detail in chapter 15, Market Options, and chapter 17, Grass-Finished Beef, this discussion of the cattle year on grass would not be complete without addressing how the *timing* of our marketing strategies can be molded to take advantage of the seasonal rhythms of a natural cattle year on grass and identifying the marketing opportunities created by calving in sync with nature's seasons.

Most producers have an excess of grass during summer, when the herd simply cannot keep up with the fast growth. Much of the grass matures, slows in growth, and loses nutritional value by late summer. By grazing your own calves as stockers through summer, each calf produces more beef before you sell it. Thus, it takes fewer cows to produce the same number of pounds of beef per acre of grass. Fewer cows means fewer mouths to feed through winter, but then the herd's grass consumption virtually doubles during summer, when grass is plentiful, because you continue to graze the stockers through the summer months instead of selling them directly after weaning, as most conventional production and marketing systems do.

The best time to wean is just as the fresh grass starts to make its appearance at the end of the long winter.

The increased grass consumption caused by retaining your stockers during the summer allows you to take advantage of the summer grass excess, but by selling the stockers in the early fall or slaughtering grass-finished animals in the late fall, you reduce your herd's grass consumption significantly in the winter. Consequently, during the winter months, the grass consumption of the remaining cows and their very young calves (which consume only a small amount of grass at this time) coincides with the significantly reduced availability of forage.

In addition, you can take advantage of wintering the calves on the cows' milk, then capitalize on the calves' compensatory gain when spring grass arrives. This allows you to produce beef at a much lower cost per pound than most other producers. By late summer, just as grass growth is slowing, the stockers reach an ideal weight to sell before the fall rush of weanlings starts to drive down fall auction prices. Or you can continue on to grass-finish and direct-market your steers and cull heifers, which will be ready for slaughter at the end of fall. If you grass-finish, you will require even fewer cows to produce the same number of pounds of beef per acre, which further reduces your production costs.

Many producers overlook the opportunities presented by their open and cull cows. The standard thinking is that it is best to sell opens (cows that failed to conceive during the breeding season) and culls (high-maintenance cows, cows with poor genetics, sickly or mean cows, and so forth, that you must remove from the herd regardless of whether they are pregnant) as soon as possible because it costs money to feed them. Their low price per pound at auction supports the misconception that cutters (a term used for culls and opens that are considered too old or of too poor quality to use for all the prime cuts and which are thus ground into hamburger by the butcher) do not have much value. In a summer-calving scenario, however, rather than immediately selling your opens and culls just before the grass flush starts, you can fatten them on the spring grass flush to take advantage of their compensatory gains before they are sold. They can gain weight far more efficiently than stocker cattle because they are neither lactating nor growing. If a cow's BCS goes from 5 to 7 in a month, she may well add 150 to 200 pounds to her market weight before you sell her! Even at cutter prices, this is a phenomenal increase to your profitability and means that your cutter cows will rival even your best stockers in net value. It also allows you to begin your direct-marketing season much earlier because young open and cull cows can be direct-marketed by the cut, and even the older cutter cows can be direct-marketed as hamburger for the summer barbecue season.

Moving the Calving Season

There are many production and marketing advantages to a natural, grass-based production system, but they can be attained only if the calving season is in sync with nature's seasonal rhythm. The calving date is the most crucial building block that makes possible the low-cost, high-profit grass-based enterprise.

Unfortunately, moving the calving season from early spring or late winter to early summer involves missing a year's income, because the calves are not sold as weanlings in fall but are held over into the next fiscal year. I don't know an easy way around this. The cautious approach would be to switch the herd slowly, to have two herds and make the transition gradually over a number of years. Theoretically, this may sound more attractive, but this approach is a financial and management nightmare because each system demands different management strategies, calving needs, equipment, and grazing programs. Each herd will impede the other's smooth operation and you will be stretched so thin, both financially and mentally, that you will not be able to realize the full benefits of your new system.

The calving date is the most crucial building block that makes possible the low-cost, high-profit grass-based enterprise.

None of the benefits of winter grazing, stress-free calving, and reduced inputs is possible with an "out-of-season" herd. Until the entire herd is switched to summer calving, you can't even sell your expensive forage and feeding equipment to pay for infrastructure changes (for example, electric fences and water sites) necessary to take full advantage of the benefits of summer calving. And what of the maintenance costs? Do you start rebuilding feed bunks and replacing forage equipment for only half the cow herd, knowing that these investments will be obsolete as soon as you complete the transition?

The best approach is to plan the transition all the way through to its completion before changing anything and then to make the transition as quickly as possible, preferably in one fell swoop. This way, you can give your new production system your full attention without being distracted by a remaining out-of-season herd. You will start enjoying the benefits and results of a single summer-calving herd much sooner, allowing you to gain confidence quickly. By committing yourself fully, you can sell off a great deal of equipment such as feeding and grain-handling equipment that is no longer necessary or applicable to your new production system: For example, you'll be able to reduce to a bare minimum — or sell altogether — your forage-making lineup. This will give you the financial cushion to survive a year without income until your first set of yearlings is ready for sale, and also allows you to make new investments in grazing infrastructure.

Fall-calving producers who need to move their herds to an earlier calving date can adjust their breeding season ahead a little each year until the cows reach their ideal calving date instead of delaying their entire breeding season for many months. Synchronizing the herd with the presence of a bull incapable of breeding or by short-duration weaning just prior to breeding will help improve conception rates during the transition period. The closer you get to the ideal early-summer calving season, the more nature will support the change, with cows coming into estrus soon after calving. Moving the breeding season ahead by three weeks to a month each year is quite safe in a forty-two-day breeding cycle because the cow's recovery period will be shortened by only one estrus cycle. Some cows undoubtedly will not get rebred, but you can take comfort in knowing that these are the less-fertile, high-maintenance animals that you'll want to cull to improve your herd's genetics.

THE JULIAN CALENDAR

Introduced by Julius Caesar in 46 BCE to match the length of the solar year and divide the year into roughly equal months, the 365-day Julian calendar is a wonderful tool to help plan your calving and breeding dates (see page 298). Each day of each month corresponds to a day on the 365-day Julian calendar, which allows you to quickly calculate when to turn out your bulls based on when you want to calve and vice versa. In chapter 23, you will find a full Julian calendar and instructions on how to use it, along with complete guidelines on how to plan your ideal cattle year.

Chapter

Grass and Grazing

THE CATTLE INDUSTRY IS UNIQUE: IT TURNS sunshine into cash. Most industries require purchased energy to turn raw resources into something that can be used and sold, but the cattle industry relies on sunlight, which, in the process of photosynthesis in grass, combines with chlorophyll to convert carbon dioxide and water into the organic materials that are used as the building blocks for plant growth. By converting the grass to beef, we make a salable product. And best of all, the cattle do all the work of harvesting, receiving as payment the joy of eating.

It sounds like an impossible dream, but this simplicity epitomizes an efficient grass-based system. Our opportunity lies in managing the cattle's grazing behavior in such a way that the grass collects as much free solar energy as possible and our bovine labor force efficiently converts this energy to salable product. We are the orchestra conductors; the grass and cattle are the "moosicians" that do all the work.

The Rumen: The Forge That Turns Grass into Beef

Grass is tough and is not easy to digest. It is so tough, in fact, that the human digestive system cannot sufficiently break down its cellulose structure to derive any nutritional benefit from it.

When grass appeared on the landscape twenty-four million years ago, new grazing species such as ruminants — four-stomach-chambered cud chewers — evolved to make use of it. To break down the

cellulose structure of grass, these creatures, such as cows, evolved an ingenious digestive system. An army of microorganisms lives inside the *rumen,* the first stomach chamber, and waits for grass to come down the hatch in order to feed on it. These microorganisms break down the tough cellulose structure of grass so calories and nutrients can be extracted in the next three stomach chambers. Without these cellulose-digesting microorganisms, grass would offer cows very little nutritional value.

As a reward for their vital activity, the microorganisms in the rumen consume their share of proteins. Only excess proteins and the protein released by the microorganisms at the end of their short life cycle pass to the other stomach chambers and become accessible to the cow. Even when the rumen environment is functioning optimally, the rumination process is not particularly efficient at extracting nutrients from the grass, which is why so much partially digested grass is returned to the soil via manure. Soil microbes rely on this partially digested manure to maintain the soil's fertility, thereby completing the self-sustaining cycle between grass, cow, and rumen and the soil that supports them.

The microorganisms responsible for fermenting and breaking down the cellulose structure of grass in the rumen are most active in an environment with a slightly acid pH of 6.4. If stomach acidity increases to a pH of 6.2 or less, these cellulose-digesting microorganisms become inactive and the nutrients in the grass remain locked inside the cellulose struc-

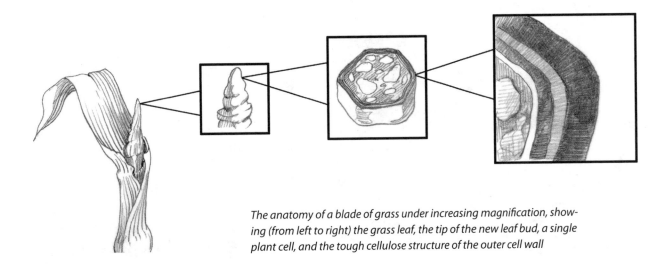

The anatomy of a blade of grass under increasing magnification, showing (from left to right) the grass leaf, the tip of the new leaf bud, a single plant cell, and the tough cellulose structure of the outer cell wall

ture. Digesting grains (described below) increases the stomach acidity inside the rumen. Thus, when a cow is fed supplemental grain as part of her diet, the valuable cellulose-digesting microorganisms in her rumen are deactivated by the increased stomach acidity, causing the grass to pass through the digestive system without the cow being able to access the nutrients within it. The small amounts of grass seeds consumed on pasture are not enough to cause a sufficient increase in stomach acidity, so cellulose-digesting microorganisms remain active.

We are the orchestra conductors;
the grass and cattle are the "moosicians"
that do all the work.

In the wild, ruminant diets consist primarily of grasses and herbs. Only a very small portion of their diet is made up of the seeds and grains that occur naturally on pasture, with a little making its way in when ruminants are forced to eat overly mature grass. Unless these seeds and grains are damaged in some way, a large proportion of them will pass through the animal undigested. Often, this process is a necessary part of the life cycle of the grass seed: The acid environment of the stomach weakens the tough seed shell, allowing it to germinate more easily when it reaches the soil. After the seed is expelled, the manure around the seed provides fertilizer that kick-starts the germination process. In agriculture, this process is mimicked by *seed scarification,* a mechanical process that breaks down the tough outer seed shell to help facilitate germination. Efficient digestion of seeds is left to other non-ruminant species such as birds, which are able to break down and mill the seeds in their digestive tracts.

The small amount of seed that inadvertently passes through a cow's digestive system during grazing is digested by a second set of microorganisms in the rumen that evolved specifically to process seed grains. These grain-digesting microorganisms are more aggressive than the cellulose-digesting ones, but the lack of grain in a typical grassland diet generally keeps them from overpowering those that digest cellulose. If a cow consumes unnaturally high amounts of grain, as do conventionally raised cattle, the grain-digesting microorganisms quickly overrun the rumen environment. Their digestive process increases the acidity of the rumen to a pH between 5.8 and 5.3, well outside the tolerance levels of cellulose-digesting microorganisms, effectively causing their digestive process to shut down. As a result, the grass passes through the cow's digestive system without the cow being able to access the nutrients locked inside the grass's tough cellulose structure. Therefore, when a cow's diet becomes too rich in grain, the grass portion of her diet will function strictly as a source of roughage (fiber), slowing down the grain

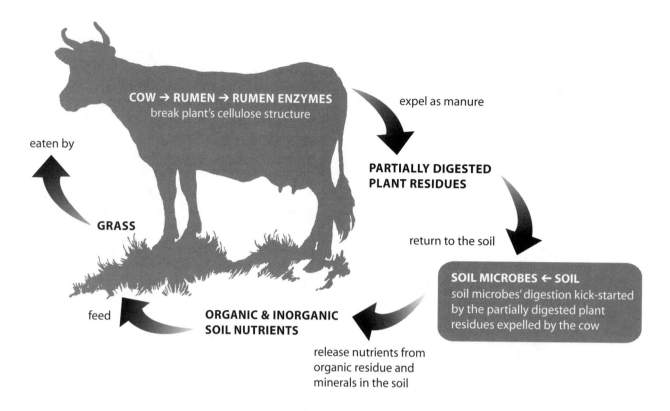

The self-sustaining cycle involving grass, cow, and rumen and the soil that supports them.

in the digestive tract so the grain microorganisms can digest them before they are expelled in manure.

Because cattle are not equipped to deal with a high-grain diet on a regular basis, their bodies cannot cope with the natural digestive toxins and high acidity of grain digestion. In feedlot animals on extremely high-grain diets, the resulting stress over the long term is so high that their livers shut down, causing them to die sometime after their second birthday if they are not slaughtered before. If we want our cattle to live naturally on grass, grain simply has no place in the equation.

Root Reserves and the Ideal Grazing Interval

A grass plant has roughly the same amount of plant material below the soil surface as is visible above it. The roots are essentially a mirror image of the green grass above ground. As the leaves and stems grow, so do the roots. When the grass is cropped, the roots die back in an effort to maintain a balance between the solar-energy-harvesting material above ground and the nutrient- and water-harvesting root system below. A large plant requires a large root system to gather nutrients and water from the ground. When a plant is grazed, the remaining aboveground portion responsible for photosynthesis is able to provide solar energy to only a small root structure. The excess root mass dies off to rebalance the energy drain on the plant. Each time a plant is grazed, dead grass roots decompose, building up the organic components of the soil.

Roots contain a plant's stored energy reserves. When drought or cold weather hits, this energy reserve allows the plant to survive; a small plant with an inadequate root system would simply die. As the dormant season approaches, a large root system beneath the soil surface gives the plant greater resistance to frost or extreme heat. After the plant becomes dormant, the root reserve remains unchanged, even if the plant material above the soil surface is removed by grazing. These stored reserves in the dormant plant's roots allow it to regrow quickly at the onset of warm spring weather or the arrival of the rainy season. If a plant is cropped too close to the ground before it becomes dormant, it will have much less energy to draw from when it resumes growth. Clearly, this has important implications for grazing management, in terms of both the timing of the harvest and the amount to harvest before the onset of winter.

As a plant grows, its root reserves must build up before maximum growth above ground is possible. Once the root reserve has reached a critical mass, plant growth above ground accelerates until it reaches the mature, seed-forming stage. If a herd leaves a tall grass residual after grazing, the root reserves will not die off below that critical mass, allowing rapid regrowth to continue without interruption. If the grass is cropped too short, however, the root reserve must rebuild to regain critical mass before accelerated grass growth can resume.

For example, if grass is cropped to a three-inch residual, the regrowth will have only a three-inch root mass from which to draw energy reserves and access soil moisture. Clearly, this is not much, particularly when the summer sun has dried out the soil and the top few inches are dusty and dry. If a minimum grass residual of ten inches is always maintained during the growing season, the roots will have a ten-inch root zone to draw on, which gives the plant a much larger energy reserve and mineral bank and a deeper moisture zone to draw on for quick regrowth. The grazing cow (or harvester) should not spend too much time in any one area or repeat its migration pattern too soon to safeguard the grass from being cut too short. Grass can be cropped much shorter during the winter season after the frost kills the vegetative portions of the plant above the soil's surface. This is explained in much greater detail during the discussion about winter grazing in chapter 7.

When grass reaches maturity, its growth slows. The plant's energy is redirected toward seed forma-

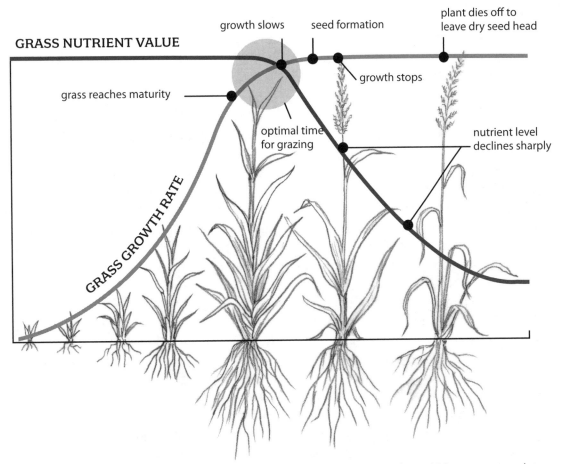

The stages of grass growth (left to right): The seedling matures to fully grown grass plant, which matures to seed formation and then dies, leaving behind a dry seed head and stalk. The grass growth rate stops altogether at the seed-formation stage. The grass nutrient value drops off sharply once grass begins seed formation.

THE S-CURVE OF GRASS PRODUCTIVITY

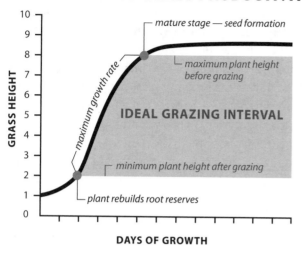

The curve of grass productivity speeds up after the root reserve has been established and slows when the plant reaches maturity. Note the ideal grazing interval (gray).

tion and nutrients from the green vegetative plant parts are drawn up into the seed head. Eventually, growth halts and the nutritional value of the leafy parts decreases significantly. To maximize grass production (solar harvesting), grazing should be timed to prevent grass from going to seed.

As grass matures and begins to die, the dead material prevents sunlight from reaching the new shoots at the base of the plant. Unless the dead material is removed, it chokes out future grass productivity. Grazing animals help keep grass healthy and productive. For a cow to have an incentive to remove it, she must have access to it before it becomes overly mature, unpalatable, and less nutritious. The grass must be in the vegetative (growth) stage for the cow to gain enough energy from the grass to make it worth her while to harvest it. The grazing rotation has to be timed carefully so the herd returns before the grass enters its mature, seed-formation stage. Striking the perfect balance here is the key to a highly productive, low-cost beef-production system.

The illustration above shows how grass production speeds up after the root reserve has been established and how it slows when the plant reaches maturity, forming an S-curve of growth. Maximum production is possible when grass is grazed just prior to reaching the seed-formation stage and when the residual is tall enough to provide an ample

root reserve so the grass can resume growing quickly after grazing. This is the ideal grazing interval.

A Recipe for Healthy Soil

Not surprisingly, there is more to keeping grass healthy and productive than just careful timing of a herd's migration. Healthy soil is also critical. The top five inches of soil is the gold mine that keeps grass healthy. It is where almost everything happens. The success of your entire grazing program depends on how well you manage the topsoil.

Oxygen penetrates only about 5 inches into the soil, slightly deeper if the soil is cultivated. Below this layer, the soil is anaerobic, making it far less active biologically. That is why at ground level a fence post rots and tips over, but below the oxygenated soil layer the wood remains sound much longer. Oxygen has to be present in sufficient quantity for soil microbes to do their job. These aerobic microbes are responsible for decomposing dead plant material and animal manure and reincorporating the nutrients and organic matter in the soil in a form that can be used by plants.

Essentially, the soil functions like a cow's rumen. Many of the microorganisms responsible for breaking down grass in the rumen are also responsible for breaking down organic material and releasing nutrients into the soil. Without a healthy microbiology to keep the nutrient cycle functioning, soil would behave like a hydroponics system in a greenhouse. The nutrients for plant growth would need to be added artificially in water-soluble form to make plants grow, but the system would not be self-sustaining. Instead, it would be sterile and unproductive, dead without artificial inputs.

Healthy soil and its biological components need oxygen to breathe and live. Organic topsoil should be composed of 25 percent air and 25 percent water. When the soil becomes compacted, air and water cannot reach the soil microbes to keep them active and the plant roots are not able to spread out to access nutrients and water. As we will see below, cattle are our most powerful tool in creating and maintaining the sustainable, healthy, well-aerated pasture soils that are essential to the survival of the vast diversity of plants, animals, and micro-

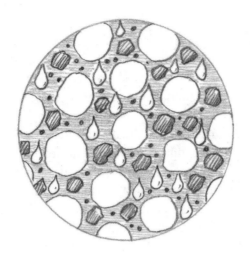
 AIR

 INORGANIC PARTICLES

WATER

ORGANICS

A healthy topsoil is composed of 25 percent air, 25 percent water, 45 percent inorganic particles (such as gravel, sand, silt, and clay), and 5 percent organic matter.

organisms that sustain valuable grass cover and determine our profitability.

Organic matter is the most important soil component, but it accounts for no more than 5 percent of the soil. Inorganic soil particles such as clay, sand, silt, and gravel comprise the remaining 45 percent. Plant nutrients are stored in the organic layer as stable organic compounds. Without this organic component, the soil microorganisms would not be able to live and the nutrients would be flushed away by groundwater. The organic component binds the soil together, giving it a porous structure that retains air, water, and nutrients. Without organic matter, the soil becomes extremely vulnerable to erosion, and without cohesiveness, wind and water can grab soil particles and sweep them away.

In a healthy grass-based pasture management system that mimics the relationship of wild herds and grassland, soil compaction is prevented in a number of ways. When a herd grazes an area, it quickly sweeps across it before continuing its migration (pasture rotation). The animals do not stay long enough to use the same shady sites and paths day after day, and because they do not linger, they do not pound and compact grazed areas. As they sweep through an area, they break up the soil with their hooves, thereby naturally aerating the soil. This is particularly effective when cattle become excited, as in the presence of a predator, for example. If they remained too long in an area, the repeated impact of their hooves on the soil would compact the soil instead of providing this beneficial aerating effect.

The herd keeps the grass healthy and, in turn,

grass roots loosen the soil as they force their way down in search of water and nutrients. As plants die back in response to grazing, dead roots provide pathways for air, water, microorganisms, and earthworms to penetrate the soil. As these organisms travel, they further aerate the soil, increasing its porous, sponge-like texture.

Hooves also break apart dead plant material and manure piles, spreading them out and pushing them down into the moist soil, where they will come into contact with microbes responsible for decomposing the dead plant and manure particles. Without this trampling action, the nutrients from the manure and dead plant materials would not be recycled (see pages 11–13 for a discussion of how decomposition varies according to climate). Instead, they would remain on the dry soil surface and slowly oxidize in the sun until they are blown away by the wind.

A cow's sharp hooves can be nature's mechanism to ensure effective plant decomposition and nutrient recycling, loosen compacted soils, and prepare the ground surface for maximum water absorption.

Particularly in dry regions, trampling hooves are the recycling mechanism that completes the nutrient cycle between soil and grass.

Grass Varieties

The world's wild grasslands contain a phenomenal variety of plant species, contrasting starkly with the monoculture approach typical of most modern farmland and pastures. This diversity is possible in the wild because nature does not need to maximize harvest yields on specific harvest dates and does not require crop-specific harvest equipment. The plants are specifically suited to the grazing needs of the great herds, which require a stable food supply that remains vegetative and nutritious over the longest possible time period during the growth season and that maintains its nutrient quality so it can be rationed successfully over the winter. A single massive yield that matures all at once would be wasted; the excesses would become unpalatable and lose their nutritional value. A monoculture grass supply cannot satisfy a grazing herd's nutrient requirements through the seasons without artificial feed supplements.

The wider the variety of food the herd can choose from, the more balanced its diet and the healthier it will be.

A variety of plant species produces a more consistent food supply over the full growing season and extends the growing season in spring and fall as grass species complement and supplement each other. This smorgasbord of grazing plants flourishes in the various environmental niches and climatic conditions of the growing season. The buffet includes frost-tolerant plants, heat-tolerant plants, plants adapted to specific soil conditions, and plants whose growth peaks at different times during the growing season. For example, a pasture in the northern prairie region might include Altai wild rye for its ability to stand up through the snowpack and provide nutrient-rich winter grazing; tall wheatgrass for its winter hardiness; reed canary grass for its tolerance to water-logged soil conditions; cicer milk vetch for its drought tolerance; crested wheatgrass for its early spring and fall grazing; Russian wild rye for its tolerance to salinity and for providing early-spring grazing; meadow fescue for its heat and drought tolerance; alfalfa for its deep roots, drought tolerance, and nitrogen-fixing ability; and alsike clover for its tolerance to spring flooding and alkaline soil conditions. Because the nutrient and mineral content of each plant differs, animals are able to pick and choose based on their individual needs. The wider the variety of food the herd can choose from, the more balanced its diet and the healthier it will be.

As plant varieties increase, so does the range over which their roots access soil nutrients, making the plant community as a whole more resilient to nutrient deficiencies and reducing the nutrient drain from any single root zone. For example, while alfalfa plants have roots that extend down twenty feet into the soil, with some reportedly going as deep as one-hundred feet, many grasses are shallow feeders, their roots staying within the top few feet of the soil surface. If deep-feeding plants such as alfalfa are part of the plant community, their roots bring up minerals from deep in the subsoil, making them available to other shallow-rooted plants that otherwise would never have access to them. In addition, the root acids of certain plants are more efficient than those of others at accessing minerals from inorganic soil particles, which helps them to introduce new minerals into the nutrient cycle. These specialized root acids are able to dissolve minerals out of inorganic soil particles to feed their nutrient requirements, thus making these new minerals part of a plant's organic molecular structure so that once the plant dies and is reincorporated into the soil, these new minerals can be recycled by other plants that are not capable of accessing minerals directly from inorganic soil particles.

Plant variety is the key to creating a resilient food supply that extends uniformly across the longest possible growing season. Each species fills a different niche in the food supply during the year.

- Cool-season plants flourish in spring and fall but slow their growth and lignify (become hard and less palatable) in summer heat above 30°C/86°F.

- Warm-season plants flourish during summer heat but are dormant during the cooler spring and fall.

- Deep-rooted plants provide access to new nutrients from deep in the subsoil.

- Some plants are able to overcome nutrient deficiencies in the soil by using their specialized root acids to access nutrients from otherwise inaccessible inorganic sand, silt, and clay particles.

- Some plants concentrate certain minerals and vitamins that are essential to livestock.

- Nitrogen-fixing plants such as legumes draw nitrogen from the air.

- Plants thrive in different moisture and soil conditions (for example, drought conditions; moisture-saturated, swampy soils; sandy, well-drained soils; acidic soils; cold soils; and frost), balancing the food supply.

- While the leaves of some plants retain their nutrients longer over the winter or dry season, the nutritional value of other plants collapses in the dormant season, and still others are more susceptible to nutrient-leaching from winter rains and wet weather. Altai wild rye, for example, becomes more palatable in winter as frost and moisture break down its hard stems. Typically, native grasses hold their nutrient value for longer periods during winter than do domesticated grass varieties.

- Strong-stemmed plants that stand up and are visible through the snow (for example, Altai wild rye and grazing corn) attract livestock; weak, frail plants are crushed by the snowpack.

Ultimately, there is a plant that thrives in almost every imaginable environmental niche. The trick to planning a grazing program is to consider the various environmental conditions on your farm and to plan for a variety of species that spread the food supply across the full spectrum of temperature, weather, moisture, and soil conditions. Wild plant species thriving in uncultivated areas on your farm such as natural meadows and fence rows are excellent indicators of which species will do well in your pasture conditions. Your local agricultural Extension service, agriculture department, or forage seed dealer will be able to provide you with publications about your local grass and legume species, describing their specific tolerances or adaptations to the various soil, grazing, and climatic conditions of your area.

Rejuvenating Old Pastures

Although it is tempting to rip up unproductive or monoculture pastures to reseed them with the most ideal grass varieties, it is far more economical to begin by reworking the grazing pattern of your herd so it efficiently takes advantage of the full potential of your current grass resource.

Changing your grazing management inevitably improves old pastures, even without reseeding or overseeding (laying grass seed onto an existing turf) because it creates an environment that makes desirable plant species more competitive and restricts competition from less-desirable grazing plants. Niches open up and new plants emerge in the pasture. Seeds that blow in or are already in the ground have an opportunity to germinate, unlike in a continuous-grazing system or an inefficient grazing rotation where hard ground and hungry mouths prevent new seedlings from taking root. Introducing new plant species into pastures with less-than-ideal conditions (for instance, overgrazed compacted soils, poor nutrient cycling, poor manure distribution, dead material) at high cost merely treats the symptoms of poor grazing management and does not translate into a long-term, sustainable increase in production.

At some point, you may want to introduce new seed to your pastures — for example, if your herd's grazing management reaches the full carrying capacity of your land, if you are trying to improve the quality of your pastures in preparation for grass finishing, or to balance out the grass availability over the course of the seasons for successful winter grazing. Overseeding established pastures is never easy because competition from existing plants makes it difficult for new seeds to germinate and establish themselves. There are several methods to counter this. One is to use "range seeder" implements designed to cut through the sod mat and plant seeds into it. Another is to harrow the pasture aggressively after broadcast seeding and then repack it to ensure good seed-to-soil contact. Many unproductive grass sods and legumes can be rejuvenated simply by disking across the pasture, which aerates the soil, loosens root-bound sod mats, and promotes tillering (new shoots that sprout from the base of the grass). Unfortunately, all of these approaches require expensive equipment and fuel and the results are highly unpredictable, depending on primarily the success of your grazing management after overseeding to ensure that new seedlings are not outcompeted by the existing grass sod.

Alternatively, you can let the cattle do the work for you. Simply mix the grass and legume seeds into their loose mineral supplements and they will distribute the seeds for you in their manure. After all, that's how nature reseeds itself in the wild. The manure provides essential nutrients and nitrogen right where the new seedlings need it, and the trampling hooves help plant the seeds into the soil. The harsh digestive environment of the cow's stomach also increases the likelihood of successful germination, because acidity helps break down tough outer seed shells. You can also periodically delay grazing until after the grasses have gone to seed so that fresh seeds can be knocked to the ground and pressed into the soil by cows' trampling hooves.

You can manipulate the timing of your grazing rotation to promote specific desirable grass species or you can graze cattle heavily when undesirable species are most vulnerable. Every plant has a weak point in its life cycle during which grazing either kills it or makes it less competitive than other plants in the pasture. Similarly, each plant has a period in its life cycle when grazing promotes tillering and seed transport or gives it a competitive edge over its neighboring plants so it can choke them out.

Given the unpredictable results of overseeding, it is rarely worth the large investment. Cultivating and reseeding a pasture from scratch will certainly produce a pasture with desired plant species, but at a considerable financial cost and with a limited life span, if you do not implement a healthy, well-managed grazing strategy. In addition, it generally takes two to three years to establish a new pasture sod, during which the pasture produces very little and the seedbed is extremely vulnerable to the harsh impact of the grazing herd. Yet by manipulating the timing and intensity of each grazing rotation and by using cattle to introduce new seeds to the pasture, you can rejuvenate old pasture stands as effectively or even more effectively than you could by mechanical means without the cost of heavy equipment and labor, without putting your current pasture productivity on hold while you wait for the new seedbed to be established, and often within a similar time frame as reseeding. A healthy, well-planned grazing strategy will continue to improve pasture productivity, robustness, and resistance to drought and other climatic challenges long after the three transitional years of introducing a healthy, well-planned grazing strategy are over. But patience and planning are key.

Note that my references to "your farm," "your pasture conditions" and "your area" are equally valid whether the land is deeded or leased, because in either case, readers will be responsible for managing the plant species and grazing strategy on the land that they are using. I prefer not to open up the topic of leasing versus buying at this point so that readers can concentrate on one concept at a time.

INFRASTRUCTURE
and
Management

Electric Fences
and Rotational Grazing

IT IS ONE ROLE OF PREDATORS TO KEEP WILD herds bunched together and migrating as a group. Shepherds mimic this role with domestic herds, but fences have largely replaced shepherds in North America and Europe.

The classic barbed-wire fence is expensive to build and highly labor-intensive to construct and maintain, and provides no flexibility for grazing rotations as the pasture growth rate changes throughout the seasons. It also tends to be very hard on wildlife and people, restrict wildlife movement, and have a short life span (averaging between ten and twenty-five years, depending on climate, livestock pressure against the fence, and how rigorously it is maintained).

The electric fence is a cheaper, more effective, longer-lasting, safer, more animal-friendly option. It is designed not as a physical barrier that restricts cattle movement, but as a psychological barrier that cattle fear to cross. Fear of that hard-hitting little electric wire enables us to use a minimal amount of material — usually just one or two strands of wire — to accomplish what rolls of barbed wire and page wire simply cannot achieve. Wildlife can easily pass over or under the wires and there are no barbs to tear the life out of an animal that gets tangled.

An electric fence is also easier and faster to put up than other fences and can be constructed at a fraction of the cost. Because it does not function as a physical barrier, it has more than twice the working life of a barbed-wire, post-and-rail, or page-wire fence, despite requiring significantly less maintenance. Portable electric-fencing gear allows you to further subdivide pastures at your convenience. You can mimic wild migration patterns by varying the size and number of pasture subdivisions over the course of the seasons to achieve an effective grazing rotation.

A combination of permanent and portable electric fencing gives you the greatest flexibility in man-

IMAGINE THE POSSIBILITIES

The ability to use electric fencing to focus your cattle's grazing impact according to your needs opens up some surprising labor-saving opportunities. With a few well-placed permanent wires to hook on to with portable wire reels, I have used cattle to mow road edges and used the bull herd to mow my lawn, though sheep would have left less of an impression. The cattle's ability to plunge through brush, graze between rocks, and clamber up steep hillsides allows you to maintain unsightly or difficult terrain at little or no cost; the cattle cheerfully grazing for free what would otherwise cost fuel, time, and wear-and-tear on equipment, if it is even possible to maintain these areas by mechanical means. Besides, it is much more fun to sit on the lawn with a glass of iced tea, watching the bulls grazing and fertilizing the lawn, than it is to push a lawn mower and breathe exhaust fumes.

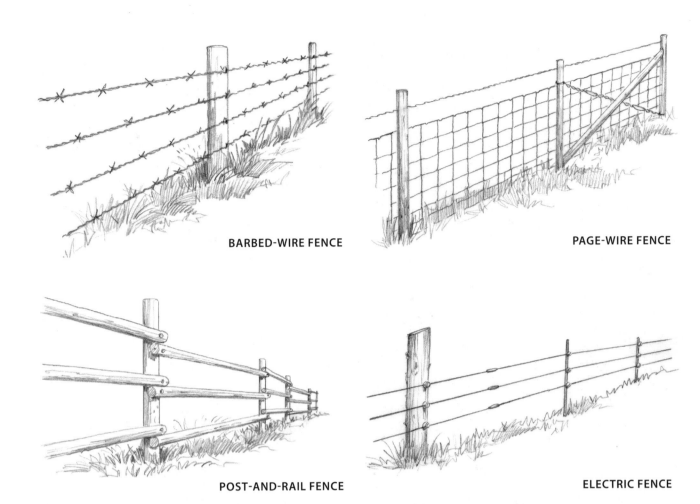

BARBED-WIRE FENCE

PAGE-WIRE FENCE

POST-AND-RAIL FENCE

ELECTRIC FENCE

aging the separate environments on your land as well as the greatest ability to respond to the changing seasonal grass so you can ration it effectively over the entire year. Chapter 24, Your Grazing Infrastructure, contains a practical, step-by-step guide to help you design your own electric fence grid for your farm. With the help of electric fences, our livestock become a versatile tool to manage a farm's ecosystems.

Factors to Consider

When designing permanent electric fences, keep portable fencing subdivisions in mind. Design with an eye toward flexibility and convenience and plan to use portable electric wires to ration the grass supply over the entire year. If permanent paddock divisions are too wide, a single reel of portable electric wire will not be long enough to span the distance, which makes setting up portable wire subdivisions frustrating and impractical. If the design of your electric fence grid requires you to use long, portable fence

subdivisions (greater than ⅛ mile), check with your local electric fence supplier to be sure that it sells portable wire spools that are long enough for your purposes.

Note that there is considerable variation in conductivity in different portable wire products on the market. All wire has some resistance to the electric current passing through it, causing a predictable voltage drop the farther you get from the energizer. The thinner the wire, the greater the resistance. Consequently, it is important to know the voltage drop of the portable electric wire type that you choose to ensure that it will retain a sufficient shock through the entire length of each fence subdivision. In order to contain cattle, you need to maintain at least 2,000 volts on the fence. Because of their wool insulation, sheep need 2,500 volts to be deterred from crossing the fence. Bison need even more voltage than sheep; horses need less voltage than cattle. Most standard polywire and polytape drop from more than 8,000 volts to less than 4,000 volts within approximately

100 yards of wire and fall below 1,000 volts in less than 500 yards of fence! Some companies sell high-grade polywire and polytape — sometimes called turbo wire or turbo tape — that have more wire strands in the braid and therefore carry more current. The turbo wire is capable of carrying current much farther, with only a 2,000- to 2,500-volt drop per 1,000 yards of fence line.

Birds are an asset to the natural grass-based management system because they eat flies and bugs, thereby reducing parasite pressure on the cattle. Many bug-eating bird species seem to prefer living where two ecosystems meet. That's why you'll find the best bird and wildlife viewing at the edge of a forest, brushy area, or body of water: The cover and shade provided by one ecosystem is in close proximity to the abundant food resources provided by the other. Most birds remain within approximately 200 feet of the brush or forest edge, so bird activity around cattle will increase dramatically if you design your fields with no more than 400 feet of open pasture between shelterbelts.

Much like birds, cattle are shade seekers and enjoy the cover of trees and brush to cool off and relax while they ruminate. By repeatedly seeking out a favorite shady spot, they will severely trample the soil around it. In addition, by repeatedly returning to the same location, they concentrate their manure piles in this area, rather than spreading them out evenly in the field. Consequently, the valuable nutrients contained in the grass they eat will not be returned back to the field as a whole, causing a significant drain to the field's fertility and productivity while overfertilizing the concentrated area. This process is known as *nutrient transfer*. To avoid nutrient transfer, separately fence open pasture and brush.

Individual trees left within a pasture will become nutrient magnets as manure is concentrated around them because the cattle continually return to them to seek out shade. Some people remove all trees from their pastures to eliminate this nutrient drain and create an even distribution of manure around the pastures, but if you design your grazing rotation to accommodate daily pasture moves, you can limit

DON'T COMBINE ELECTRIC AND BARBED WIRE

It is tempting to string electric wire on or along an old fence to extend its life or because you need electric power along an existing fence line. Fencing companies sell offset insulators that hold electric wires away from the old fence, but I strongly recommend against combining barbed wire and electric wire.

Cattle grazing along an old fence that's been electrified are not able to reach under the wire from both sides of the fence, so the grass tends to grow up onto the fence and ground out the electric wire. If the cattle cannot keep the electric fence free of growing grass, time-consuming mowing will be required to maintain it.

In addition, the combination of electric and barbed wire can be deadly for cattle, wildlife, and even people. Any animal that accidentally gets caught in the wires is in for a horrible experience. So many closely spaced, barbed and electrified wires make a trap that entangles and slowly electrocutes and tears to shreds whatever creature it catches as the animal struggles desperately to get free.

Equally dangerous is the practice of electrifying one of the barbed wires on an existing fence, another practice used to extend the life of sagging barbed-wire fences. Likewise, barbed wire should never be used to carry an electric current when building electric fences because its barbs and the softness of the wire make it very easy for an animal to become entangled when it is shocked by the constant pulse of the electricity. Use only high-tensile electric wire or portable electric wire specifically designed for this purpose. If an animal gets tangled in it, it will break much easier than barbed wire, whereby the current is instantly cut off. In the case of high-tensile wire, if it breaks, its stiffness causes it to coil into large, open loops, which are much less likely to wrap around an animal than the soft, malleable barbed-wire coils.

Part of the beauty of an electric fence is the ease with which both you and wildlife can move under or over it. In a combination fence, ease of movement is lost. Accidental entanglement of man or beast is the stuff of nightmares.

Cattle are forced to lie farther from a tree's base and move more frequently to benefit from shade when the tree's lower branches are removed. This minimizes manure overload at the tree and reduces nutrient transfer from surrounding pasture.

access to each tree to just one day during the grazing cycle, thereby reducing nutrient transfer. If the cattle stay longer, they will return to the same tree day after day and draw nutrients from a much larger area. To further reduce the nutrient load at the base of each tree, cut off the lower branches; this causes the shade to migrate around the tree without reaching the base as the angle of the sun changes throughout the day. Thus, the cattle have to lie farther from the tree and move more frequently to stay in the shade.

Energizers

The heart and soul of an effective electric fence is the energizer. Its strength (measured in joules, which essentially reflect how far an energizer can push a current through a wire against resistance) must be sized based on the length of the fence and the number of wires used, the soil type, and how moist the ground is. Also important is how much vegetation will grow up against the wire, because vegetation bleeds electricity out of the fence and reduces its shock. If you choose a large energizer, 10 joules or more, any plants touching it will die back or burn

off as they come into contact with the fence, thereby keeping the fence relatively free of vegetation. A good rule of thumb is to calculate 1 joule/6 miles of electrified wire. For example, if you build 12 miles of a two-strand electric fence, your energizer should have a minimum rating of 4 joules/24 miles of wire in 12 miles of two-strand fence.

Grounding an energizer properly is crucial. Flawed or insufficient ground systems are responsible for more than 90 percent of all problems with electric fences. The electric-fence system is an incomplete electric circuit powered by the energizer. The circuit is completed when something touches the fence, allowing the electrical charge (more specifically, the electrons) in the wire to flow into the earth. In order to complete the circuit when electrons are lost to the ground through the feet of an unfortunate cow, the energizer must quickly replace the electrons in the wire by collecting electrons from the earth through ground rods that are connected to the energizer.

The shock delivered to the animal when it touches the wire can only be as great as the number of electrons the energizer can draw up from the earth in that same instant. If the ground system does not have enough ground rods to collect electrons or if the contact is poor between the ground rod and the soil, the shock will be significantly reduced no matter how well designed the rest of the electric-fence system may be. I have seen more than a few farmers give up on electric fencing after spending huge amounts of money replacing or fixing energizers in the mistaken belief that their soil was too dry for the electric fence to work, when in fact a few extra ground rods would have solved the problem.

A good rule of thumb is to use one ground rod for every 2 joules of stored energy. If the soil is particularly dry or is a poor conductor (as is the case with sandy, rocky, and low-mineral soils), more ground rods may be needed. In a worst-case scenario, you can use a more specialized grounding system in which some of the soil around the ground rods is replaced with a conductive clay medium. Placing ground rods where moisture can reach them such as along the drip line of a roof or periodically watering the soil around the ground rods improves the grounding of the energizer.

Ground rods must be made of a material that will not be resistant to electrical flow. Metal loses its conductivity as it rusts (the rust acts as an insulator), and even thinly electroplated rods or painted rods will lose their conductivity as they rust, so galvanized ground rods are the best option. Use a heavy-gauge, insulated lead-out cable (a thick, highly conductive, insulated wire used to connect the energizer to the ground rods and/or the electric fence if a connection must travel through buildings, underground, or across gateways) to connect the ground rods to the energizer wherever it touches buildings, soil, or

A cow can complete the circuit of an electric fence. The shock she receives is caused by the electrical charge (electrons) traveling through her body to reach the ground.

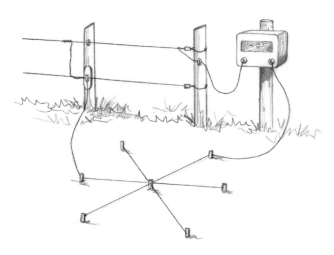

A crow's-foot design attached to a lightning diverter is the most effective ground-rod pattern for dispersing a lightning strike. Space ground rods approximately 10 feet apart.

water pipes to insulate them from electrical charges. In addition, be sure to ground the energizer some distance from water lines and milking sheds so the flow of electrons around the ground rods does not induce an electric charge in buildings where it will affect milking cows or your home plumbing.

To protect the expensive energizer if lightning strikes the fence, attach ground rods to a lightning diverter. You can use the same grounding system for the energizer and the lightning diverter. In high lightning-strike areas, add lightning diverters to your fences in spots where they can be ground in damp soil. The most effective ground-rod pattern to disperse a lightning strike is a crow's-foot design. The ideal spacing between ground rods is 10 feet.

Test the quality of the ground system this way: Short out the fence to less than 2 kilovolts at least 300 feet from the energizer so the short does not induce false readings on the ground rods. Then place an independent ground rod 3 feet from the last ground rod hooked to the energizer and push it at least 8 inches into the soil. Connect this independent rod to the ground rods via a digital voltmeter. The voltmeter should read less than 200 volts; if it doesn't, add more ground rods to the ground system. At no time should you feel an electric shock if you touch a ground rod or the soil around it with your hands.

Plug-in energizers are the least trouble-prone and the most maintenance-free, but portable energizers that run off deep-cycle marine batteries (car batteries are not designed for repeated and almost complete draw-down) provide flexibility in areas without power and offer quick backup during a power outage. Solar-powered fences recharge such batteries, but during the shortest winter days and during long periods of overcast weather, they may need to be switched occasionally with a second battery for recharging.

Permanent Wires

Use only 12.5-gauge or thicker wire for permanent fencing because it carries current significantly farther than lighter-gauge wire. A 16-gauge wire has three times more electrical resistance than a 12.5-gauge wire.

The height and number of wires required for effective livestock control depends on the class of livestock being contained and the fence application. A single wire placed thirty-two inches from the ground is sufficient to contain cow/calf pairs, yearlings, and bulls on all interior fence divisions. This height minimizes losses in current caused by tall grass and allows cattle to easily graze under the wire to prevent grass from growing over the fence line. If you want your cattle to graze under the wire to keep it clean, the grass should not carry a current, which makes it unpalatable. This height allows calves to graze ahead of the cows by walking under the wire, providing them with access to the best nutrition. Their mothers' milk and the security of the herd prevent calves from straying.

Two strands of wire placed 18 and 36 inches above the ground prevent calves from escaping and even contain bulls that can see and smell cows, as long as they are not in adjacent pastures. This is also the ideal fence for alleys and other high-pressure areas.

For property-boundary fences, I prefer two wires spaced at 18 and 36 inches or three wires spaced at 12, 24, and 36 inches. Should the power be cut off temporarily, this provides a greater physical barrier and prevents calves from running onto roadways or onto the neighbor's land. This wire spacing also discourages unwanted dogs from entering the property and harassing your cattle because of the high likelihood that they will receive a painful and memorable shock while sniffing at or attempting to cross under the fence. Water-site corrals are ideally suited to two- or three-strand fences, particularly if you use them as capture and collection sites for treating and sorting cattle. I prefer to use wood or unelectrified wire around permanent water troughs, however, to prevent electricity from bleeding into the water. Last, weaning fences are ideally constructed with two or three strands of electrified wire (see chapter 3 for a detailed discussion of weaning across an electric fence).

Some particularly dry, rocky, and frozen soils do not conduct sufficient electricity to shock cattle when they touch the fence, even if the energizer is grounded properly. In such cases, you may need to use an earth-ground return system in which you hang a ground or earth wire onto your electric fence, approximately 10 inches below the live wire, or between live wires if you are constructing a multiple-strand fence. When animals lean into the electric fence, they will receive the full effect of the electric shock when they make a connection between the live and ground wires because the ground wire completes the circuit by carrying the electrical charge directly back to the energizer, rather than the circuit being completed through the soil.

This ground wire does not require insulators where it is attached to the posts but does need to be a separate continuous wire connected back to the ground rods at the energizer. It should also be grounded to additional ground rods every ¾ mile to safeguard against potential lightning strikes and to drain away any induced current that builds up on the ground wire through the process known as *electrostatic induction* (an electrical charge is produced in the ground wire due to the proximity of the electrical charge in the wire carrying the current, without the wires actually touching). This type of fence requires much more work to install and maintain because of the potential for shorts between ground and live wires.

A cow completes the electric circuit in an earth-ground return system. The bottom wire is not electrified but connects directly to the grounding system attached to the energizer. When the cow touches both the top wire and the ground wire below it, the electric charge in the top wire is transferred to the lower wire through the cow's body, causing the cow to experience a shock.

To revolutionize your experience of working with high-tensile wire, learn to tie wire knots by hand. Using crimping tools and wire cutters can be a discouraging and frustrating experience, but once you learn to tie wire knots and break wire by hand, working with high-tensile wire becomes a pleasure. Some electric-fence suppliers teach these knots and host electric-fencing workshops, which are well worth attending. Practical lessons are best, but if they're not available in your area, here's an overview of how to break a wire by hand, how to tie wire directly to a strain post and then break off the excess wire by hand, how to tie a strain insulator to a strain post by hand, and how to make two different wire joins that allow you to tie two wire ends together by hand.

Cutting a Wire by Breaking It

Hold the wire in both hands, with your hands about a foot apart. Bring your hands together; the wire will naturally form a loop. Pull your hands apart without letting the loop open, and the loop will tighten until it pinches or kinks.

Give it a good hard kink, force the loop open again by straightening the wire, and then rekink it. With a couple of kinks the wire will break. The more rigid the wire, the faster it will break.

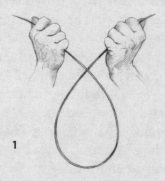

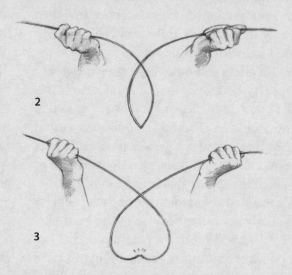

Breaking a wire: 1) Form a loop, 2) kink the wire, and 3) open the kink. Repeat until the wire breaks.

Tying Off to a Post

This knot (see top of page 73) is used when tying a wire directly to a post. When making wire wraps around the strain end of the wire (the end that will hold the tension and pulls against the post when the fence is complete), kink the tail end of the wire (the end with which you tie the knot) into a handle, then use this handle as a crank to tighten the wraps around the strain end. The handle also allows you to easily break off excess wire. When you have made several wraps around the strain end with the wire handle, the wire is ready to be broken off using the following wire-break method.

To break off the tail (handle), bend the handle back on itself so it passes back over the strain end as though unwrapping, except that the length of the handle should remain parallel to the strain end as it passes over the wire to create a tight kink (see middle illustration at top of page

73), instead of remaining perpendicular to the strain wire, which would actually cause the tight wire wraps around the strain end to begin unwrapping. Once you have created the kink shown in the middle illustration, the handle can be cranked alongside the knot until the wire tail (handle) breaks off, creating the completed self-locking knot shown in the illustration on the right.

The combination of kinking the wire in the manner described above and then cranking the handle will break off the wire tail so close to the knot that you will be able to run your hand over the knot without feeling a barb. That's better than cutting off the excess wire with wire cutters, which leaves a little barb that livestock and people can get caught on when they get shocked by the fence. Using wire cutters is also slower and more strenuous than breaking the wire by hand.

A self-locking, tie-off wire knot for a strainer post. The tail end is bent in a "handle" shape to facilitate wire wrapping: Wrap the tie end around the strain end with the handle (left), kink the handle back on itself before cranking the wire handle alongside the knot to break off the excess wire (middle), and complete the knot (right).

Tying on a Strain Insulator

Tie a strain insulator to the wire by making several wraps around the wire with the tie end. Again, make a little crank to facilitate wire wrapping. Break off the tail with the same reverse-direction kink-and-crank motion.

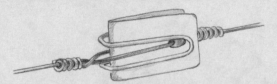

A strain insulator. Note the series of wire wraps used to attach the electric wire.

Tying a Wire Join

There are two effective wire joins. A wire join must create as much contact as possible from one wire end to the other. A join that has only a single point of contact, shown below, will be prone to shorts, and provides too little contact between the two joined wires to transfer the full current from one wire to the other, much like a kink in a garden hose will reduce the flow of water beyond the kink. To ensure that there is sufficient contact between the two wires you are tying together, use a reef knot or a figure-eight knot (see below).

In both wire joins, the excess wire ends are wrapped around the wire and broken off after creating a reef or figure-eight knot. Of the two joins, the figure-eight knot has more wire contact. Another unusual and beneficial feature of this knot is that the slack gained when the wire knot is

tightened will be squeezed out toward the tail ends of the wire knot instead of slackening the tension in the joined wire, causing the tail ends to get longer while the length of the joined wire being tensioned remains exactly the same. When joining a wire break, the slack required to tie the knot can usually be gathered by loosening the permanent wire tightener (a small device used to provide tension to an electric wire by wrapping the slack onto the device with the help of another tool, called a permanent wire-tightener handle). If you do this, move the wire tightener to a new location on the wire. Retightening the wire tightener in an old kink weakens the wire and risks breaking it. The kinked wire in the old location will straighten out when the wire is retightened and will not be weakened.

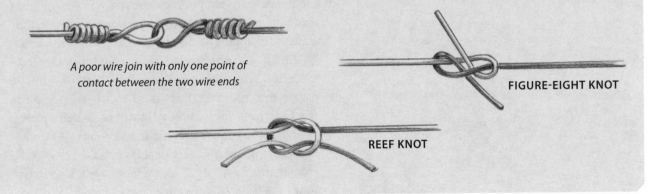

A poor wire join with only one point of contact between the two wire ends

FIGURE-EIGHT KNOT

REEF KNOT

Posts and Braces

The wires of an electric fence do not require a lot of tension because they do not act as physical barriers. But there must be sufficient tension to hold the wires up off the ground so they provide a psychological barrier and carry the shock at the proper height above the ground. Thus, electric fences can be built with far fewer posts and much lighter braces than a standard fence, and the fence line does not need to be as straight. Obstacles such as deep gullies, rough terrain, and steep ridges can be circumnavigated within reason as long as the insulators and wire are always placed on the outside of the curve so the wire pulls into the post rather than pulling the insulators away from the posts.

The traditional H-brace (top) is necessary only in soft ground, on high-tension fences, or if you are using multiple wires with very long spans between braces. An angle brace can be used for medium-tension fences. A four-foot stay block (a piece of timber that is placed into the ground to brace against the pressure of a post or angle stay — also known as a bedlog) should be dug horizontally into the ground against which the angle stay (the angled post used to support a post against the tension pulling on the post) is braced, as shown here (middle illustration). The fence post is notched to prevent the angle stay from slipping up, and the angle stay is rammed into place between the block and the post.

The bedlog brace is the simplest and quickest brace to build. It is ideal for most corner and strainer posts (end posts in a fence line that hold the tension, or strain, of the wires on the fence) on low- to medium-tension fences. All that is required is a 4-foot-long stay block or bedlog, which is dug horizontally into the ground at the base of the strainer post on the side of the tension and packed firmly into place, as shown below. The top of the bedlog should sit just beneath ground level.

Post spacing varies according to topography, to allow the wire to follow the contours of the ground. On flat ground, one post every 50 to 60 feet is sufficient; alternatively, posts can be spaced 100 feet apart and a fiberglass spacer used between each to eliminate wire sagging and to hold apart multiple wires.

Livestock producers considering organic certification for their land and livestock should be aware that the chemicals used to treat fence posts are becoming increasingly tightly regulated or even prohibited by most organic associations. If you are considering organic certification, please refer to the information on page 75 on approved alternative fence posts for organic producers.

TRADITIONAL H-BRACE

ANGLE BRACE

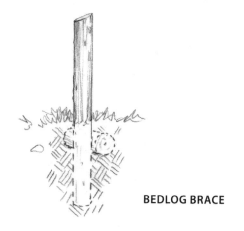

BEDLOG BRACE

Insulators

Unless you use plastic or Insultimber posts, you will require some form of insulator. (Note that Insultimber posts are made of high-density, nonconductive eucalyptus or acacia wood that is so hard that it does not require chemical preservatives, hence their popularity among certified organic producers. They are a product of Gallagher Animal Management Systems, which has a comprehensive worldwide network of electric-fence dealers.) I strongly recommend using high-quality insulators because they are more resistant to ultraviolet degradation and so will last much longer. The wires should be able to move freely through the insulator to absorb shocks in the event of an impact on the wire from a fallen tree or cattle running into it.

Some electric-fencing companies have come out with a new insulator style called a swivel-lock insulator (sold as PowerLock Insulators by Gallagher Animal Management Systems) that gives the farmer more flexibility. By swiveling the wire-holding mechanism in this insulator, you can easily remove wire while the fence is "hot" and under tension. Pin-lock insulators, sold by most electric fence companies, also allow you to remove the wire from the insulator, but with some difficulty and often not without receiving a shock from the wire. The swivel-locking mechanism of insulators like the PowerLock allows you to drop the wire to the ground on interior

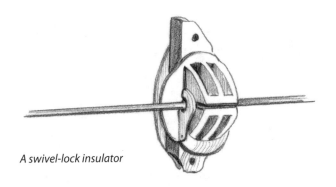

A swivel-lock insulator

single-strand electric fences so you can drive over them with equipment.

Alternatively, you can prop up a single wire with a long stick or notched PVC pipe to create a temporary, makeshift gate anywhere you need one. If the diameter of the PVC pipe is greater than the diameter of the post, you can slip the bottom end of the pipe over the post after lifting the wire to hold the pipe upright, creating a very stable temporary gatepost tall enough to enable cattle and equipment to pass underneath it. This gives your pasture rotations an added degree of flexibility and allows you to sort out or add cattle to the herd even if a permanent gate is not in a convenient location. If your cattle are trained to follow your ATV (discussed in detail in Training a Herd to Pasture Moves, this chapter; and Teaching a Herd to Follow in chapter 9), simply drive under the propped-up wire and call them through the temporary gate to show them its location. With this method you don't even have to touch the wires or turn off the power to create a temporary gate.

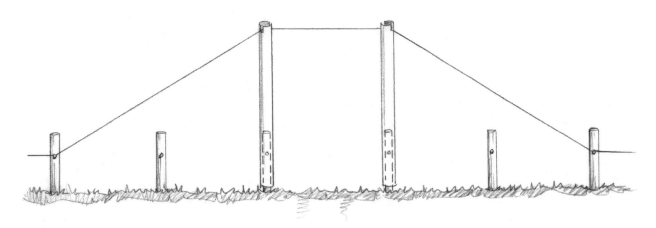

Create a temporary gate in an electric fence line by removing the wire from its insulators along a short section and propping up the wire with notched PVC pipes that fit over two of the posts.

A simple pair of insulated rubber gloves from the hardware store, such as the type sold for handling fuel, is essential for working around electric fences. When the gloves are dry, they insulate against electric shock, allowing you to lift the wire over your ATV (assuming you are using single-strand wire). The gloves come in handy for putting up portable fencing if you choose to work with live wires instead of connecting them afterward, as you might if you have a wire break and want to stop the cattle from traveling any farther.

Insulated rubber gloves also give you the option to repair live fences without disconnecting the power beyond the section you need to fix or modify. Wrap the hand grips of your permanent wire-tightener handle with heavy-duty electrical tape for added protection so you can loosen any wire sections that need work.

Swivel-locking insulators also allow you to access water alleys from anywhere in the pasture by simply propping up the wire; you don't have to build a multitude of permanent gates for each paddock subdivision. They are particularly useful when using portable electric fencing to change the geometry and size of your paddock rotations in response to grass growth rates and seasonal weather conditions.

Permanent Gates

Permanent gates built into your permanent electric fences should be constructed with cattle movements, not vehicles, in mind. Locate gates where cattle will naturally travel, such as in the corner of a pasture and in the bottom of a gully. A gate in the middle of a straight fence line or at the top of a hill causes cattle (who are lazy by nature) to scatter sideways instead of going through the gate. Put gates in locations that naturally draw your sight line so you can see from a distance whether they are open or closed. A gate placed just in front of a 90-degree bend in a road, facing a grove of trees, or with a second fence running parallel to it confuses cattle because they cannot see whether the gate is open. Instead of being drawn forward by a clear visual path, they mill around, bunch together, or scatter.

Insulate gates at their hinge end so they are live only when closed. A live open gate not only loves to bite the person trying to untangle it, but also snaps loudly if it's on the ground, scaring the cattle and stopping them from passing through.

When you connect the wires on either side of a gate, you have two options: overhead connections and insulated buried connections. In areas that get severe ground frost, the insulating plastic layer around underground wires will crack over time because of the expanding and contracting ground. In all climates, rodents are prone to gnaw through the insulating layer of underground wires. If voltage begins to drop in a fence line, underground wires in gateways are often the cause. Burying the insulated underground wire in a short section of polypipe or garden hose can help extend its life and protect it from ground compaction caused by vehicles driving over it. Overhead wires require less maintenance, unless someone snags them with tall machinery.

Portable Electric Fencing

A single strand of portable wire is sufficient for most portable electric-fencing divisions. If you design the permanent fence grid with portable fencing in mind, it will be narrow enough for one reel of portable wire to span the permanent wires. Insulated geared reels are a most useful invention because they allow you to swiftly wind up live wires while protecting you from any risk of electric shock.

Most polywire and polytape come in orange or white. Although white is somewhat more visible during summer and at night, it becomes invisible in snow, making late-fall and winter grazing a nightmare. Orange polywire, which is easily seen in snow and fairly visible during the rest of the year, is a better choice. I prefer polywire to polytape because it does not twist as much and rolls up more compactly.

Until recently, fiberglass posts were the most commonly used portable fence posts. They allow you to adjust the height of the insulator on a post, and they fit easily into a quiver (see box at right) for quick, portable-fence moves. But if anything hits the wire, the insulators have a tendency to pull off the posts and disappear. To avoid shattering the posts when you pound them, you have to use a driver cap and hammer, which you also must carry with you. In addition, ultraviolet solar radiation causes fiberglass to deteriorate, limiting its life span and making it brittle and splintery.

Step-in insulated portable metal posts are gaining in popularity. They have insulated hooks or loops to hold the wire and no tools are needed to put them in the ground; a quick push with your boot heel is all it takes. This makes them quite versatile, although their extra weight and slightly bulky design makes them heavier than fiberglass. Their wire loops and tread-in design are prone to entanglement if carried in a quiver, though they do fine in an ATV box.

Portable fencing works best as cross-fencing between permanent wires; the permanent wires form corridors subdivided by portable wires like the rungs on a ladder. One wire is set up behind the cattle and one in front. I also like to set up the next day's grazing subdivision so there are always two wires protecting the ungrazed grass from trampling cattle. Putting up the next day's wire ensures against a failed portable wire costing you many days' worth of carefully rationed grass and prevents an escaped herd from spreading manure (a.k.a. fly hatcheries) across future grazing divisions. It is not necessary to put up a second fence behind the cattle because the grass has already been eaten on that side. If the por-

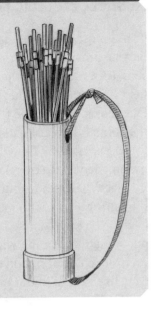

A quiver for your portable electric fence posts is easily constructed by gluing a 4-inch sewer cap to one end of a 3-foot section of 4-inch-diameter PVC pipe. Then cut a couple of holes through which to tie a rope or strap for carrying.

table fence fails, there is less incentive for the cattle to go back and eat the grass because it won't be as fresh as the food in their current paddock. In addition, the grass behind the cattle has fresh manure piles scattered throughout, which will begin hatching flies in just a few days. With all their biting and buzzing, the flies are irritating and so function as a sort of secondary back fence.

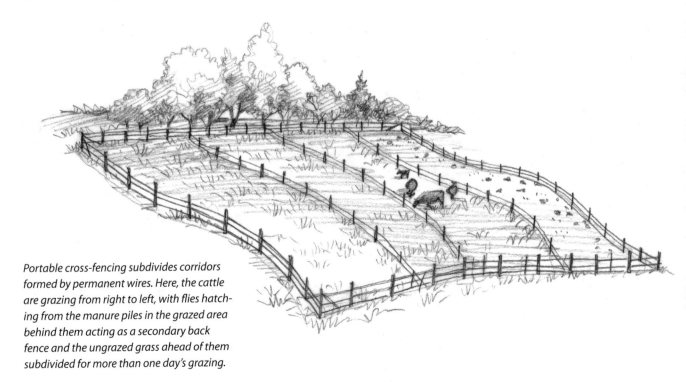

Portable cross-fencing subdivides corridors formed by permanent wires. Here, the cattle are grazing from right to left, with flies hatching from the manure piles in the grazed area behind them acting as a secondary back fence and the ungrazed grass ahead of them subdivided for more than one day's grazing.

Training Cattle to Respect Electric Fences

Cattle must be trained to respect the psychological barrier provided by an electric fence. The place to start is in a contained pen rather than out in a big field. To train a herd that is unaccustomed to electric fences or to train cattle that are new to the farm, put the animals into a situation where they are all forced to get shocked by the wire. A strong multiple-wire training fence in a corral or training pen works best.

Place the fence through the middle of the pen so the cattle are forced to make an inconvenient detour to get from their food to their water. For added effect, put something on the wire to attract the cattle to it, such as a colorful ribbon or something aromatic and tasty like peanut butter or molasses.

It will be obvious when the animals learn to respect the wires and maintain a safe distance from it, and within a few days they will be ready to leave the training pen. Once they learn this lesson, they will never forget it, no matter how flimsy and thin a wire may look to you. The occasional accidental tickle they receive when grazing underneath the electric fence seems to serve as a sufficient reminder of the fence's unpleasant bite. I have seen several instances where a malfunctioning energizer or electrical short cut all the electrical current to a fence, but it went unnoticed by the cattle for weeks before any of the animals tested the wire and crossed the single unelectrified wire containing them.

A HARD LESSON

The first time I turned cattle loose behind an electric fence, it was a catastrophe. I had carefully divided a huge hay field with a long network of portable electric wires to ration the grass. Quite proudly I turned out 300 uninitiated cows in the first pasture division. Untrained to calm pasture moves and completely unaware of the psychological barrier this electric fence was supposed to present, they rapidly stampeded through the entire hay field in a flurry of excitement and wire-busting enthusiasm. I spent the next two days collecting polywire, portable posts, and insulators from every corner of the trampled hay field that I was attempting to ration. It wasn't the cows' fault, though I'm sure my language at the time would have suggested otherwise. I had forgotten the most important lesson of electric fencing in my enthusiasm over its effectiveness and flexibility: Train the cattle. They had not learned to be afraid of the powerful bite of the little orange and silver wires.

This training pen is partially subdivided by an electric fence to train cattle to respect the psychological barrier of such a fence. Notice the flags on the fence to attract curious cows.

Understanding the Goals of Pasture Rotation

Your grazing philosophies during summer and winter are opposite because you are trying to achieve a different goal in each season. In winter, grass is food for the cattle, but in summer the cattle are your tool for managing the grass. The goal of summer grazing is to create the tallest possible grazing reserve without any of the grass becoming overly mature so it can be rationed during winter.

Calculate how many days it will take before your cattle need to regraze your summer pastures to prevent the grass from getting too mature. This period is the length of the grazing rotation. Because pasture-recovery time changes throughout the year, it is best to create a written, week-by-week grazing plan for the entire year. This helps you to determine how many acres you need to graze each day and, therefore, how many days you will spend in each pasture division. You can't simply eyeball it every morning when you go into the pasture. (See chapter 25 to learn how to calculate an appropriate grazing schedule and to learn how to calculate the amount of grass in the grazing reserve so you can adjust your herd numbers to fit the size of the grazing reserve on your farm.)

FENCES VERSUS SHEPHERDS

Until recently, we used shepherds to keep our domestic herds bunched together and migrating as a group. Although shepherds have largely been replaced by fences in North America and Europe, they continue to be used elsewhere, and in North America they are making a comeback, managing extremely large herds and low-stock-density rangeland. A single shepherd can easily eliminate all the fence and water-site infrastructure investments and maintenance costs related to fencing while still maintaining an intensive rotational-grazing program. Shepherding allows far more flexibility in managing the migration patterns of cattle because paddock divisions do not restrict movement. (See chapter 9 for a more detailed discussion of shepherding.)

The beauty of using portable electric fence inside a permanent electric-fence grid is that the portable wires allow you to change the size of the paddocks over the course of the grazing season as the grass speeds up its growth, slows down, or becomes dormant; one size of paddock does not have to accommodate every condition throughout the year. During the growth peak, it may take up to thirty days to make a complete rotation around your entire land to prevent the grass from getting too mature, but by fall the rotation may slow to about sixty days. After the dormant season begins, the grass has to be rationed daily until the next season's growth starts, so you may need more than 160 paddock subdivisions for thorough rationing.

> **TIP**
> The more pasture divisions you create, the less grass reserve is lost to trampling. A cow's feet consume five times more grass than her mouth.

For example, a forty-acre pasture may feed a small herd of cattle for two weeks if the herd grazes as a single unit, but it may last a month or more if you ration it out in daily "slices." More pasture divisions allow you to maintain a consistent grass quality from day to day during the dormant season, because each day the cattle begin with a fresh slice of high-quality grass. This is very important to your herd's health, fertility, body condition scores, and metabolic rates because it allows nutritional intake to remain consistent from day to day. If the grass is not rationed daily, the cattle will top-graze the best grass first and the remaining grass will steadily decrease in quality until the herd is ready to move to the next pasture. Also, because daily rotations force the cattle to graze each slice thoroughly in a single grazing session, in winter far less grass will remain buried and wasted under the snow after the herd moves on to the next slice, thereby allowing your grass supply to last longer.

Portable electric fencing enables you to realize your changing grazing goals if your livestock-watering options permit it. But you won't need 160 water sites to make it through the winter because you don't need a back fence to protect the regrowth from cattle mouths and feet. You simply ration your way outward from the water sites, and the livestock walk back across what has already been grazed to get to water.

Managing Summer's Grass Excess

In an ideal world where grass grows at a consistent rate and cows graze with your grazing calculations in mind, careful planning should be enough to create the best winter-grazing reserve and to prevent grass from becoming overly mature during summer. But life in the real world is rarely this simple. Even the cow's increasing nutritional demands during summer cannot compensate for the incredible grass flush that occurs then.

In order to manage this flush without ending up with an excess of overly mature grass, some producers make hay. It is far more economical, however, to vary livestock numbers on your farm in response to the seasons: more animals in summer and fewer in winter. You can accomplish this without buying off-farm stockers by keeping your yearlings through the following summer, which effectively doubles the mouths that feed during the growing season.

For example, if you have a herd of 100 brood cows, you will have only 100 cow/calf pairs to graze through winter on your grazing reserve, but you will have 100 cows and 100 yearlings during the summer growing season, and you will also have your open and cull cows to refatten on the spring grass flush before you take them to auction (or direct-market them as hamburger). Furthermore, selling your calves as yearlings the following fall (especially if you grass-finish and direct-market them by winter) allows you to produce the same pounds of beef per acre of grass with fewer brood cows, so winter herd size is even further reduced without affecting your beef output.

Pasture Rotation: Herd Migration in Your Own Backyard

Daily pasture moves are the key to managing a herd's impact on the land. Cattle and grass evolved together as a consequence of the herd's migration, which impacts the grass and soil briefly but intensely before the herd moves on. As few as twenty-four grazing subdivisions (also knows as cells) in a pasture rota-

tion will prevent nutrients from being exported from the pasture to water sites, shady spots, and favorite bedding sites. Although the manure will be evenly distributed across the fields, the frequency of pasture moves in a twenty-four-cell grazing rotation is still not enough to prevent pasture deterioration and an onslaught of weeds. It takes daily pasture moves to mimic the natural migration pattern and create self-sustaining and self-improving pastures.

Daily pasture moves are the key to managing a herd's impact on the land.

In a daily pasture-rotation schedule, the cattle herd changes its grazing habits. When cattle have access to large pastures for extended periods of time, they spread out and graze independently. But in a pasture rotation where space is limited and the herd moves frequently, the cattle begin to behave as a group, much like a flock of migrating birds or caribou. Members of the herd become very competitive to secure access to fresh grass each day. They stay together, grazing around the pasture as a group, neck to neck, shoulder to shoulder. Their selectiveness disappears. Rather than idly searching for the best mouthful, they grab what is in front of their noses and move on to keep pace with the herd. All grass species are grazed evenly, and inedible plants that would have been carefully avoided before are indiscriminately trampled instead. Manure falls evenly behind the herd as it grazes, much to the benefit of the grass underfoot.

Moving the cattle also becomes easier. As the cattle grow accustomed to grazing as a group, the herd becomes their place of security. During a herd move, the cattle eagerly follow each other to avoid being left behind, whereas if cattle are accustomed to grazing independently, they are less concerned about keeping up with the herd during pasture moves.

Training the Herd for Pasture Moves

Conventional search-and-gather pasture moves typically involve droves of cowboys, the bulk of a day, and a big dose of patience to get all of the cattle through the gate. But cows are remarkably trainable and thrive on the routine of frequent pasture moves. They learn to anticipate a move and become eager to pass through the gate to fresh grass.

Much as a dog salivates and comes running in response to his rattling food bowl, cattle are equally trainable to certain cues. A *come-cow* call that you use every time you call your cattle through the gate will excite the herd to follow you to new pastures. During the move, stay ahead of them, either on foot or on an ATV, and they will quickly learn to associate following you with reaching their next grazing location. Don't let them pass you. Try to slow them down until you train them to walk behind you to the next pasture so you can control the direction and they can calmly filter through a tight gate without flattening it in their excitement.

When you train them to follow you, don't just take them through the gate before you stop calling. Take them to the best, most lush part of the pasture so they get a really good reward for coming, and keep calling until everyone makes it through the gate.

When all the animals are accounted for, give them an all-okay call, such as *pau*, which sounds very different from the *come-cow* call. After a while they will know this means that the herd is complete, everyone is safe, and they have reached their destination. This trains them to continue to pay attention to you until they hear your all-okay signal. Consistency is key: If you are predictable, they learn to obey your signals because they know what the signals mean.

Over time, they will learn to follow you until you reach whatever destination you have picked for them, such as a faraway paddock a mile up the highway or the center of the sort corral on processing day. Just don't disappoint them at their destination. A treat at the center of the sort corral, like a few flakes of hay, will reinforce that your calls were worth following. Food is a powerful motivator. This training is also the foundation of shepherding, in which cattle are trained to follow a cue to the next unfenced grazing area and back and forth to a central water site once per day, even if it is a long distance away. (For more on shepherding, see chapter 9.)

TRAINING FOR PASTURE MOVES: A CASE IN POINT

To illustrate how effective this training can be, I'll share an experience I had with a small herd of 150 cows just 3 weeks after I started using *come-cow* calls in combination with a "follow me" training cue. When it was time for the herd to move, I attached the mineral feeder to the back of the ATV and pulled it to the next pasture while calling to the cattle.

(I like using the mineral feeder dragged behind an ATV as the "follow-me" cue because it prevents the cattle from associating you or the ATV with the cue — so you don't get mobbed every time you enter the pasture — and it is a visible, familiar object that remains with the herd at all times. After all, the mineral feeder is central to the cattle's lives in every pasture they use, so why not train the cattle

to accompany it to the next pasture when it is time to move?)

On this particular occasion, the herd had to leave a large, 80-acre pasture of bushland through a narrow, complicated gate configuration at the top of a long slope at the far end of the pasture. The cattle then had to walk uphill through knee-deep lush grass and cross almost a mile of lush grass fields to get to their next grazing paddock in the rotation. It was as complicated as it sounds!

With only one other person available to help, we were anticipating a long, frustrating day of old-fashioned cattle moving. But once the mineral feeder started rattling behind the ATV and I started calling, our freshly trained herd of cattle followed me up the hill, through all the complicated

gates, and across some of the most delicious-looking grass this side of heaven, all without stopping. They just plodded along until I stopped to unhook the mineral feeder.

The fellow who was helping closed each gate behind the herd and walked along, feeling let down by the lack of old-fashioned rodeo excitement. What stood out for me was how willing the cattle were to follow my training cue despite distractions after only a few short weeks of this routine. It also illustrates how training cues and a calm, predictable training routine can compensate for even the most rudimentary and poorly designed fencing infrastructure. Cattle are creatures of habit, but they certainly are not dumb.

Managing the Calving Season on Pasture

Like deer, cows have the habit of hiding their newborn calves in the grass to conceal them from predators until the calves are strong enough to follow the herd. Although calving on grass during the warmth of summer increases the vigor of calves and good herd selection eliminates cows with poor mothering instincts, you nevertheless risk some calves being left behind during daily pasture moves. What do you do when the pasture rotation causes newborn calves to be left behind or cows to break through gates to retrieve their young? And what if you find calves starving to death while they patiently wait for their mothers after the herd moves on or being eaten by predators after being left?

Calving during an ongoing grazing rotation would seem to spell disaster for newborn calves hiding in the grass. Yet everything we have learned from the herd regarding grass management and animal health suggests that the migration must continue; it cannot be suspended for the forty-two-day calving season. The herd must move on to avoid overgrazing and the migration around the pastures must continue to keep mature grass from forming in pastures that are not reached.

Time your grazing rotation so cows will calve on the lushest open pastures on your farm.

Likewise, keeping the grass in its S-curve of maximum grass productivity (see page 60) and preparing the winter-grazing reserve both depend on the constant movement of the herd. The herd must move on to avoid nutrient transfer, parasites, and diseases. Even the newborn calves, which fare so well when calving on grass, become prone to disease through oral-fecal transfers and flies hatching from manure piles if the migration stalls. A balance must be struck to successfully manage calving on grass.

Wild herds manage their calving seasons in a number of ways. First, their calving seasons are usually much shorter than our typical forty-two-day or sixty-day calving season. They also tend to calve in the same areas each year: areas that are fairly open so they can spot predators and where the grass is very lush so the migration can slow and only small distances must be traveled each day for fresh grass.

Most cows calve during the middle of the day while the herd grazes and ruminates. By the time the herd moves on the next day, the newborn calves have had time to gain their strength and suckle, and are strong enough to follow the herd. As soon as calving season is over, the migration rebounds to full speed, and the herd leaves behind its calving grounds, manure piles, afterbirths, flies, predators, and the predators' victims.

Shortening our domestic calving seasons to forty-two days or shorter will help make this period more manageable. To strengthen the bond between cow and calf so the calf will be more likely to follow its mother during pasture moves, do not tag or process calves during the calving season. Also, rigorously cull cows with poor maternal instincts to improve the maternal instincts of your herd.

The location you choose as your calving area is extremely important. Time your grazing rotation so cows will calve on the lushest open pastures on your farm. That way, calves cannot hide in the brush, and the herd requires less space to calve, which makes it easier to find calves that are left behind.

A consistent, rigorous, daily pasture rotation will help establish a routine that the cows can count on (moves occur at the same time each day; don't cheat and move cattle every other day). Invariably, when the herd is bunched tightly in a predictable daily pasture rotation, the majority of births occur *after* the daily pasture move, during the grazing period or the rumination, settle-down period that follows grazing. In this way, the calves have almost a full day to bond with their mothers and gain their strength before they are disrupted and the excitement of a herd move tests their attachment.

In a tightly bunched grazing scenario, the cattle identify strongly with the herd for security and fresh grass keeps curious cows occupied so they don't steal calves, which cows approaching labor are prone to do if they do not have sufficient space and too much time on their hands — common conditions in a conventional calving scenario. As a result, calving cows seek to calve in the middle of the herd for protection, rather than at the herd's periphery. Calving at the herd's periphery is reserved for continuous-grazing, winter-calving, and unhealthy cows (predator decoys) that are pushed there by the other cows. In the wild, a calf born outside of the herd is gradually introduced to the herd by its mother at a later date, when the calf has gained some strength. Neither the calf nor the herd identifies with one another until after the introduction. A calf born within the boundaries of the herd is automatically accepted by the herd and immediately identifies the herd as its security zone. When it feels threatened (such as during herd moves), it is more likely to follow the herd than hide.

Three Management Options

We have three options for successfully managing our grazing rotation during the calving season. The least desirable option is to suspend the daily pasture rotation for the duration of the calving season by providing the herd with enough pasture to last the entire calving season. The second-best option is to continue providing the herd with a fresh slice of pasture every day without moving the back fence, so the pasture space gets larger every day. The best option is to manage the cattle in such a way as to enable you to continue the daily pasture rotation with the back fence following behind, exactly as during other times of the year.

Suspend the Daily Pasture Rotation

In the first and least desirable option, you designate a large calving area that contains enough grass to accommodate the herd's grazing needs for the duration of the calving season, eliminating the herd rotation for this period of time. By avoiding herd moves during calving, you avoid accidentally leaving calves behind and you prevent the pandemonium of cows losing their calves in the confusion of a complicated pasture move.

To prevent unnecessary nutrient transfer, choose a site that is as flat as possible without trees or bushes or any other cover so there are no shade magnets that will continually attract cattle and accumulate manure. Prepare the site for calving by having

it grazed briefly at the very beginning of the spring green-up so it can produce a significant volume of forage prior to calving without becoming overly mature. Also, have the grazing herd sweep across areas that will be excluded from the calving grounds to try to prevent the grass in these areas from becoming overly mature by the time the rotation resumes. At the end of calving season, you have a big roundup and resume the rotation.

Despite the best plans and preparations, this option inevitably leads to a significant disruption in the effectiveness of your grazing management; causes significant nutrient transfer to favorite cattle hangouts, shady areas, and water sites; increases parasite pressure from flies hatching out of manure piles; and puts the calves at risk of disease from oral-fecal contamination, particularly if a period of wet weather occurs while the grazing rotation is suspended. An outbreak of scours or coccidiosis on grass is almost always caused by oral-fecal contamination due to a stalled grazing rotation, especially if there is wet weather to incubate the disease.

Continue the Daily Pasture Rotation, but Do Not Move the Back Fence

The second-best option for managing your calving season on grass is significantly better because the grass in your chosen calving area is rationed out daily, slice by slice. Although you give a fresh slice of grass to the cattle every day, however, the back fence is not moved so the cows have access to the entire previously grazed pasture in case they leave their calves behind. Flies hatching out of the manure piles act as a back fence, and fresh grass is the motivator for cattle to remain with the herd.

Each new slice must be adjacent to the previous slice, so a natural progression keeps the animals off the previously grazed, manure-laden areas and the cows can easily retrieve their calves. If the progression works so that the already-grazed grass slices get farther and farther away from each new day's grazing slice, it will become progressively more tedious for cattle to go back to the old slices through the hatching flies. Calves usually stop their hiding behavior within five days as they gain strength to follow their mothers, so unless you see cows calve in previously grazed areas, you can move the back fence to follow the grazing rotation about a week behind the lead fence.

Again, fresh food at the leading edge of the rotation motivates the herd to stay together, but only if the water-access point moves along with the leading edge of fresh grass. Focus the cattle's water needs at the leading edge of the grazing rotation by using either a portable water trough placed inside each fresh daily slice of grass or an access point (gate or raised wire) opening into a water alley from the fresh grass slice. Water sites, mineral licks, shade, and anything else that draws cattle back into previously grazed areas will prevent you from closing the back fence behind the cattle because of the risk of cows depositing their calves in these areas.

Slowing down the grazing rotation in an effort to create smaller pastures so abandoned calves can be spotted more easily creates an abundance of overly mature grass at the leading edge of the grazing rotation. The combination of overgrazed and undergrazed grasses limits the productivity of your grass and hurts your winter-grazing reserve. Tall, seedy grass irritates cattle's eyes, creating an ideal environment for pinkeye, and the abundance of tall, mature, unpalatable grass clumps that young calves can hide in makes moving the back fence even more difficult. The grazing rotation during calving season must continue to be matched to the speed of grass re-growth, just as during the rest of the year.

Continue the Daily Pasture Rotation, Including Moving the Back Fence Daily

In the third and best option for calving during the grazing rotation, the cattle are managed in such a way that the grass in the designated calving area can be rationed out daily in adjacent slices and access to water accompanies each day's fresh grass slice. The back fence is moved with the cattle as it is during the normal grazing rotation for the rest of the year.

The herd must be moved at the same time each day, without fail, to promote calving within the herd and to encourage cows to calve *after* the daily pasture move. Because each new slice is adjacent to the previous day's slice, just across a single thirty-two-inch-high portable wire, the mothers are never separated from their calves by any great distance, and calves can run underneath the wire to rejoin them.

In daily moves to adjacent paddocks, the trip is not difficult for the calves and the cows do not feel the panic of having to run to keep up with the herd. A cow knows the routine so she knows the herd will not abandon her. This sense of security ensures that she either will wait for her calf or will return for it within the first few minutes after the pasture move. By contrast, longer pasture moves risk a cow's temporarily abandoning her calf to ensure her own place in the herd. If a cow starts bawling or pacing along the back fence line with a full udder a few hours after the pasture move, you'll know that a calf has

This optimal layout for calving on pasture enables daily pasture moves into immediately adjacent pasture slices and facilitates moving the back fence to accompany the cattle's daily pasture rotation.

been left behind. Opening the gate is usually all that's needed for the cow to retrieve her calf and rejoin the herd. Implicit to the success of this approach, however, is that you observe your herd's behavior very carefully throughout the day so you will notice any behavioral clues that indicate that a calf has been left behind. You cannot simply move the herd and then walk away for the remainder of the day, as you would during the rest of the year's pasture moves. Any cow that does not have a calf at the end of the calving season should be culled. This method, therefore, is also a system of improving maternal instincts.

The layout of the pasture divisions in the calving area plays a big role in the success or failure of an ongoing pasture rotation during calving with a moving back fence. If the terrain is flat and divided into narrow adjacent slices of fresh grass that run parallel to one another (see the illustration on page 84), the herd is never visually separated during a pasture move. But if the paddock layout creates a situation where the next slice is a long distance from the previous slice or if the topography can hide a cow from the herd, she will choose the herd before the calf. In the wild, the calf that does not follow the herd jeopardizes its own survival as well as its mother's. If the cow is visually separated from the security of the herd, she is predisposed to abandon a calf that does not follow. This is also why it is extremely important to train the herd to move calmly at the sound of the *come-cow* signal; it maintains a sense of security in the herd. If the herd rushes headlong through the gate to reach fresh pasture, the excitement of the moment will panic a new mother into thinking she needs to hurry to follow the herd. If she panics, she is more likely to abandon her calf than coax it to follow a calm migration.

Livestock Water

An effective pasture rotation hinges on cattle having access to an adequate supply of clean water. Water is required to digest food and flush out waste toxins produced during digestion and is a significant component of weight gain: 70 percent or more of cattle's total body mass is water. If water intake is restricted, appetite decreases to maintain the balance between food intake and waste processing so all digestive waste can be flushed from the system. Water also allows the immune system to function properly by flushing from the body stress and disease toxins and other by-products of a functioning healthy immune system. In other words, in cattle, weight gain and good health are tied to the amount and quality of water they consume.

How Much Water Do Cattle Need to Drink?

The quantity of water required by cattle (in gallons per day or gpd) changes dramatically throughout the season depending on growth rates, moisture in the food, lactation, summer heat, access to shade, and humidity.

Peak Water Supply

Although the daily requirement of water for your cattle determines the amount of water they will consume over a twenty-four-hour period, it has to be

DAILY WATER CONSUMPTION BY LIVESTOCK

Livestock class	Average gallons per day (gpd) per animal
Dry cows and heifers	6–15
Lactating cows	11–18
Bulls	7–19
Growing cattle	
400 lbs	3.5–10
600 lbs	5–14.5
800 lbs	6–17.5
Finishing cattle	
1,000 lbs	8.5–20.5
1,200 lbs	9.5–23
Dairy cattle	10–25
Bison	*See* Cattle
Horses	8–15
Chickens (for every 100 chickens)	4–10
Sheep (for every 100 lbs)	1–4
Pigs (for every 100 lbs)	2–5

Let me tell you about a water system I spent many sleepless nights trying to repair. One part of the family farm I was managing was a 600-acre summer pasture some distance from the main farm. It was divided into a wagon-wheel grazing rotation around a single central water corral, which was supplied by one mile of underground pipe from an artesian well at the property boundary. The grazing cell had just been switched from a 120-cow and calf pair summer range, with part of the land set aside for grain production, to be used exclusively as a summer stocker grazing operation for more than 350 stockers.

The stockers' weight gains on this property were disappointing, however, and were getting worse. The grass was remarkably lush, the soil showed no nutrient deficiencies, no fault could be found in the mineral program, and there was an abundance of unused grass — so what was the problem?

The water trough was always full and seemed to be working properly, but one day I happened to arrive for a pasture check when the whole herd was drinking at the water corral. There was quite a bit of pushing and shoving around the trough, which the cattle had drunk to the bottom. Although water was still flowing into it, the sides were bone dry, telling me that the cattle's thirst dramatically exceeded the water supply's flow rate. This realization was the prover-bial lightbulb moment during which all of the cattle's behavior and weight-gain issues became clear to me.

The water was flowing into the trough at 3.5 gallons per minute (gpm). The cattle had to wait at the trough all day to get a turn at the water, and even then, it was simply not enough to meet the demands of high weight gain. They simply could not drink enough water to eat more.

The flow from the well had slowed. After I cleaned out the well and added a pump, the flow returned to its original volume of 6 gpm at the trough, which had been adequate for the previous, smaller herd. Still the problem persisted. The stockers continued to drain the trough faster than the well could refill it, so the cattle spent their whole day in the water corral, pushing and shoving to get their turn, instead of grazing in the pasture. The corral was turning into a muddy, manure-filled fly magnet and all of the valuable nutrients from the field were quickly accumulating around the water trough.

Although pumping increased the well's flow, only a fraction of the well's capacity was actually reaching the trough. The underground lines had been sized for the smaller herd. The narrow pipe diameter created tremendous friction losses to the water pressure (see the discussion on piping water on page 90) over the long distances that the water was flowing. Replacing the underground lines over that great distance to meet the peak water demands was not feasible, for it would have required digging up a public road separating the well from the rest of the property. The only other option I had was to create water storage at the corral.

The problem was solved by installing a giant aboveground water tank that gravity-fed water to the trough at over 40 gpm. The well might take all day to fill the tank, but when the cattle came to the trough to drink, every last one of them got their fill in minutes.

The change in the behavior of the cattle was remarkable. Water ceased to be a limited resource. Their weight gains skyrocketed and the dominant animals no longer felt the need to control access to the water trough because water was no longer in limited supply. Soon the stockers began to drift into the corral one at a time to drink, after which they immediately turned around and left. They no longer felt the need to rush to the trough together to ensure that the first arrivals didn't deplete the water. The corral dried up, the flies disappeared, the manure stayed in the fields where it belonged, and health concerns such as pinkeye and foot rot all but disappeared because the animals no longer spent the day surrounded by flies and with wet muddy feet.

This process taught me the difference between a clean water supply and an *adequate* clean water supply capable of meeting the cattle's demands.

MINIMUM PEAK FLOW RATES USING THE DAILY WATER REQUIREMENT

Daily requirement (gpd)	Minimum peak flow rate (gpm)	Daily requirement (gpd)	Minimum peak flow rate (gpm)
1,000	8	7,000	39
1,500	12	8,000	42
2,000	16	9,000	45
3,000	24	10,000	48
4,000	28	12,000	50
5,000	32	15,000	55
6,000	36		

From Lance Brown, ed., *BC Livestock Watering Manual*, 1990: 2–4. BC Ministry of Agriculture and Fisheries, Abbottsford, BC

delivered within a much shorter time period to prevent it from becoming a limited resource, which creates stress among the cattle. They will fight to control it, with the more dominant ones monopolizing the trough and the weaker ones getting pushed aside. The dominant cows will then spend a good portion of the day guarding the supply from others in the herd.

In addition, if water is delivered too slowly, valuable grazing and ruminating time is wasted while the animals wait for the supply to catch up to their drinking rate. The water must flow at a rate that can meet the peak demand and fulfill the drinking needs of multiple animals at one time. The above chart shows the minimum peak flow rates required to meet the daily water needs of a herd of cattle. The daily requirement is the maximum amount each animal in the herd should drink (see the chart on page 86) multiplied by the number of animals in the herd. For example, during the summer heat, 350 stockers weighing 1,000 pounds each require:

20 gpd × 350 stockers = 7,000 gpd

According to the chart above, that means a peak flow rate of 39 gallons per minute (gpm). Cattle are unpredictable in how often they drink. Some drink two or more times a day; others drink once a day or even every other day. Thus, the peak flow rate should be sustainable over a six-hour period so the entire herd can drink at once, even if some cattle come to water only every second day. If the water supply cannot provide or sustain this rate, water should be stored on-site to meet the demand. Ideally, a water storage facility will contain at least two days' worth of water to supply the sustained peak flow rate.

I like to periodically test the water flow rate at the trough with a 5-gallon bucket. Well production may decrease during a summer drought, nozzles may get plugged, and lines may accumulate sediment, but you would never know it by the sound of the splashing water. You will, however, notice later when the livestock are weighed as they come off pasture. Flow rate is calculated by determining how long it takes to fill a 5-gallon bucket and converting that time to gallons per minute (gpm). For example, if a 5-gallon buckets takes 15 seconds to fill, your water flow rate is:

5 gallons ÷ 15 seconds =
0.33 gallon per second × 60 seconds =
19.99 or 20 gpm

(or divide 60 seconds by the number of seconds required to fill a 5-gallon pail, then multiply by 5; [60÷15] x 5 = 20 gpm).

Water Quality

The amount of water that cattle will drink and the water's ability to regulate their digestive and immune systems also depends on the water's quality and taste. This may seem obvious, but when you look at the kind of water many cattle drink, it bears mentioning: Livestock water should be the same quality as our own drinking water. Besides the obvious pollutants such as pesticides, parasites, salinity, rust, and chemicals, there are a number of additional contaminants that can be present in a water supply. I am amazed at how often I see cattle drinking water that is obviously contaminated with their feces. Cattle feces are not just *E. coli* hazards to humans; drinking manure-contaminated water exposes cattle to all sorts of diseases. Cattle will not drink from contaminated water by choice.

Having said that, many of us have seen a cow drink from manure- and urine-laden puddles right in front of the water trough, but this is usually a sign that an animal's health is compromised or that it is suffering from a nutrient deficiency. It may be seeking urea to supplement its diet, for example.

Blue-green algae are also a concern in slow or stagnant water. Although green algae are not harmful, the blue-green variety produces a toxin that kills livestock, in some cases very quickly. It is important to clean the troughs periodically.

To prevent contamination from drowned birds and small mammals, install a little ramp in the trough that creatures can use to climb out if they've fallen in. If the ramp is designed to provide easy access to water and not just to function as an escape route, it will attract many birds, which will return the favor by harvesting bugs and reducing the parasite load on the cattle.

Direct Access to Open Water

Cattle have messy drinking habits. Something about all that water rushing down into a cow's belly stimulates her urge to drop a pile of manure, usually right where she is standing. And if she can wade out into the water to drink, she will, particularly if there are other cattle crowding the water's edge. This puts a great deal of manure into the water, which is undesirable and even illegal in many areas because of water-quality issues. The churning feet of cattle can also rapidly destroy the water's edge, so that eventually the nutrients in the manure create algal blooms in slow-moving or stagnant water and cattle are forced to wade through a death trap of muck at the edge to reach clean drinking water.

If a body of open water such as a lake, stream, or dugout (artificial pond) must be accessed directly by cattle (rather than having water pumped from one of these into a trough), it should be fenced off and access points should be constructed. At each access point, the slope to the water should be relatively shallow. If the ground is not rocky, then gravel should be hauled in to make it firm. If the soil in

A water trough with a ramp installed to prevent birds and other small animals from drowning. A safe water source attracts many birds, which will repay you by eating bugs and flies and picking parasites off cattle.

the access point is muddy, particularly soft, or high in organic matter, the added gravel will be pushed down into the underlying soil by the cattle's hooves. Prevent this by laying under the gravel a geogrid (a liner or mesh made of plastic or polymers that was developed by the road-building industry for building on soft ground; the mesh locks the gravel into place).

The illustration shows an access point to a watercourse or lake. It provides just enough space for the cattle to get their noses and front feet wet and prevents them from wading into the water. The access point is quite narrow, only a little wider than the length of a cow, so she will have to back away from the water source after drinking if she shares the access point with other cattle. This discourages the cattle from turning around in the water and planting their manure piles in it. If the water level changes throughout the season, constructing an adjustable gate across the front of the access point using panels that are tied onto the main fence will allow you to move the gate up and down the bank to accommodate this changing level.

Once the temperature drops below freezing in winter, however, these access points become treach-erous. The ice buildup on the slope makes for a dangerous slide down to the water's edge. Although the water will be kept relatively open by the livestock traffic and will require only minimal chopping, water splashing on the leading edge of the bank will create a giant step of ice that the cattle have to negotiate into the cold water or which they have to kneel on to reach the water below. In addition, overflow and ice buildup in a stream can move the water away from the access point at a moment's notice. If you witness cattle trying gingerly to maneuver themselves in and out of an access point like this to avoid breaking a leg or falling onto their pregnant bellies, you will likely come to hate these access points in winter as much as the cows seem to.

Pumping Water

Unfortunately, not every water site will have a well as a water source, nor is electricity, diesel, or gas always available to power a pump. But there are a number of innovative pumping solutions addressed in detail

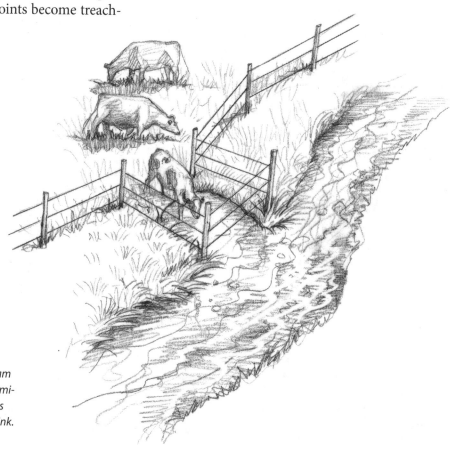

A properly constructed livestock water access point protects stream banks, minimizes manure contamination of the water, and provides the cattle with a safe place to drink.

below that utilize the energy provided by the sun, the wind, the weight of the water, the current of the water, gravity, or the animals themselves to allow you to pump water from a lake, stream, ditch, dugout, or swamp directly into a trough or a storage facility, from which it can then be gravity-fed into the trough. Many of these alternative pumping methods are mobile, allowing the pump and trough to serve multiple locations or to be moved along the water's edge alongside a pasture rotation. If the banks are steep, muddy, or extremely overgrown, the cattle will prefer to use a convenient trough full of clean water on the bank rather than make an inconvenient journey to the water's edge, even if the stream banks are not fenced. All of these pumping methods are limited by the "head," or height the water has to be pumped to reach the trough or storage facility, and by the volume required by the herd.

If you use a continuous pumping method, the trough can be constructed with an overflow line to drain unused water back to the water source, or a float can be mounted in the water trough to shut off the flow when the water reaches a certain level. Traditional float systems require protection from cattle to safeguard the brittle float arm and shut-off valve, but now there are float systems that tie to a shut-off valve with a short piece of rope whose length adjusts the water level in the trough. The float can be pushed around by cattle without harm and the shut-off valve can be mounted underwater at the base of the trough, out of the cattle's reach.

Solar-Powered Pump

A solar-powered pump is a fairly reliable way to pump water. Its main drawback is the relatively high initial investment in solar panels, though the system is fairly maintenance-free and has a long life expectancy. It is also relatively mobile, so you can combine it with a portable water trough that travels along a creek, drainage ditch, or lakeshore. If a solar pump is combined with a water-storage facility, water can be supplied to the cattle through the night, even when the solar pump is inactive.

Wind-Powered Pump

In consistently windy areas, a wind-powered pump such as a windmill is another option. It should be combined with a large multiple-day storage facility so water will be available when the air is calm.

Hydraulic Ram Pump

A hydraulic ram pump can be used wherever there is a sufficient stream gradient to create an artificial water gradient in a pipe (called the drive pipe) to provide the energy to drive the pump. The speed and weight of the water accumulating in the drive pipe power the pump much like an old-fashioned sewing machine is driven by pumping a foot pedal: Apply weight on the downstroke, release weight for the upstroke of the pump piston (or sewing machine needle). Stream gradients that create as little as 1½ feet of drop in the drive pipe are sufficient. The speed and weight of the water rushing down the drive pipe to the pump cause a valve to slam shut, like wind slamming a door. This "water hammer" drives a smaller set of pistons that pump out water through a smaller pipe controlled by a one-way check valve

An underwater flow valve operated by a float system attached to the valve by a rope regulates the water level in the trough. Unlike traditional float systems, this setup is extremely resistant to damage from cattle.

that prevents water from rushing back to the pump between each water hammer. This allows the pump to push water quite a considerable distance uphill (up to 600 vertical feet, depending on the water volume pumped, the supply head or drop in the supply line that drives the water hammer, and the manufacturer and size of the pump used), making it a versatile pumping unit. For example, depending on the supply head on the drive pipe, a large Glockemann water-powered pump ranges from lifting up to 5,800 gallons per day to a height of 60 feet to reach the trough to 690 gallons per day lifted to a height of 600 feet.

Stream-Driven Pump

A stream-driven pump can be used if the stream has enough current. The current spins a propeller on the pump suspended in the current to push water up the bank to a trough or water-storage facility.

Animal-Driven Pump

An animal-driven pump uses the livestock themselves for power. Small water bowls are fitted with obstructions to block the cattle from reaching the water in the bowls. The cattle push the obstructions out of the way with their noses, thereby working the pump. With each stroke, the water bowls are refilled. When the cattle have drunk all the water from the bowls, they pull back. As the obstructions slide back into place, they push fresh water back into the bowls. The cattle sniff it and push aside the obstructions to drink, repeating the process. One nose pump will water 20 to 50 animals and can lift water 20 feet above a water source.

A stream-driven water pump uses current to spin a propeller, which then powers the pump.

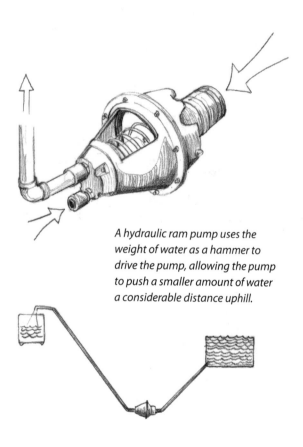

A hydraulic ram pump uses the weight of water as a hammer to drive the pump, allowing the pump to push a smaller amount of water a considerable distance uphill.

An animal-driven pump is powered by cattle when they push aside an obstruction to reach water inside the drinking bowl.

Gravity-Fed Lines

You can also use gravity to provide water flow if you're able to tap into a water source on high ground. A screened pipe intake can be hung above the bottom of the water source to avoid debris or from a floating pontoon or barrel to allow the water intake to compensate for changing water levels. Gravity-fed water lines will effectively siphon water over any ridges or high spots between the water source and the trough.

When these lines follow undulating ground, the high spots are prone to collecting air bubbles. These counteract the flow of water, which has to push the air bubbles down through the next dip in the line. Prevent this air locking by installing air vents on the high points in the line. A little ball float seals the line from water leakage but allows air pockets to bleed out when they displace the water and break the seal.

Water Storage

Unfortunately, it is not always possible to have a sufficient water supply that is capable of satisfying the peak water demands needed at the trough. High water volumes quickly become very costly to pump because they require high-pressure systems or expensive large-diameter pipes to transfer water from the source to the trough (see the discussion on piping water on page 96). Or the water source may simply not be capable of supplying water fast enough, as in the case of a low-volume well. Or perhaps there is no readily available water source where it is needed so an artificial water source has to be created. In any of these cases, some form of water storage is needed, ranging from the smallest of tanks used to temporarily boost water-flow rates to meet peak demands to enormous water-containment structures capable of storing or even collecting a sufficient water supply to last an entire season.

Storage Tanks

In a cost-per-gallon comparison with the other water-storage methods discussed below, storage tanks are the most expensive form and are therefore typically cost effective only when a relatively small water volume needs to be stored, such as when water flow must be temporarily boosted to meet peak water demands at the trough. When used as such, by setting the storage tank aboveground, water can be gravity-fed into a trough, thereby eliminating the additional cost of a second pump to transfer water from the tank to the trough. The taller the tank, the higher the water pressure at the trough (every additional foot of vertical height above the water trough will increase water pressure by 2.31 psi). This allows a trough to be filled at high peak flow rates even if the supply line to the tank carries only enough flow to satisfy the minimum daily drinking requirement.

An aboveground storage tank allows you to gravity-feed water into a trough at a high delivery rate, even if the tank is filled very slowly from the water supply.

The tank can be filled from a remote location and the water supply coming from the source can be turned on and off by attaching a float valve inside the storage tank. Thus, no power is necessary at the trough. Bear in mind that the water in aboveground storage tanks is prone to freezing. A discussion of watering livestock during the winter and how to effectively winterize water sites is found toward the end of this chapter.

Grain Bins

A cheaper option for storing a somewhat larger water volume is to convert a grain bin for water storage by lining it with a special vinyl liner. The bin should be covered, however, to prevent rodents and birds from falling into it and contaminating the water supply. A solid roof that blocks sunlight will prevent evaporation losses, reduce water warming, and slow algal growth. Converted grain bin rings cannot be built very high because the water constitutes a much heavier load than the grain-storage design intended.

Ponds and Dugouts

Ponds and dugouts can be built at a much lower cost per gallon than can storage tanks and grain bins, making them suitable for storing much larger volumes. They can also be lined with a variety of materials to prevent seepage.

Gleization

Gleization is the cheapest method of lining a pond or dugout. The bottom of the dugout is lined with a thick organic layer such as straw, which is then covered with a layer of clay to seal the decomposing organic layer. As the straw decomposes underwater, the lack of oxygen causes a natural chemical alteration to the mixture of clay and organic material, creating an impermeable bluish gray sticky soil called *gley*. Clay can be used by itself, but a twelve-inch-thick layer is needed to create an effective seal, and it won't work if the clay has silt in it.

Polyethylene and Rubber Liners

Liners made from polyethylene or rubber are commercially available. Cover the liner with a thin layer of soil to prevent breakdown by ultraviolet radiation and punctures by sharp objects.

Bentonite Clay, Soil Cement, and Chemical Sealants

Bentonite clay can be used as a sealant: It swells up to twelve times its dry volume and forms an impermeable, gelatinous layer at the bottom of a pond. Soil cement, a compacted mixture of soil, portland cement (the most common type of cement in general usage), and water, is another viable option for sealing a pond floor. Chemical sealants such as sodium

Water can be stored in a converted grain bin lined with vinyl and covered with a solid roof to block sunlight (preventing algal growth) and to prevent birds and rodents from drowning.

carbonate that interact with the clay particles already present in the pond soils can also be used, but chemical methods usually must be repeated to maintain their effectiveness and they may negatively affect the quality of the drinking water.

Water Harvesters

A water harvester allows you to collect rainwater in areas where there is no other available water source. It can be constructed in a number of ways, but the most common is to contour the landscape as a catch basin so the runoff funnels into a smaller storage facility such as a dugout. Another common method is to surround a storage container with a rooflike structure, such as sheet metal, to catch and direct water into the storage facility, much the way that a rain barrel collects runoff from the roof of a building.

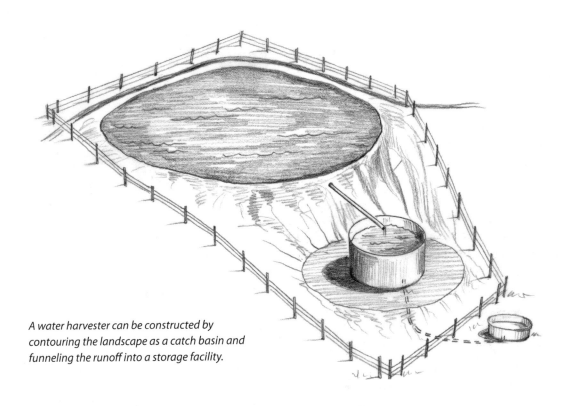

A water harvester can be constructed by contouring the landscape as a catch basin and funneling the runoff into a storage facility.

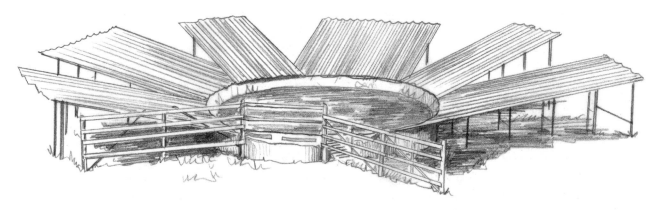

A water harvester can also be constructed by surrounding a large storage container with a rooflike structure made of sheet metal to funnel rainwater into the container.

Piping Water

When water must be transported any distance, friction in the pipe reduces the pressure and consequently limits the flow of water. Pumping uphill increases pressure and flow, and pumping downhill decreases pressure and flow. To design a water system that guarantees a certain pressure and flow (gpm) at the trough or storage facility, all pressure gains and losses must be converted into a *pressure head,* which is a way of expressing pressure in feet, corresponding to the height of a vertical column of water that would produce the equivalent pressure at its base through the force of gravity (1 psi is equivalent to a 2.31-foot-high vertical column of water). The practice of calculating water pressure as a pressure head developed as a simple tool to determine gravity-related pressure increases or decreases in the water system caused by vertical rises or drops throughout the water system. (To learn more about pressure heads, see chapter 24)

The minimum flow rate required at the trough or water-storage facility determines how you will build your water system. To guarantee a particular flow, you must size the pump (if you are using one) appropriately and use pipe of proper diameter to minimize friction losses. By adjusting pipe diameter, you can manipulate friction losses in the supply line and keep the pumping pressure within the safe working range for which the pipe is rated, preventing it from bursting. To learn how to size the pump, choose the proper pipe diameter, and assess potential pipe friction loss, see chapter 24.

Cattle Behavior at the Water Trough

As we learned in the discussion of peak water supply, when there isn't enough water for the cattle's daily drinking needs or when the flow is insufficient to quickly water the entire herd, water becomes a limited resource. When this happens, the social hierarchy becomes important and the cattle struggle to assert dominance and gain first access to the water.

After a short time, the watering site becomes the main hangout for the cattle, not only to wait their turn, but also because the dominant cattle attempt to keep the other cattle away from the limited resource. Manure and urine thus accumulate around the watering site instead of out in the field where they belong, and the site quickly turns into a muddy, messy fly hatchery.

After sleeping, ruminating, and resting in the fly-infested muck, the cattle become prone to foot and eye diseases, not to mention a stressed immune system and slower weight gains caused by drinking insufficient water. They limit their grass intake to get back to the water trough sooner instead of eating their maximum fill; as soon as one cow starts moving toward the water trough, the whole herd follows.

Grazing rotations must be designed around water-access points so that each division is accessible to water.

When water ceases to be a limited resource, the herd no longer feels the need to hang around the water site and compete for access. The cattle go to water individually as they become thirsty rather than going together all at once. They drink their fill, then immediately return to the pasture to eat and regain the companionship of the herd, and manure and urine transport to the corral is virtually eliminated. When water ceases to be a limited resource, the corral dries up, provided it is in a well-drained location and the mud-prone soils are covered with gravel or sawdust to lessen the effects of the wet season.

The distance between the water site and the grazing area also affects watering behavior. If the herd can see the water site from where it grazes, animals that go to water alone do not feel separated from the herd. But if the herd is not visible from the water site, individual animals feel vulnerable to predators and will travel to the water site only in a group.

Terrain influences how far cattle are willing to travel for water. The rule of thumb for the maximum distance between the grazing area and the water site is ¼ to ½ mile on rough terrain, ⅜ to ¾ mile on rolling terrain, and ¾ to 1 mile on level terrain.

Rotational Grazing Patterns and Water

Grazing rotations must be designed around water-access points so that each division is accessible to water. We'll look at three options: the wagon-wheel configuration, water alleys, and water lines along electric fences, listed in order from least to most effective livestock-watering system to facilitate the most flexible grazing management, uniform grazing impact, and efficient nutrient recycling.

Wagon-Wheel Configuration

For this design, the water site is positioned at the center of the pasture. Fences radiate out from it like spokes of a wagon wheel, with each wedge having access to the water site and each typically having a gate midway along the fence to facilitate cattle rotations.

This design is often used for dividing up large tracts of land, but it has several disadvantages. The cattle graze the areas closest to the corral quite heavily while areas farthest from the water site are not grazed enough. Because of this, the narrow end of the wedge closest to the corral is prone to compaction from repeated cattle crossings. Also, this graz-

ing pattern is difficult to subdivide further with portable electric fencing, which makes it challenging to vary the size and number of paddock subdivisions in response to the seasons and grass growth rates.

Water Alleys

In this design, alleys are used to provide access to a central water site from all parts of a grazing rotation. Each pasture division in this design is grazed more uniformly than is possible in the wagon-wheel configuration. Within each pasture "run," permanent or portable wire can be used to create ladder-rung-style pasture subdivisions. Portable wire divisions make it easy to vary the size and number of pasture divisions in response to the seasons and weather conditions. This design also lends itself well to a pasture-calving scenario, the back fence following the herd daily or approximately five days after the lead fence. (See chapter 5 for more on pasture calving.)

A water alley is accessed via permanent gates or is constructed with swivel-locking insulators so the wire can be removed from the posts and lifted to create a temporary gate anywhere. The alleys can be seeded with a grass species that is resistant to traffic and high-grazing pressure, such as the lawn-grass

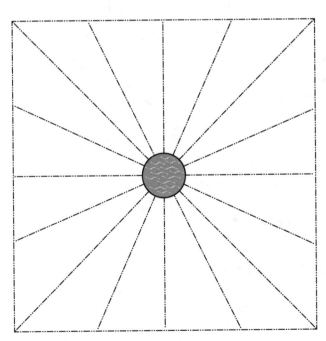

A "wagon-wheel" rotational grazing layout contains a water site at the center of the pasture with pastures radiating outward from it like the spokes of a wheel.

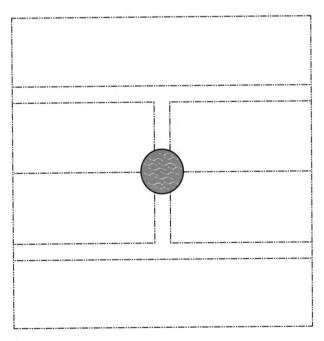

A "water-alley" design for rotational grazing connects all pasture divisions to a central water site through a series of alleys.

varieties, or they can be graveled or covered with sawdust, as are many grass-dairy alleys.

Alley width is important. Each should be wide enough to accommodate some bunching during cattle moves and to prevent problems at difficult junctions. The more cattle in the herd, the wider the alleys should be, but even if the herd is relatively small, they should be at least 30 feet wide.

Like gates, alleys should be constructed with clear visual cues that help the cattle flow through without balking. Tight 90-degree bends look similar to closed gates from a distance, causing the cattle movement through an alley to stall. In corners tighter than 90 degrees, the lead cattle move in a direction opposite the rear of the herd after they have rounded the bend, which causes the animals at the rear to balk, reverse direction, and bunch up in the alleys as they attempt to follow the direction of the herd leaders. Therefore, wherever possible, bends should be wider than 90 degrees to facilitate cattle movement in the right direction.

Water Lines along Electric Fences

This configuration is similar to the "water-alleys" design, but the water is brought to the cattle rather than the cattle having to follow the alleys to reach the water. Water lines (typically PE pipe) are laid out along every second fence line. Quick-connecting couplers are installed periodically so a water line can be attached to a portable water tub that follows the cattle as they are moved to different pastures. These couplers allow water to be accessed from both sides of the pipeline when the water tub is attached to a flexible hose. The water lines can be laid aboveground or directly under the fence line, or buried. The electric fence protects the aboveground water lines from being trampled. Bury the pipe under gates so equipment can pass through and cattle can cross from one division to the next without trampling it.

This configuration allows you to disperse manure buildup over multiple water sites, thus further reducing nutrient transfer. The convenience of a water tub in the middle of a grazing area also helps to spread

CONFUSION CAUSED BY TIGHT ALLEY CORNERS

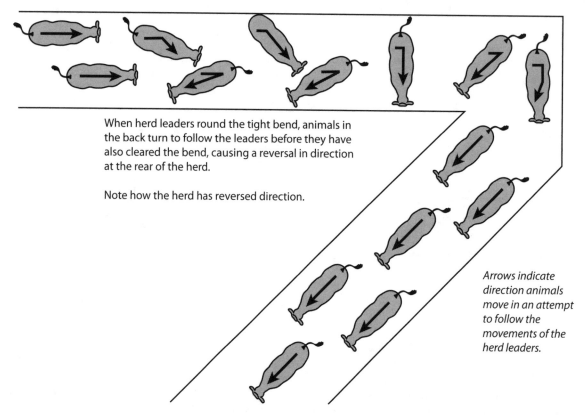

When herd leaders round the tight bend, animals in the back turn to follow the leaders before they have also cleared the bend, causing a reversal in direction at the rear of the herd.

Note how the herd has reversed direction.

Arrows indicate direction animals move in an attempt to follow the movements of the herd leaders.

Tight alley corners cause confusion in cattle.

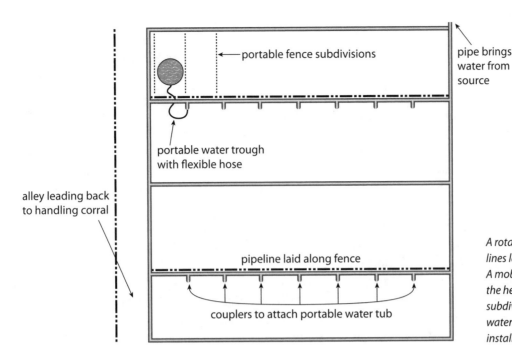

- portable fence subdivisions
- pipe brings water from source
- portable water trough with flexible hose
- alley leading back to handling corral
- pipeline laid along fence
- couplers to attach portable water tub

A rotational grazing grid with water lines laid under the electric fence lines. A mobile water tub moves along with the herd through each of the pasture subdivisions and is attached to the water lines by quick-release couplers installed along the lines.

the drinking demands of the cattle over the entire day. Consequently, the cattle do not mob the water tub as heavily and therefore the water line typically does not have to supply the full peak water-supply rate, as long as the tub contains several hundred gallons of water to meet the drinking demands of smaller groups. If the water tub is not visible from all areas of the grazing division, however, the cattle will begin mobbing the water tub again and the peak supply rate may be necessary.

Large plastic or rubber tubs containing several hundred gallons of water will withstand the abuse of daily tub moves much better than galvanized water tubs, and they are much lighter to move. I am particularly fond of the six-foot-diameter, round, plastic poly-tubs that hold up to 360 gallons of water for larger groups of cattle. The easiest way to move a tub is with a flat-deck ATV trailer and a few stakes to hold the tub in place.

Some electric-fencing companies (for example, Gallagher) sell plastic quick-couplers that reseal themselves automatically when a hose-attachment insert is removed, making it convenient to relocate the water tub; these eliminate the need for large numbers of taps and also prevent rusting hose threads. The tub should have a large drain with a rubber bung installed along its side so it can drain quickly

when the cattle are moved to their next grazing rotation and will have time to fill before the first grazing animals are ready to drink at their new location. The first groups of animals seem to come for water within a half hour after a pasture move, even if the grass is lush, simply because the water is so convenient. If the thirsty animals arrive at the tub before it has had time to fill, there will not be enough water to satiate them and the whole herd will soon be fighting over the tub. It is remarkable how quickly a herd notices that water has become a limited resource.

This design requires frequent pasture moves so the cattle traffic does not damage the grass around the water tub. Using a long flexible hose to connect the tub to the water line under the fence allows you

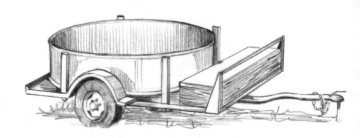

A flat-deck ATV trailer is fitted with pocket stakes to transport a mobile water tub between pastures as cattle move.

to place the tub in a different location with each passing rotation. In especially wet weather, trampling hooves can easily damage the grass, so plan the grazing rotation so that it is on firm ground during the wet season. In extreme cases, where pasture soils are particularly prone to livestock damage around the water tub, overseed with a sod-forming traffic-resistant grass variety or place a load of gravel near each quick-coupler connection to prevent excessive trampling damage.

The convenience of this grazing pattern makes it easy to coordinate daily pasture moves from one adjacent division to the next. It is advisable, however, to create an alley along one side of the grazing unit. The alley should run perpendicular to the permanent divisions so the cattle can be moved easily from one end of the farm's rotation to the other or moved efficiently to the handling corral (see the illustration on page 99). An alley also makes it convenient to remove individual animals to the handling facility for treatment, culling, or slaughter without having to bring the entire herd.

Laying Pipe

Aboveground water lines should always be PE pipe, which does not become brittle when exposed to ultraviolet light over a period of time, is flexible, and is easy to join and repair with pipe inserts and hose clamps. PVC pipe cannot be used aboveground because it becomes brittle over time when exposed to sunlight. In addition, short PVC pipe lengths are time-consuming to glue together, and because they are not flexible, they must always be laid straight in pre-dug trenches, a nearly impossible task on undu-lating terrain or when following fence lines that are not perfectly straight. Thus, PE pipe is by far the better choice for most pasture water lines, both above- and belowground.

PE pipe also expands and contracts with changing temperatures, even when laid belowground. When installing it, give the pipe plenty of slack so joints will not be pulled apart by cooling temperatures, which cause the pipe to contract.

PE pipe is black, which makes it susceptible to solar heating. If water use is insufficient or if cattle wait long periods between drinking, the water may become so hot that it is undrinkable. After the line has been on the ground for a while, it will be covered with debris, which will help shade it from direct sunlight, thereby reducing excessive heating. A tall grass residual will also help shade the pipe. If, on a hot day, you observe your cattle refusing to drink from the water tub, you should suspect that the water is hot. Solve the problem by flushing the lines until cool water from the water source reaches the trough.

PE pipe comes in long rolls usually wrapped around wooden frames. When laying the pipe aboveground, you can carry it on a steel A-frame with a removable bar on which the roll can spin freely. Placing the A-frame in an ATV trailer makes it quite portable and easy to unroll along the fence line. If you tie the end to a post, the pipe will unroll as fast as you dare to drive.

When the PE pipe is installed belowground, you can weld a curved metal pipe along the back side of a subsoiler shank or other ground-ripping shank that you pull behind a tractor. Feeding the PE pipe by hand down through the curved metal pipe as the shank rips through the soil allows you to lay the pipe

For easy installation of aboveground PE water lines, carry the PE pipe roll on your ATV trailer on a steel frame that allows the roll to spin freely. For installation, tie the end of the PE pipe to a post and drive along the fence line as the pipe unrolls behind you.

PE pipe can be installed belowground by feeding the pipe through a hollow tube welded along the back of a ripper shank pulled behind a tractor.

fairly quickly without digging open trenches. When using this method, it is particularly important to provide enough slack in the pipe to accommodate seasonal pipe cooling.

PE pipe placed belowground should always be laid early in the morning or on a cold day to minimize heat stretching the pipe. In most cases, it is not necessary to install drain valves for frostproofing the water system because the entire system can be blown out with an air compressor prior to winter and PE pipe is flexible enough to accommodate small puddles that freeze inside the lines.

Winter-Grazing Patterns and Water

In winter, dormant grass does not need to be protected from cattle after it has been grazed because there is no regrowth. This allows you to graze progressively outward, away from a winterized water site, so the cattle migrate back across vast stretches of grazed frozen ground to drink. Therefore, only a few water sites need to be winterized. Here are several winter-watering options.

Eating Snow

Like bison and other wildlife, cattle are able to satisfy their water requirements by eating snow. Their food intake must increase somewhat to supply the added energy needed to melt snow in their mouths, but because melting small mouthfuls of snow is such a slow process, eating snow will not chill their core body temperature.

By contrast, drinking very cold water during winter can be harmful to a cow. Her core body tempera-

ture drops significantly as she tries to heat the giant volume of cold water she has consumed in a short period of time. This quick cooling of the core body temperature can cause a great deal of stress to the cow's body and weakens her defenses against disease and extreme winter temperatures. Eating snow does not compromise the health of the cow in this way.

Do not expect a herd of untrained cattle to know how to eat snow if you turn off the water; you must train them. The easiest way to accomplish this is to make their trips to water increasingly tedious and inconvenient. The farther they need to walk to get from food to water, the greater their motivation will be to replace their liquid intake with melted snow. As the distance between food and water increases, the tracks of cattle going to water will dwindle until one day there will be none. At that point, you'll know that it's safe to turn off the water.

Open Water Holes

If the cattle are given a small access point from which to drink in a creek or lake, their continuing presence will keep ice buildup to a minimum, allowing you to keep a water hole open, with minimal effort, using an ax or a saw.

I don't like this option because ice-cold water is stressful to the cattle's core body temperature. In addition, open water holes are prone to icing along the edges, creating a tall, dangerous, slippery ice step that the cattle have to kneel on or step over to reach the water. Animals may slip or be pushed into the cold water, and broken legs and aborted fetuses may result from bad falls. The gradual slope to water will also ice over, creating a rock-hard slide down to the icy pool below.

This type of watering scenario encourages cattle to eat snow, particularly if they have to walk any distance to reach the open water, but that is the only positive thing I have to say about it.

Continuous Flow

A water trough with a continuous flow of water has a fairly good chance of remaining open during winter. The water can be moved by gravity or pumped continuously (using an electric, hydraulic ram, or solar pump) or redirected from a flowing stream with a pipe (an alternative to an open water hole in a stream).

Remove the float valve from the trough and install an overflow pipe near the top of the trough to carry overflow away from the trough's immediate vicinity or back to the original water source so water does not splash out and ice up the trough's periphery. Insulate the walls and floor of the trough to make better use of the warmth of the flowing water. A large-capacity trough is more resistant to freezing than a small trough because more water has to cool before it can freeze. Such a trough can be fitted with an insulated cover having a small access hole from which the cattle drink, thereby reducing heat loss from the surface of the water.

Ice-Preventer Valve

An *ice-preventer valve,* or thermostat-controlled bypass valve, can be used to regulate when the water in the trough must be flushed to prevent freezing. Install it just ahead of the water trough's flow valve. When it senses that the water is near freezing, it allows water to bypass it into the trough to flush out the cold water with fresh, warmer water. Excess water is carried out of the trough through an overflow pipe. There is sufficient latent warmth in water that is 48°F/9°C to keep a trough ice-free using an ice-preventer valve. If the water supplying the trough drops below 48°F/9°C, the ice-preventer valve will become increasingly ineffective at keeping the trough free, depending on the trough set-up, how much insulation is used around the trough, the number of cattle drinking from it, and how cold the surrounding air temperature is. Thus, below 48°F/9°C it remains a matter of experimentation to determine if an ice-preventer valve will achieve its purpose in your individual circumstances.

Heated Trough

If power is available, a water source can easily be kept ice-free. Insulated heated troughs are available commercially. As an alternative, heating coils can be installed in most troughs to heat the water, as long as the supply pipe is insulated or heat-traced (wrapped with electrically powered heating tape). Unfortunately, electricity is not always an option, and generators, propane heaters, and woodstoves placed underneath or built directly into the walls of water troughs are very labor-intensive to maintain, despite their effectiveness at keeping a trough frost-free.

Energy-Free Trough

An energy-free trough uses the water's latent warmth to keep ice at bay. These commercially available troughs come in many different sizes, rated according to the maximum number of cattle each energy-free trough model can supply and the minimum number of cattle required to ensure sufficient water replacement to keep the trough ice-free. They are heavily insulated and small holes at the top provide water access. Each drinking opening is sealed by

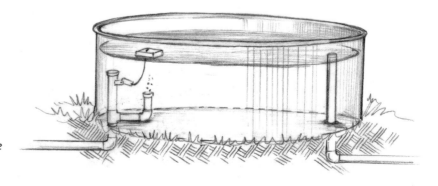

An ice-preventer valve in use: The valve is installed inside the water trough on the supply pipe ahead of the flow valve.

a large floating ball that the cattle must push aside in order to drink. It is often useful to install two smaller troughs rather than a single large trough so that one of the troughs can be shut off and drained during cold weather to accommodate smaller herd sizes in the winter season. A single large trough can freeze if the pasture is used by smaller herds that cannot guarantee the minimum water usage required to keep the trough ice-free.

An energy-free trough can also use heat from the ground to keep the trough from freezing. Two tall cylindrical tanks, joined at the top and the bottom to allow the water to circulate between them, are dug into the ground (at least half of the tanks extend below the maximum frost depth). Because cold water is denser than warm water, it drops to the bottom of the tanks, where it is warmed by the ground. As it warms, it rises to the surface, where it again cools, causing the water to circulate in the tanks. The top of one tank is fitted with an insulated drinking-bowl lid; the other remains sealed.

The warmth of the ground also can be used to keep a supply pipe frost-free. Bury a large-diameter riser pipe through which the supply line rises to a trough. The purpose of the riser pipe is simply to provide a large air space around the supply pipe. The air in the large-diameter riser pipe is warmed by the ground below frost level. It begins to circulate as it cools at ground level but is warmed from below. The top of the large pipe must extend under the lid of the water trough and remain open to the air space under the lid. The large-diameter riser pipe and the

AN ENERGY-FREE TROUGH

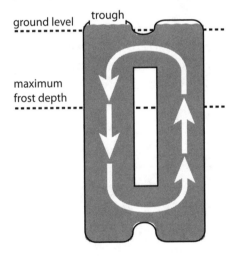

An energy-free trough uses the ground's latent heat to warm the circulating water and keep the trough ice-free.

HEAT RISER PIPE UNDER A WATERER (not to scale)

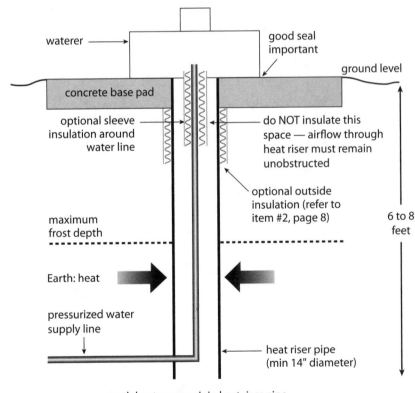

Warmth from the ground can keep the air inside a heat riser pipe warm enough to prevent the water line inside the pipe from freezing.

supply pipe should both be wrapped with insulation to below the maximum frost level, but be careful not to seal the air space between the riser pipe and the supply pipe with this insulation because if this space between the riser pipe and the supply pipe is blocked, air convection will not be possible and the supply pipe will freeze.

Gas Bubbler

A gas bubbler allows a small amount of gas to leak out when attached to a gas source such as a propane tank. It is laid under a water hole or at the bottom of a water trough so the gas will bubble up where the cattle drink. The bubbling keeps the water in motion to prevent ice buildup, and the motion helps circulate warmer water from the bottom of the trough or water hole to the surface.

Watershed Conservation: Managing Riparian Areas

In recent years, watershed management has become a significant environmental concern. There has been a strong backlash against cattle drinking directly from water sources such as streams, ponds, and lakes

because of *E. coli* bacteria contaminating the water and causing serious health risks to humans. (*E. coli* is shed naturally by cattle in their manure. A discussion on how grass-fed cattle reduce the human health risks caused by *E. coli* bacteria can be found in chapter 17.) In addition, there is concern about the extensive damage trampling and manure have caused such watersheds. The most popular solution has been to remove the cattle and permanently fence off riparian areas. Certainly, this eliminates the risk of *E. coli* bacteria entering the water source, reduces algal blooms, and prevents trampling at the water's edge, but lack of grazing along stream banks has created an entirely different, less obvious problem: erosion.

If grazing along stream banks is prevented, grass matures and dies without being harvested. The dry dead grasses mat, block out the sun, and choke out new grass growth. Without the active root mass created by periodic grazing, the soil along the water's edge loses integrity and becomes prone to severe erosion.

More than any other part of a pasture, the water's edge is susceptible to erosion by fast-moving floodwaters and mudslides on overly steep, unstable banks. In a healthy riparian area, lush, vibrant

A gas bubbler in action: The rising bubbles keep the water in motion and circulate warmer water up from below, thereby keeping the water hole from freezing.

grasses (without any bare soil showing) grow to the water's edge to protect the soil from rising floodwaters. A healthy sod mat and root system protect the stream banks from becoming unstable and sloughing into the water. The simplest and most cost-effective way to manage these riparian areas, therefore, is to use livestock to keep the grass healthy.

Unlimited access to the water would perpetuate the problem of trampling and water pollution, but fencing off the watercourse in wide corridors provides a simple solution. Corridors allow the livestock to graze the watercourse periodically in a very quick rotation, ensuring that their exposure to the water is brief. Supplying a trough away from the water's edge further decreases livestock pressure: Placed on firm ground level with the grazing area and adjacent to the creek, drinking from it is more convenient for cattle than descending down a steep stream bank or navigating muddy shores to reach fresh water. If livestock are used in a deliberate way such as this,

stream banks will support healthy plant life and will resist erosion, thereby reducing or eliminating our heavy reliance on other expensive artificial streambank management strategies.

Cattle, sheep, and goats are becoming increasingly popular tools for managing seasonal watercourses and flood channels because their grazing effectively creates stable soils and erosion-resistant sod covers. These grazing animals are also incredibly efficient at removing dead grasses and clearing underbrush. Otherwise, this scrub must be removed mechanically to prevent flood channels from becoming choked with debris that would obstruct floodwater runoff, causing flooding that would damage property and put communities at risk. In some areas of California, in fact, ranchers are paid to graze their goats in municipal flood channels to keep them erosion-resistant and free of brush and other channel-choking debris (a free grazing opportunity for innovative graziers).

A healthy riparian area (foreground) and an unhealthy riparian area (background) are separated by a fence. In the healthy area, grass sod grows right to the water's edge and the stream banks are stable with no bare soil. In the background, the stream banks are bare and show heavy signs of erosion. Grazing management can keep a riparian area healthy.

Planning for Winter Grazing

A YEAR-ROUND GRAZING STRATEGY IS ONE of the most important management practices that determine the success and financial viability of grass-based beef enterprises. But year-round grazing is not a haphazard activity. It requires meticulous preparation of your grazing reserve during the growing season, a carefully thought-out winter grazing plan, rigorous year-round monitoring of your livestock's fluctuating body condition scores (BCS), systematic measuring of the nutrient and mineral values of your grass reserve throughout the year, and a carefully designed mineral and feed supplementation program to compensate for specific nutritional shortfalls in the grass reserve. Only then can you maintain your livestock's good health, ensure their fertility in the next breeding season, and meet their nutritional requirements in a year-round grazing program. In this chapter, I discuss each of these factors along with several low-cost options for making and feeding stored feed during the transitional years of learning how to extend your grazing season and teaching your herd to graze year-round.

Winter Grazing: The Twelve-Month Plan

Winter feeding is the largest expense for most cattle operations. Keeping cattle fed during the growing season is a negligible expense, but the dormant season sucks the profit margin from many beef operations. The bulk of many cattle farmers' expenses revolves around equipment used to make, haul, store, and feed stored forages and to fertilize hay fields that haven't benefited from a grazing rotation because they are being saved for the baler or silage chopper.

Such effort runs up an impressive maintenance, fuel, and labor bill, and, unfortunately, the bulk of time is spent operating equipment, a job that has the least direct impact on the finished beef product. Jobs that make the biggest difference to a farmer's bottom line, such as management, planning, and marketing, are fit in whenever time can be found for them. A chief executive officer of a company can earn a quarter of a million dollars a year while his equipment operators make only a fraction of that wage not because the equipment operator doesn't work as hard or puts in fewer hours, but rather because the CEO's contribution — management, planning, and marketing strategy — has a huge effect on the profitability of the company.

Although beef farms typically tend to be bloated with equipment-operator and maintenance jobs, and management, planning, and marketing strategies tend to be severely neglected, change is possible. The key is altering the way we approach the most expensive time of year. We need to implement a winter-grazing program that eliminates stored forage as well as the hidden costs of making and feeding it.

Wildlife does not have the benefit of stored winter feed or human handouts. Yet millions of bison thrived on the far northern prairie for thousands of years in bitter cold and deep snow, digging through hardened, windswept snowdrifts for frozen grass while hungry predators followed close on their heels. How did they thrive in some of the most challenging conditions on the planet, and how can we prepare our cattle to do the same? Cattle should be able to graze year-round, with or without snow, while your expensive machinery depreciates in someone else's yard.

It Begins with the Growing Season

Winter grazing does not just happen. You can't turn out your cattle into any old field of dead grass (with or without snow), say a few magic words passed down by the bison, and sit back to watch your winter-feeding costs disappear. You need a plan, you need your cattle to work for you, and during the growing season you need to meticulously prepare the grass reserve for the onset of winter.

Summer Calving

All the planning in the world will not allow brood cows to winter-graze profitably unless the herd calves on grass during summer. Although I have already discussed summer calving at length in chapters 3 and 5, I mention it again because you must match the yearly nutritional cycle of the cow to the yearly grass cycle if you expect (1) the nutritional value of the grass reserve to sustain your cattle through the entire dormant winter season; (2) the cattle to supplement their nutritional intake with their own fat reserves without having their health affected; and (3) the cattle to maintain their conception rates without requiring significant feed supplements prior to calving.

Grass Quality and Motivation

A cow will continue to forage for food during the dormant season only as long as the grass she finds is able to meet her nutritional requirements. Her taste buds are just as sensitive as any forage analysis (see page 114). If the energy she must expend to find and digest the grass is greater than what the grass is able to give her, she will not eat it. As grass quality deteriorates, she will not dig as deep or walk as far to find it.

A cow searching for good-quality grass under the snow will submerge her head to eye level.

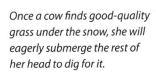

Once a cow finds good-quality grass under the snow, she will eagerly submerge the rest of her head to dig for it.

Tall grass is easiest to find under the snow. When searching for grass under snow, a cow submerges her head to eye level, but once she finds good-quality grass, she eagerly submerges the rest of her head to dig for it, even using her head like a plow, much as bison do.

Forcing cattle to continue grazing in a tight daily pasture rotation as the snowpack increases is the easiest way to train them to dig for the grass reserve. But even in midwinter, new animals can be trained to dig for grass if they are provided with a pasture full of highly nutritious grass that is tall enough to emerge above the snowpack. Seeing grass motivates the cattle to begin digging as you gradually cut back their stored feed. Initially, the cattle will complain and just pick at the grass tips visible above the snow, but within a few weeks, as hunger increases and their digging is rewarded, they will all be rooting for grass under the snow. Each successive generation of cattle will become better at winter grazing.

Grass retains its highest nutritional value if it goes dormant while in its vegetative stage (the growth stage of a plant's life cycle during which leaves and stalks grow, following germination and emergence). Plants in this stage are most resistant to leaching by rain and meltwater and contain more nutrients (and are more palatable) than mature grass that has gone to seed. The taller the grass is as it goes into the dormant season, the larger the root reserve. Thus, in spring, at the end of the dormant season, it greens up sooner and regrows faster.

Variety of Grasses

A pasture with a variety of grasses provides a more consistent food supply through the dormant season than does a monoculture pasture. Because some grasses continue to grow much later in fall and others resume growth earlier in spring, the overall dormant season is shorter. Some species, including many of the native grass varieties, retain their nutritional quality through more of the dormant season than other grasses. You will notice that these grass species are typically eaten first by cows grazing through the snow. Many of the tall, high-volume domestic forage varieties most popular for summer grazing and hay production are especially prone to nutrient leaching, making them the least desirable grazing varieties during the winter months. In a pasture of diverse grasses, because every grass variety contains a different balance of nutrients, the cattle can choose from them to meet their constantly changing nutritional needs during winter.

Some grass varieties that are less palatable during the summer growing season actually become more palatable during winter, as moisture and frost soften their hard structure. Although Altai wild ryegrass is particularly unpalatable during summer due to its coarseness, it becomes quite palatable during winter and retains its nutrient value even after it goes to seed. Its coarse, tall stems are ideal for winter grazing because they are strong enough to stand up through the heavy snowpack and they encourage cattle to dig below the snow to find other buried grasses. Grazing lush, highly digestible, protein-rich leaves of alfalfa while it is growing or immediately after a killing frost, before the alfalfa plant has dried out, poses to cattle a potentially fatal risk of bloating (see chapter 17). But fields of pure alfalfa can be safely grazed during winter without risk of bloat, providing grass-finishing-quality grazing.

Altai wild rye grass provides excellent winter grazing.

Grass Height and Solar Wicking

Grass that sticks up through the surface of the snow helps to prevent snow crusting by functioning as a thermal conduit, concentrating solar energy down through the crust around the grass stem. If you walk through a snow-encrusted field, even one in which the crust is hard enough to support your weight, but step in a spot where the grass shows through the snow, you will break through the ice, as will a cow's nose during grazing. This is why backcountry skiers avoid *tree wells,* spaces around trees. The solar energy concentrated by a tree recrystallizes the snow around the trunk into loose granular crystals that remain unconsolidated right down to ground level, even in mountain areas that receive as much as twenty-five feet of snow. But solar wicking is possible only when the grass is taller than the snowpack. If snow covers the top of the grass, the snow will compact.

If you plan your summer-grazing rotation to maintain a tall-grass reserve so that at least some grass is taller than the maximum average snowfall, cattle will be able to graze through even the deepest and crustiest snow conditions.

Snow and freezing temperatures are advantageous to the winter grazer. The cold freeze-dries the grass reserve so it is significantly more resistant to leaching and deterioration over the long winter months. The snow also provides an insulating cover

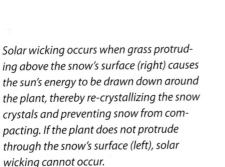

Solar wicking occurs when grass protruding above the snow's surface (right) causes the sun's energy to be drawn down around the plant, thereby re-crystallizing the snow crystals and preventing snow from compacting. If the plant does not protrude through the snow's surface (left), solar wicking cannot occur.

LESSONS FROM THE HERD

I remember one autumn when my soil-sample collecting was delayed until long after fall freeze-up. The snowfall had not yet begun, so the grass and bare soil were exposed to intense freezing temperatures. The crop fields were frozen so hard that I had to use a large hammer to pound in the soil probe to get my samples. Surprisingly, the ground in the pastures was not frozen nearly as deep; the frost had merely formed a thin crust on the surface of the pasture soil.

I also remember a pasture that had a particularly tall grass reserve compared to the other pastures I had sampled (the grass was over 2 feet tall before the frosts stopped its growth) and which had been managed with only organic fertilizer inputs and organic production methods to preserve the soil organisms. Despite the bitter cold, I could easily push the soil probe through the loose soil. The frost had not even penetrated entirely through the thick grass cover. Below the frosted grass tips, the grass

retained 10 to 12 inches of green material.

By contrast, in the short-cropped pastures with the thin frost crust, which had been grazed to within 6 to 8 inches of the soil, the frost had made the grass crisp and brown to ground level, despite being immediately adjacent to the greener pasture. The only difference between them was that the relatively frost-free pasture had a much taller remaining grass reserve and had been managed to promote a vibrant soil organism community.

to protect the underlying grass reserve. If you dig down through the snow, you will almost always find that a portion of the grass remains green, appetizing, and nutritious throughout the winter. The taller the grass before it enters the dormant season, the more resistant it will be to deterioration and frost damage, assuming it hasn't entered the seed-formation stage.

Every living thing generates heat, and the larger it is, the more heat it generates. The heat produced by a vegetative-stage plant with a large root reserve reduces the amount of frost in the ground around the plant and prevents the plant from freezing as deep. Added to this: As the grazing rotation improves soil health, the soil is able to support more biological organisms, which generate additional heat. When pastures are managed to maximize grass height before the frosts arrive and to promote a healthy population of plant life and soil organisms, the soil does not freeze as hard and the grass retains a higher nutrient value throughout the winter-grazing season.

Winter-Grazing Preparations

Managing the pasture-grazing rotation during the dormant season is much simpler than in the growing season because there is no regrowth to worry about. This is the season when you get to relax and concentrate on rationing, slice by slice like a jealously guarded pie, the grass reserve that you carefully created during summer.

Before the growing season ends, you must determine how much grass reserve (cow grazing days, or CGD) is left in your pastures (to calculate CGD, see chapter 25). Then you must adjust the size of the herd to a number that the remaining grass will support without shortages throughout winter.

By adjusting your herd numbers in fall, you will have to sell fewer animals to correct for grass shortages than you would if you waited until later in winter. And because your surplus animals are in prime condition, they are worth more at auction and can also be direct-marketed as grass-finished beef.

If you wait to sell animals until the cows' body condition begins to drop as grass quality diminishes, the surplus animals will be lighter and worth less at auction. In addition, you will not have the option of slaughtering them as grass-finished beef because

they must be gaining weight at the time of slaughter for their meat to be tender (see the discussion in chapter 17 on how to create consistent, high-quality, grass-finished beef). Another factor to consider: If you run into a grass shortage during winter, the body condition of the entire herd could drop below a BCS of 4 (see chapter 3 for more information on BCS), which would prompt a significant drop in the herd's conception rates the following breeding season unless you resort to expensive stored feed and feed supplements.

Winter-Grazing Strategy

Your grazing strategy changes dramatically after the grass goes dormant in fall. Once the frost kills the grass from the soil up, grazing can no longer affect the dormant roots, so you'll want to thoroughly graze each pasture division right down to a two- to four-inch residual. Just be sure to leave enough grass residual and debris to protect the soil from erosion and excessive moisture loss. The herd should completely graze the entire grass reserve within each grazing slice on the first pass because it cannot be grazed a second time after it has been knocked down, trampled, soiled by cattle feces, and covered by additional snowfall.

Ration Daily

The most effective winter-grazing strategy is to ration out the grass reserve to the cattle on a daily basis. Less frequent rotations will cause the cattle's feed intake to fluctuate too greatly from the first to the last day in each pasture subdivision, and the extra space will allow the animals to trample and destroy too much grass. Remember, the tighter you ration your grass reserve, the longer it will last.

Because there is no regrowth, a back fence is less important during winter. Consequently, you can graze your herd progressively outward from a single winterized water site, allowing the cattle to traverse back across portions of the previously grazed areas to return to the water site. In order to distribute the manure evenly across the land, however, their route to the water site should be restricted to prevent cows' favorite sheltered areas from becoming manure magnets.

Provide Access to Natural Bedding and Shelter

Some producers like to provide bedding or a shelter of some sort during the winter season, but this concentrates the cattle in a location that soon collects manure and becomes disease-prone. In addition, you would have to use diesel fuel and tractors to transport bedding to the sites and, later, to compost the bedding and spread it back out on the fields. If you calve during the summer grass flush, however, you will not require these sheltered bedding sites during winter. By the time the snows arrive, the calves are old enough to sleep in the snow. The daily paddock rotation ensures that they will lie in a fresh patch of snow each day, so manure and disease cannot catch up to them.

Brush and forest areas make the best natural shelters for cattle during the coldest part of the winter. They provide protection from the wind, and the trees create a cover that helps keep these areas significantly warmer than open fields. Tree-sheltered areas will also be less prone to snow drifting and snow crusting (see Grass Height and Solar Wicking on page 109).

Time Rotations to Assist Cattle

By calculating the cow grazing days, or CGD, in each pasture, you can time your grazing rotation to save the best-quality grasses and the shelter of trees and brush for the coldest and windiest period of the winter, when the cattle will have the highest energy requirements to produce enough heat to keep warm.

Save windswept ridges as your defense against days when the snow gets really deep. Ridge tops and south-facing slopes will be the first to lose their snow in spring and can be grazed while the remaining land is still under a snowpack.

In spring, there is typically a wet, muddy season during which your cattle's hooves will damage many pastures. Again, plan for this and save well-drained soils, ridges, or gravelly ground for this time. Spring may also be a good time to graze back across the winter-grazing areas to pick up grass that was lost under the snow or inaccessible to the cattle in areas of drifted snow. Make a map of the farm and plan your winter-grazing season so you know where you will be during all predictable weather conditions from the beginning to the end of the dormant season.

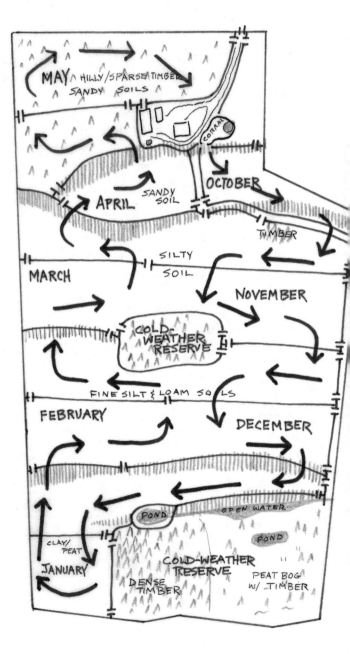

A winter grazing map for a farm with a dormant season lasting from October (first snowfall) to mid-May (new grass growth). Grazing is managed to avoid soft soils during the spring thaw and to take advantage of boggy areas during the midwinter freeze. Also note the cold-weather reserves, which provide shelter and grazing during the coldest weather and offer several months of additional winter-grazing reserve.

Continuing the grazing rotation into the winter months poses its own unique set of challenges, particularly the challenge of using portable electric fence posts that must be pushed into the ground when the earth is frozen, the challenge of snow recrystallization and frost movement in the soil in the presence of livestock traffic, and the challenge of overcoming the reduced effectiveness of electric fences caused by frozen ground and the insulating properties of snow.

How do you move portable fence posts during the dormant season when the ground is frozen? The hardness of the ice crystals in the soil depends on their temperature. As the temperature drops, their molecular bonds become tighter and stronger. Just below freezing, ice is about the same hardness as your fingernails, but by −40°F/−40°C, ice becomes almost as hard as glass.

The combination of snow and tall grass insulates the soil so it does not freeze as hard. Further, a vibrant, heat-producing community of soil organisms and a large root and plant reserve at the beginning of winter generate enough warmth to keep the soil from freezing as solidly (remember the story of soil sampling in winter). In many cases, you can still pound fiberglass posts into the ground without shattering them if you protect the posts from the hammer with wooden driver caps (try using a 6-inch piece of broom handle with a small hole drilled into it to fit over the top of a post). Furthermore, by grazing each area only once, you can pound your posts into soil that is covered by an untouched insulating layer of snow.

Once the snow has been disturbed, it recrystallizes, crusts over, and sets as hard as concrete, preventing the cattle from digging through it and you from pounding fiberglass posts in the same area a second time. Livestock traffic will drive the frost deeper into the ground, making it harder to pound portable fence posts into the soil. Putting up the next day's portable fence division a day early prevents losing more than a day of grass to trampling if the cattle break down a fence and ensures that you are always pounding your posts through fresh, undisturbed snow.

If pounding fiberglass posts is not possible, you can use a tripod to hold the portable wire off the ground. Three fiberglass posts firmly tied together at the top and stuck into the snow will work quite well. Plastic buckets half filled with water will also work if you freeze the base of the portable posts into them, although they are a bit more cumbersome to move each day.

I have heard many farmers complain that electric fencing does not work in the winter because the cattle are insulated from the shock by the dry frozen ground and cold snow.

First of all, make sure that your electric fence has an effective grounding system so the wire retains a good kick; this is your most important line of defense against poor conductivity during winter (see page 69 for more information on how to set up an effective ground system for your electric fence). If the cattle are trained to respect the fence during summer, the portable wire will still work remarkably well as a psychological barrier during winter, even as the voltage on the fence begins to drop. Cattle still fear the sound of the electrical pulse in the wire, and the smallest zap will continue to serve as a reminder.

If all else fails, this is the time when you may need an earth-ground return system, in which a ground wire hanging below the live wire causes the cattle to make a connection between the two wires and get a shock (see page 71 for more information on setting up an earth-ground return system).

Even during subzero weather in Canada's notoriously dry Yukon Territory, however, I found single-wire fences were still sufficiently effective that I did not require a ground wire to contain my cattle.

An earth-ground return system can be used to maintain an effective electric fence in non-conductive soils or if cold snow or frozen ground insulates the cattle from the shock of an electric fence.

Swath Grazing and Crop Residues

Swath grazing is gaining popularity as a winter-grazing tool, although it requires the use of equipment and fuel, which makes it more expensive than grazing a grass reserve. The fields are swathed (a mower called a swather cuts the grass and bunches it together in narrow strips called *swaths,* or *windrows*) late in fall when temperatures have cooled enough that the crop won't rot in the swath. The cattle are given access to one or two swaths at a time. A portable electric wire is used to ration the swaths so that nutritional intake stays consistent from day to day and the swaths are not wasted by trampling or sleeping cattle. The cattle soon learn to dig into the swaths for food because the grass has been concentrated to make it easier to find.

One popular crop for swath grazing is oats, which are deliberately planted late so they do not mature before they are swathed in fall. Swathing stops the oats from drying out further during winter; the swath becomes a green, freeze-dried source of feed. In grain-producing areas, straw swaths and chaff piles can also be swath-grazed.

Crop areas are a wonderful source of cheap, additional feed for cattle. Not only are crop residues easy to come by, but so are crops that were weather damaged or not harvested before the onset of winter. Commodity market cycles can make some crops uneconomic to harvest; thus, astute graziers can provide a needed service to crop farmers by removing plant debris and recycling the nutrients back into the ground with their cattle.

Grazing Annual Crops

Annual crops for winter grazing are another alternative, though, like swath grazing, they are labor- and machinery-intensive. The most important application of both swath grazing and annual grazing crops is to extend the grass-finishing season deep into winter so you can continue to direct-market beef without using grain-finishing rations long after the pasture grass reserve deteriorates. In southern areas, winter annuals such as fall rye, turnips, and kale can provide finishing-quality grazing right through the winter months, when most grasses are dormant. This same land can then be used for cash-crop production during the summer months. Some annuals providing winter grazing can actually be planted directly into dormant pasture sod with a range seeder.

Grazing corn can be used during summer as a finishing pasture: The cattle graze it periodically while it is in its vegetative state (in this case, meaning anytime before cob formation begins). It is also an excellent winter-grazing and grass-finishing crop because it is highly nutritious and is able to stand up to a heavy snowpack. To use it for winter grazing, plant the corn late in spring to prevent it from maturing in fall. Planted too early, the corn will mature before the killing frosts arrive, causing it to wilt (dry down) and redirect its nutrients from the leaves to the cob (seed head), just as all other grasses do (corn is a specialized grass variety).

Unlike forage and grain-corn crops, which we stop irrigating by early fall to trigger wilting and seed formation, you should continue irrigating grazing corn until frost to maintain vegetative growth as long as possible. If the corn becomes overly mature before winter, its high sugar and starch content makes it susceptible to molding, particularly if the crop is exposed to wet weather during fall. Some of these molds are toxic to cattle if they are ingested.

For winter grazing, the goal is to maximize leaf matter rather than cob production. Look for grazing-corn varieties that are leafier and mature later, such as Baldridge grazing maize, so the corn will be frost-killed before cob formation is complete.

Grazing corn should be rationed daily with portable electric fences. You can drive an ATV through the crop to flatten a strip on which you can set up the fence. The cattle will clean up the whole plant once they get accustomed to grazing it. Despite the prohibitive fertilizer, tillage, and equipment costs of growing grazing corn, it has a place in some operations to extend the grass-finishing season and guarantee 12-month grazing in areas of extremely high snowfall.

Fortunately, when corn is grazed, the bulk of its nutrients are recycled directly back to the ground through manure, which makes it less of a nutrient drain on the land than corn harvested for grain or silage.

Supplementation and Forage Analyses

During winter, the dormant grass reserve is pounded by rain, frost, meltwater, sunlight, and wind. As time passes, the protein and energy content of the reserve is slowly leached out of the plants. At some point the grass will no longer contain enough protein or energy to meet all the nutritional needs of a cow and she will begin losing weight. If the deficiency becomes too great, she will lose more weight (body condition) than she can hope to regain in the spring before calving, resulting in reduced fertility at breeding or even more serious health concerns related to emaciation.

This does not mean that the grass reserve should be abandoned and that you have to resort to feeding expensive hay or silage. Instead, you can intervene by supplying the cattle with small amounts of feed supplements to make up for individual protein or energy deficiencies. These supplements will allow the cattle to continue to meet the bulk of their nutritional needs from your grass reserve, which is significantly cheaper than feeding hay.

To know what is deficient in the grass supply, when to begin supplementing, and how much sup-plementation is required to prevent the cattle from dropping below a BCS of 5 during winter, you must map out the seasonal changes in the nutritional value of your grass supply with a series of monthly forage analyses. These should provide a record of the changing grass quality over the entire year, not just the winter months.

The winter supplementation program hinges on knowing the quality of the grass during the summer so the nutritionist you choose to work with can calculate how quickly the cattle will regain weight in the spring, how much weight they will gain during the summer, and how long the forage quality and fat reserves gained during the summer will sustain the cattle into the winter before they require supplementation. From this forage map, the nutritionist can calculate which months your cattle will need supplementation, how much supplement to feed, and the most cost-effective supplement program.

The illustration below shows how the nutritionist compares the forage supply data to the yearly nutritional requirements of the summer-calving cow. Keep in mind that the supplementation program must only prevent the cow from losing more body condition than she is able to regain on the spring grass flush in order to reach a BCS of 7 in time for

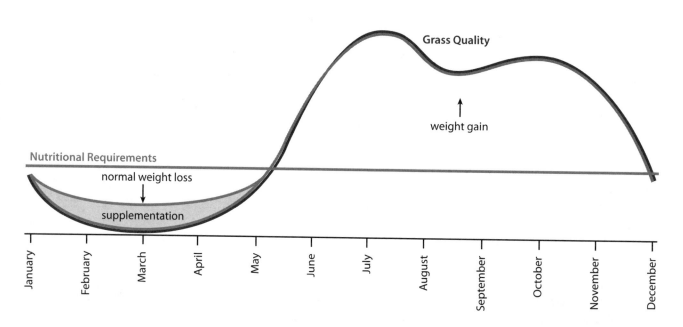

Seasonal grass supply (curved line) plotted against nutritional requirements (straight line) of a summer-calving cow. When the grass quality exceeds the cow's nutritional requirements, she will gain weight; when the grass quality falls below the line, she will lose weight.

calving. A supplement program designed for a cow calving out of season or for one whose body condition is not allowed to fluctuate between a BCS of 5 and 7 will not be cost-effective. Only by calving in the early summer on grass and by allowing the BCS to fluctuate from 7 down to 5 during the winter, as the cows supplement themselves from the fat reserves they have stored during the summer months, can you utilize your grass reserve year-round at a very small supplementation cost.

Monthly Forage Analyses

In order to build a reliable forecasting map of the protein and energy provided by your yearly grass supply, it is important to take monthly forage analyses over a two- to three-year period. I schedule the first Monday of every month for analysis to avoid accidentally postponing it. The date must be set in stone so meetings, chores, and other responsibilities do not compromise the accuracy and reliability of your forage forecasting database. The nutritionist will tell you which elements to test for and how much material is required for each sample every month.

Locating Representative Samples

Your sample should represent what the cattle are eating each month, not the entire plant community in the pasture. Observe the cattle grazing and then try to take the same mixture of grass species and only the plant heights that the cattle are eating. Perhaps certain species are favorites at certain times of the year while other plants are avoided altogether. Your sample should represent this selectiveness. It is not important to sample the entire farm each month; you need to sample only the grass from pastures in which your cattle are eating on that day. There will be more nutritional variation from month to month than from pasture to pasture, and your rotation will be similar from year to year anyway. After a few years you may want to resample to check your database in case your soils (and therefore the nutritional value of your grass) change over time. You will also want to resample if you add any fertilizer to the soil.

Collecting Samples

Do not use your bare hands to collect the forage samples. Even a trace amount of sweat on your hands can dramatically affect the sodium content of your sample. Using a clean pair of gloves, place each sample in a new plastic bag to avoid contamination; avoid reusing bags such as plastic bread bags, even if they look clean, because tiny crumb traces can affect the lab analysis. This is particularly important if you are analyzing the mineral content of your forage samples at the same time to create a database for your mineral supplement program.

Supplements to Avoid

There are two sources of feed supplements that I discourage you from using: grain and urea. In chapter 3 we have already discussed the negative impact of grain on the rumen. In addition, grain will limit or even reverse many of the dietary benefits of grass-finished beef (see chapter 17 for a full discussion of this topic). For example, grain feeding will lower the meat's beneficial omega-3 fatty acids while raising undesirable omega-6 fatty acids. If your animals are destined for a health-conscious or a natural or organic meat market, using grain as a supplement simply does not fit into the equation.

Although urea is a common supplement used in conventional feed rations and lick tanks because it is such an efficient source of protein, it should not be fed to animals destined for a natural or organic beef market. Certified organic associations do not allow it. Urea is produced naturally inside a cow's digestive tract and this natural form is regulated by the cow's own body. Synthetic urea, however, is extremely toxic to the cow's liver. Over time, urea supplementation will create ugly lesions and abscesses in the liver that either will earn you an undesirable reputation at the slaughterhouse or, even worse, will be fatal to your animals due to ammonia toxicity. Natural protein supplements such as alfalfa pellets and soybean meal are far more desirable because the cattle can regulate their own urea levels without adversely affecting their health or the quality or nutritional value of the finished meat product.

The amount of supplements required to compensate for the grass reserve's nutritional deficiencies is typically so small that they can be mixed into the cattle's loose mineral supplement that is offered to the cattle free-choice year-round. The feed supplement should be added to the loose mineral mix at the appropriate ratio of daily required supplement intake to average daily mineral intake. For example, if a cow requires 1 ounce of kelp/mineral mix + 1 ounce salt + 2 ounces alfalfa pellets per day to meet all her needs, the three supplements should be mixed in the mineral feeder in a 1:1:2 ratio by weight. Both the daily supplement intake and the daily mineral intake should be calculated by your nutritionist based on nutritional shortfalls in the grass reserve and your cattle's specific mineral and feed requirements as they relate to their fluctuating needs over the course of the year. The salt content of the loose mineral mixture will limit the intake of the feed supplement mixed into the mineral supplement so the cattle don't gorge on it and so each animal receives the correct amount.

I am not a fan of minerals fed in block form, even if they are tailored to your livestock's needs. Mineral blocks do not allow you to change the mineral ratios in the supplement or to add loose feed supplements free-choice. Without the salt in the loose mineral mix to limit the cattle's intake of the feed supplement, they would gorge themselves on it, preventing you from rationing it to the animals in the correct amounts.

Natural versus Synthetic Mineral Supplements

When natural sources of vitamins and minerals are used to supplement cattle, a cow's digestive tract can absorb what it needs each day and discard the unused portion. But when synthetic vitamins and minerals are used, the body is forced to absorb them entirely, regardless of its needs. Therefore, a synthetic mineral ration must be carefully calculated to change over the course of the year as the mineral content of the grass supply varies. If the minerals in the food supply change unexpectedly or if the animals' needs change in response to challenges to their health, however, the ration will be prone to errors.

The molecular structures of synthetic vitamins and minerals are microscopically different from those of their natural equivalents. Many of the natural types are chemically unstable in their pure form. If they are isolated from the plants that contain them, they are easily oxidized and evaporate into the air. The slightly different chemical structure of the synthetic products allows them to remain stable in their pure form. Vitamin C is a perfect exam-

ple: Synthetic vitamin C remains stable for a very long time in commercially prepared orange juice or vitamin C tablets. Although natural vitamin C has only a single atomic bond that differs from the synthetic molecular structure, it is very unstable when it is exposed to the atmosphere. If you squeeze an orange, the vitamin C will evaporate out of the juice in your glass in as little as half an hour. In addition, if you supplement yourself with synthetic vitamin C tablets, you need to give your body a break from time to time so it does not become accustomed to the synthetic vitamin and ignore its benefits, because despite the seemingly minor difference between synthetic and natural vitamin C, the synthetic version is a foreign substance to the body that will be tolerated for only so long before the body rejects it. By contrast, natural vitamin C from oranges will continue to benefit you day after day because the body recognizes the chemical structure of the natural vitamin: The organs will continue to absorb it whenever it is needed or discard it if it is not needed.

A natural mineral source allows cattle to selectively extract what they need to correct imbalances and discard the rest, maintaining an ideal mineral ratio.

An ideal ratio is often even more important than the individual numeric values of each mineral element in the body. If the ratio is incorrect, the chemical behavior of one mineral may inhibit the body's absorption of another or the chemical interaction between imbalanced minerals may actually inhibit a mineral's functions, *even if it is present in sufficient amounts in the body.* For example, a calcium-deficient animal will not be able to absorb some other minerals in sufficient quantities, even if the other minerals are present in the grass or mineral ration that the animal is eating. A natural mineral source allows cattle to selectively extract what they need to correct imbalances and discard the rest, maintaining an ideal mineral ratio. By contrast, synthetic minerals are absorbed in their entirety so the body must indiscriminately take in whatever mineral ratio has been prepared in the lab, regardless of mineral changes in the food supply (for instance, from the soils from different areas of your pasture rotation) or changes to the cattle's mineral requirements, such as health challenges, gestation, calving, and lactation.

Kelp Meal and Natural Rock Salt

As a natural mineral and vitamin source, I have been quite happy with kelp meal mixed with natural rock salt in a 1:1 ratio *by weight,* not volume. This loose kelp/salt mixture can be fed free-choice in portable mineral feeders that follow the cattle from pasture to pasture. Kelp contains more than fifty natural vitamins and minerals absorbed from the ocean waters in which it grows. Air-dried kelp has higher concentrations of minerals than kelp that has been heated in a dryer because the dryer's high temperatures cause many vitamins and minerals to volatilize and escape in much the same way that they escape during cooking.

The salt that is mixed with the kelp should not be iodized, because the kelp already contains an extremely high amount of natural iodine. The high dosage of iodine in kelp would be dangerous to cattle if it were synthetic, but again, the natural form allows the cattle to utilize what they need and discard the rest. Natural iodine is extremely beneficial in combating pinkeye; a kelp mineral ration increases the cattle's resistance to the disease. Kelp meal also contains some protein (approximately 6 percent), which will help increase the cattle's feed efficiency during the winter. Feeding kelp has other benefits as well: The meat from kelp-supplemented animals is noticeably more tender (even butchers notice it) and kelp supplementation seems to improve the meat's flavor.

There is a debate about whether the low selenium content of kelp will cause a deficiency in cattle in areas where the soil is deficient in selenium. Although selenium can be supplemented separately by providing access to selenium salt blocks along with the kelp mineral mix, the ideal way to prevent a deficiency in your cattle is to fertilize your pastures to correct for low amounts of selenium in the soil (selenium fertilizer should be applied to your pastures only according to the recommendations of comprehensive pasture soil analyses, discussed at length in chapter 11).

In this way, the grass supplies the selenium needs of the cattle. Either way, it is wise to routinely take blood samples from a representative cross-section of your cattle a couple of times a year to monitor for any mineral deficiencies in them. Doing so will give you the additional assurance that your mineral program is working properly and that all your cattle's mineral requirements are being met.

When switching to free-choice kelp, the cattle will typically go through a period of one or two months during which they will gorge on the kelp/salt mix to rebalance their mineral ratio, after which their kelp consumption will level off to affordable levels. The average kelp consumption is 1 ounce of kelp per yearling or small cow per day, slightly more for larger cows and bulls, but this will vary with the seasons, the mineral content of the grass, the animal's health, and even the weather.

Sometimes the cattle will go through periods of gorging on the kelp, while at other times they seem to ignore it altogether. Don't be surprised if it takes the cattle a few days to develop a taste for the kelp after you introduce them to it. Once they catch on, they will leave synthetic mineral sources untouched, even if they are offered both side by side in the same mineral feeder. If animals have access to urea, however, either as a protein supplement in a feed ration or in a mineral lick tank, they will not like the kelp. The urea seems to affect their taste buds and they will continue to prefer synthetic mineral sources, much as children who have been exposed to sweets come to prefer the taste of candy bars and soft drinks over the taste of real fruit to satisfy a sweet tooth.

Portable Mineral Feeder

By feeding your mineral and supplement mix in portable mineral feeders that travel along with the cattle's grazing migration each day, you gain the flexibility of using the feeder location to distribute manure more evenly across your pastures. A portable mineral feeder eliminates the high manure buildup and soil compaction that typically occur when minerals are fed from a stationary feeder located near the water site. Because the portable mineral feeder is a magnet attracting cattle wherever you place it for the day, you can use it to draw animals into weed or brush patches that need trampling or into areas that may be

A portable mineral feeder on pasture. The roof is essential, protecting the supplement from the elements.

nutrient deficient, such as dry ridge tops and bare soil patches in your pastures. You can even add grass and legume seeds to the loose mineral mix so the cattle can distribute the seed for you in their manure. This is an inexpensive way to overseed pastures and bare soils; you don't even have to start a tractor.

The portable mineral feeder quickly becomes a central part of the cattle's daily life. It is a familiar object they are all drawn to and they quickly become accustomed to the predictable routine of the feeder following them to fresh pasture each day. You can use this to your advantage when training your cattle to pasture moves: Make a habit of pulling the mineral feeder ahead of the oncoming cattle. Moving the feeder will become their cue; they will learn to follow it and not stop until the feeder stops. If you do this consistently, the moving feeder becomes a visual magnet to pull them through even the most enticing distractions along a pasture move, and your ATV will not be mobbed in anticipation of moving to fresh grass every time you enter the field.

Low-Cost Winter Feed for the Transition to Winter Grazing

Winter feeding is never as inexpensive as grazing. Stored feeds are expensive to make or buy and they require a huge investment in depreciating machinery that requires fuel and maintenance. And don't forget all the hidden costs of stored feed: the

opportunity cost of capital investments tied up in equipment, maintenance bills, labor, and — most important — your time away from planning, management, and marketing.

But replacing winter feeding with winter grazing requires meticulous planning and preparation. Although a year-round grazing strategy is vital to the success and financial viability of a grass-based beef enterprise, it takes time to learn the art of winter grazing, develop a solid year-round grazing plan, prepare your grazing reserve, and make the necessary changes to the remainder of your farm's production plan in anticipation of this change (see part 4, Putting Plans into Practice, for advice on how to most efficiently structure this transition period). As you progressively extend the grazing season into the winter, you will learn to overcome your specific environmental and climatic challenges and gain the necessary experience to allow you to complete the transition to year-round grazing. During this transitional phase, you can minimize the cost of winter feeding through several innovative methods to make and feed hay and silage, which are described below. Just remember that even these innovative approaches are temporary solutions that must be eliminated altogether and replaced by year-round grazing if your natural grass-based beef enterprise is to be truly profitable.

Self-Feeding Hay

The cost of using hay as a stored winter feed can be significantly reduced by allowing the cattle to self-feed from carefully constructed haystacks. With a bit of planning and deliberate preparation during summer haying season and the judicious use of electric fencing to carefully ration the hay during the winter, cattle can self-feed from these haystacks at a fraction of the labor costs normally associated with winter feeding and without the support of tractors and other equipment traditionally required for feeding hay.

The technique for self-feeding hay during the winter is particularly suited to round bales, though it is easily adapted to large square bales (upwards of 800 lbs) or the now rarely used hay loaves. Small square bales (50–120 lbs) are not as suited to this method because they are not cost-effective for feeding larger numbers of cattle, and are used primarily to feed very small cattle herds and other livestock (horses, for example) kept in small numbers; they are much more labor-intensive to make, store, and feed; if they are made and stored in large quantities, they require a high equipment investment; and they require more precisely controlled storage environments to protect them from the elements; and they are more vulnerable to spoilage.

Preparations for and Techniques of Self-Feeding Hay

When making hay, store the round bales (or large square bales, or hay loaves) in clusters (single-tiered stacks) directly in the field so you can ration them using portable electric fences, which allow the cattle to self-feed. Locate them in nutrient-deficient areas that could use a bit of extra manure and are not prone to flooding, such as ridges, if you have them. Select sites that will not turn to mud due to heavy livestock traffic if you anticipate feeding during wet weather, and keep in mind that you should make multiple small stacks to spread out the manure over a number of areas. Thus, similar to your winter-grazing plan, you can plan to feed specific stacks in different pastures at specific times during the winter, depending on the weather, the vulnerability of the feeding site to spring mud, and the access to trees or other forms of shelter during the coldest part of the season. The cattle can therefore continue to rotate around the pastures on your farm as they move from haystack to haystack during the winter.

Round bales should not be stacked; they should be laid out on the ground in perfect rows, tight against one another, to form a big square or rectangular cluster of bales. They will then need to be fenced with electric fencing along both sides to form a corridor. Do this during the summer to protect the hay from livestock and wildlife and save you from pounding electric fence posts in frozen soil during the winter. Finally, string a single portable electric wire across the vertical face of the haystack between the two electric fences of the corridor in the method described below, with portable gate handles on either end of the portable wire to attach to the elec-

tric fence corridor. The fourth side (the opposite side of the corridor) is also fenced, either simply to prevent access to the haystack or with another portable wire used to self-feed the hay from both sides of the stack simultaneously.

When it is time to self-feed the hay, the portable electric wire allows the cattle to access the hay bales by reaching over or under it. As the cattle eat away at the hay face, the portable gate handles allow you to slide the portable wire along the corridor to keep up with the rate of hay consumption. The wire also limits how much hay the cattle can access and prevents the cattle from spoiling the remainder of the hay with their manure or by trampling it or using it as bedding.

To support the portable electric wire along the vertical face of the hay bales, push portable fiberglass posts horizontally into the face of the hay bales, as shown in the illustration on page 120. Attach fiberglass insulators to the post ends protruding from the hay bales, leaving approximately six inches between the wire and the bales. As the cattle eat the hay, push the fiberglass posts farther into the bales each day or twice a day (in extremely cold weather when the cattle consume more) to maintain the six-inch separation between the wire and the bales. In this manner, as the cattle eat their way through the bales, the portable electric fence slowly moves ahead through the rows of bales to keep up with hay consumption until the stack has been eaten completely.

> **SAFETY TIP**
> Constantly check the fiberglass posts for fraying; if a cow ingests one, a fiberglass splinter can cause an abscess or intestinal damage.

The ideal height of the portable electric wire across the vertical face of the hay bales will vary depending on the size of your cattle and whether or not they are horned, though most adult cattle do well with the wire set three feet from the ground. If you want the cattle to clean up more hay at ground level, raise the posts; if they are leaving a great deal above the wire, drop it down so they can reach over the top. Very little hay is wasted this way, because you let the

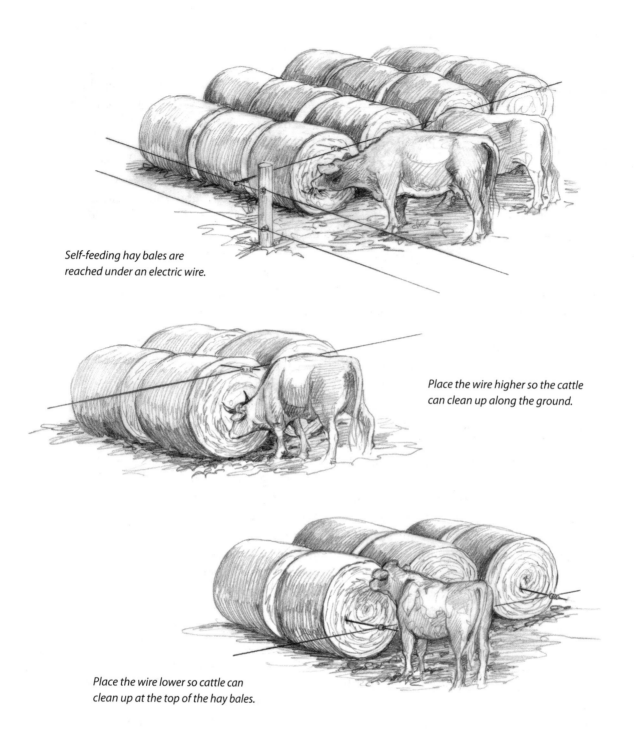

Self-feeding hay bales are reached under an electric wire.

Place the wire higher so the cattle can clean up along the ground.

Place the wire lower so cattle can clean up at the top of the hay bales.

cattle clean up what they are eating before you push the posts farther into the bales.

Remove the string from the bales before feeding, and provide enough access along the face of the bales so each cow has 1½ to 2 feet of space. When laying out the haystacks in the summer in preparation for winter feeding, it is important to ensure that the width across the vertical face of a haystack will be wide enough to accommodate the number of cattle you plan to graze on that particular stack dur-

ing the winter. Both ends of the bale cluster can be "grazed" at once to increase access.

The moisture from the animals' bodies at the hay face makes the ground damp enough that a ground wire is typically not needed. In the event that the ground isn't moist enough, string a ground wire parallel to the portable electric wire using the same horizontal fiberglass posts that already hold the portable electric wire in place across the face of the haystack. Put a second fiberglass insulator onto each fiberglass post so the ground wire is hung at the same height as the live wire but is separated from it by several inches, with the live wire closer to the hay. Oral-fecal transfer is minimized because the cattle are always moving forward, eating under the wire, where the manure cannot reach.

Buying Hay

Instead of making your own hay, consider buying all of it so you do not have to carry the depreciating investment of machinery. After all, if you are already reducing the length of your feed season by extending your grazing season, the little bit of hay you may need probably does not justify the high cost of retaining equipment. Selling off all equipment allows you to pay for further electric-fencing infrastructure, more land, or more cows.

If you decide to buy your hay, have it delivered and laid out in clusters in the locations where you will self-feed it with your electric wires during winter. Buying hay and selling off all your equipment will further increase your incentive to eliminate stored feed altogether and resort to a well-planned winter-grazing program because the convenience of making a bit of spare hay is no longer at your fingertips. It is easy to forget what it costs to fire up a tractor and make a few bales until you have to pay for hay in one big check instead of through an accumulation of small bills, time, maintenance, and hidden opportunity costs.

Feeding Pasture Silage

Silage is an alternative way of storing forage. Instead of drying grasses to preserve them as hay, the grasses (or legumes, corn, and other vegetable matter) are stored at high moisture by ensiling them — that is, the high-moisture forage is preserved through fermentation and stored in the absence of air to prevent rotting. This is traditionally accomplished by storing the forage (now silage) in silos, frequently used by the dairy industry, or more commonly in large pits or bunkers with a cement floor and walls on three sides. The air is compressed out of the silage, usually by packing it by driving a tractor back and forth across the pile, and then the pit is covered with weighted plastic tarps to seal the silage from the air. To feed the silage in the winter, the tarps on one side

of the pit are opened to expose a small sliver of feed and the silage is loaded into a feed wagon using a tractor/loader and transported to the cattle. The tarp is folded back or cut away only as quickly as the pit face advances through the bunker, while the remainder of the silage remains sealed away from oxygen under the plastic tarp to prevent spoilage.

Instead of making your own hay, consider buying all of it.

By contrast, pasture silage is made by piling the high-moisture forage directly in the pasture where it was cut, either on flat concrete pads or directly on the ground. Air is compressed out of the silage by packing it with a tractor, and then the pile is covered with weighted plastic tarps to seal the silage from the air. This style of storing silage is typically called a *clamp*. Much as hay can be self-fed using the innovative method described above, pasture silage clamps can also be self-fed to cattle using portable electric wires, thereby eliminating the need for any feeding equipment during winter and allowing you to bring the cattle to the silage, not the other way around so typical in the conventional beef industry. As for the self-fed hay stacks described above, the sites used for pasture clamps should be well-drained, nutrient-deficient locations, ideally ridges, to avoid excessive mud, particularly because the heat from the silage will prevent the ground from freezing. Ground that slopes away from the clamp face, along which the cattle will be feeding, will further reduce mud buildup and pasture damage. Pasture silage can usually be made at two-thirds the cost of making hay. Self-feeding pasture silage is described in further detail below.

The Opportunities of Long-Cut Pasture Silage

Because all silage must be compressed to squeeze the air from it to prevent spoilage, it becomes a very dense, tightly packed material that is quite difficult to remove from the pit face with a tractor/loader. Consequently, the tractor/loader used to feed bunker silage is typically fitted with a grapple attachment that functions like a sharp claw to cut or rip large chunks of compact silage out of the vertical pit face. Despite the aid of the grapple attachment, however, bunker silage must be chopped quite short to facilitate removing it from a pit face.

Long-cut silage packs together much more tightly than short-cut silage so that the long, compacted grass material becomes exceedingly resistant to the cutting action of the tractor/loader grapple. Short-cut silage, however, is relatively easy to cut and rip apart with a grapple. The drawback of short-chop bunker silage is that the finely chopped material is so small that it passes through the cow's digestive system too quickly and the cow has difficulty regurgitating the material to chew her cud. Consequently, much like grain-fed cattle (as discussed in chapter 4), cattle fed short-cut silage usually require that some hay be added to their diet to function as roughage, which slows the passage of the silage through the digestive system.

From a financial point of view,
long-cut silage is less expensive to make.

When cattle are self-fed along the pit face of a pasture silage clamp with portable electric fences, however, the short-chop length is no longer required because the silage does not need to be removed from the bunker with the aid of a tractor/loader grapple. The cattle do this themselves. From the cow's point of view, long-cut silage is the preferred form of silage because it is essentially just long, wet, pickled grass. The long grass pieces are coarse enough to be easily regurgitated for cud chewing, and because the silage material is coarse enough to pass through the cow's digestive tract slowly (at the same speed as fresh grass does), the long-cut silage does not need to be supplemented by any roughage. Finally, as compared to short-cut silage, the long, stringy material of long-cut silage is easier for the cattle to grasp with their mouths when "grazing" the self-feeding clamp face.

From a financial point of view, long-cut silage is less expensive to make. The long chop means using fewer knives inside the silage chopper unit and requires simpler, less precisely manufactured chop-

ping equipment for harvesting. This translates into less horsepower, cheaper equipment, lower maintenance costs, and lower fuel costs. Both the nutritional and financial benefits of long-cut silage are so significant that whenever possible, the long-chop silage should be used.

Making Pasture Silage Clamps

The pasture clamps for silage should be built between 5 and 6 feet high so the cattle can reach all of the face and so no silage overhang develops that can collapse on the cattle or onto the electric wire during self-feeding. At harvesttime, each clamp should be sealed after only 2 to 3 days' worth of chopping to prevent spoilage. By making multiple small pasture clamps and sealing them immediately (read about the ensiling process on page 123), you can create better-quality silage than that stored in bunkers or large clamps, which must be constantly reopened to add more harvested silage.

Multiple small clamps spread out across the farm will also feed out with less spoilage than a large clamp that is open over an extended period of time, will distribute manure more widely, and will prevent excessive damage to the fields from prolonged traffic. Clamps should be built directly in the field where the silage is harvested to eliminate traveling distances between the silage chopper and the clamp. This also ensures that nutrients are not transported out of the field.

Harvesting Long-Chop Pasture Silage

To reduce moisture seepage and nutrient leaching from the silage pit, some farmers like to let their silage wilt before chopping it. Although this works in theory, not only do you risk overdrying the windrows or having the wilting silage leached by an unexpected rain, but it also requires additional time and machinery.

A long chop length will prevent moisture leakage from the clamp (there are fewer exposed cuts in the grass from which to lose moisture), allowing you to direct-cut the silage with more consistent quality. Just avoid direct cutting during or immediately after a rain or heavy morning dew to lessen the amount of moisture in the silage. You should also always leave a 5- to 6-inch grass residual in the pasture after cutting so the root reserves are healthy enough to spur quick regrowth.

The Ensiling Process

The ensiling process is an anaerobic fermentation process. To avoid spoilage, the pasture clamps should be covered with a plastic tarp to seal out oxygen. Laying tires on top of the clamp to keep the tarp tight against the silage will prevent air pockets from forming between the silage and the tarp and prevent the wind from lifting the tarp and letting in air. Old tires or earth piled around the edges of the tarp works well to complete the seal around the perimeter of the clamp.

The effectiveness of the ensiling process depends on the quality of the grass. Old, dry, mature grass will not ensile easily, will not pack tight, and may require chemical or enzyme additives to facilitate the ensiling process, which may potentially require more energy to digest than it will give to the animals. Further, if the grass in the clamp is too dry to ensile properly, you risk developing toxic molds in the spoiled silage. More frequent harvests of shorter grass are quicker to cut and will prevent the ensiling of overly mature, dry grass in the clamp.

The ensiling process utilizes as part of the fermentation process the carbohydrates (sugars) in the grass that are produced during photosynthesis. These will peak during the day and then decrease as the plant utilizes the stored energy overnight. Therefore, it is best to wait until mid-morning, when enough carbohydrates have been stored in the plant again, before starting to make silage. Avoid extremely cloudy days, when the lack of sunshine lowers the carbohydrate levels in the grass and thereby reduces the effectiveness of the fermentation process. The small size of pasture clamps gives you the flexibility to avoid making silage on overcast days because a clamp can simply be sealed with the onset of bad weather and a new clamp started in another location when good weather returns. By contrast, once you begin putting silage into a large silage bunker, harvesting must continue until the bunker is full, regardless of the weather, because the silage in it will spoil if it is left unsealed for any length of time while you wait for good weather to return. It is also labor-intensive to seal and reopen large silage bunkers between periods of good harvesting weather, and the ensiling process is disturbed by reopening the seal to add more silage after it has begun. The layer between the old and the newly added silage usually will not ensile properly, risking spoilage and the development of potentially dangerous molds.

Direct-Cut Silage Clamps

You can make direct-cut silage clamps with a minimal equipment investment. Because of the short hauling distances (ideally within the pasture being harvested), direct cutting can be a one-person operation.

Self-Unloading Wagon Pulled Behind a Direct-Cut Flail Chopper

The most commonly used method to make direct-cut, long-chop pasture silage is to pull a self-unloading (rear-unloading) wagon behind a direct-cut flail chopper. This wagon can be used in a number of ways. One option is to unhook the wagon and ferry it separately to the clamp, where you unload it by driving over the silage clamp from end to end, dragging the wagon, which is unloading from the rear. This spreads out the silage as you go, and the weight of the tractor and wagon packs the clamp at the same time. Remember that long-cut silage packs much more tightly in the clamp than conventional short-chop silage, which requires almost continual packing as it is layered into a pit, so very little additional packing is necessary. Usually one good pack at the end of the day will suffice.

An even more efficient way to use this system is to convert the PTO-powered (Power Take Off–powered) unloading mechanism on the wagon to operate using the tractor's hydraulics and drive the whole unit (tractor, silage chopper, and wagon) along the front of the clamp, where you'll dump the silage. This approach eliminates the need to unhook the wagon to access the unloading mechanism. Unload the silage as tight "loaves" next to one another rather than spreading out the material as you drive. After accumulating ten or fifteen loaves, unhook the tractor and pack the loaves together with a front-end loader.

To convert the unloading mechanism to operate via the tractor hydraulics, remove the PTO drive assembly from the wagon and install a small hydraulic orbital motor on one of the drive sprockets with a set of hydraulic hose extensions to reach the tractor.

At one time, flail choppers and self-unloading wagons were extremely popular in the dairy industry for making green feed, but they have fallen out of favor; thus, used flail choppers and self-unloading wagons can usually be acquired for a song. Even new, they are not particularly expensive to buy because they are not precision instruments.

Their horsepower requirements are significantly lower than precision short-cut silage choppers so even the size of your tractor can be downgraded.

Typically a 45- to 75-horsepower tractor will be sufficient, depending on the cutting width and whether or not the direct-cut assembly utilizes a slip clutch.

To speed up the process, a second person and tractor can be added to ferry the silage between the chopper and the clamp. In this case, a high-dump wagon pulled behind the flail mower allows the chopper unit to dump into a self-unloading wagon pulled behind the second tractor. The clamp can again be built by driving over the pile with the wagon in tow, unloading as you drive. The extra person and dump wagon will double the clamp-building rate. Once again, however, this method accumulates machin-

A high-dump wagon can be pulled behind a flail mower (top). Direct-cut, long-chop forage can be dumped from a high-dump wagon into a self-unloading wagon pulled behind a second tractor (bottom).

ery and forces you to spend your time on equipment operation rather than on planning winter-grazing management.

Self-Contained, Self-Unloading Wagon with Direct-Cut Attachment

Alternatively, a self-unloading (rear-unloading) wagon with a direct-cut chopping unit attached to the front can also be used to make direct-cut silage clamps. This style of wagon is fitted with a pickup head/silage chopper attachment built right into the front of it. Some of these can direct-cut silage right out of the field; others need to have the crop swathed (not chopped) ahead of the wagon, but then can pick up and chop the crop into the wagon with the chopping pickup head. In either case, the wagon is pulled directly behind the tractor and does not require a separate silage chopper unit to be pulled in front of it. Unload the wagon by dragging it over the top of the pile behind the tractor in the method described above. Unfortunately, despite its convenience to the one-man silage-making operation, because it is no longer made for the mainstream North American beef and dairy industry, this style of chopping, self-unloading wagon has become extremely rare and can be found only in used condition.

Fencing the Clamp

Once the clamp is built, protect it with an electric fence to keep away livestock and deer. It is remarkable how attractive the sweet-smelling clamp can be — deer will climb onto the top of the clamp and paw through the plastic if given the opportunity. A double layer of plastic tarps will also help to discourage birds from pecking through the top of the plastic to get at the sweet, fermenting silage. The clamps are fed out during the winter with portable fencing (see pages 118–121).

As when rationing hay, fence two sides and move portable electric wires along the open pit face between the wires. Both ends can be opened at once to increase access to the clamp. Push fiberglass posts into the pit face horizontally, approximately 3 feet above the ground. The wire should sit about 6 inches or less from the silage wall. Provide a minimum of 8 inches of silage for each cow along the clamp face and a minimum of 6 inches for each replacement heifer. Thus, when building clamps during the harvest season, the width of the clamps will be determined by the number (and classification) of cattle you plan to feed with each clamp during the winter.

Each day you feed from the clamp, cut or fold back a new strip of plastic and push the horizontal

A silage clamp can be made self-feeding with portable electric fencing.

fiberglass posts a little farther into the silage wall. At least six inches of pit face should be eaten every day to avoid spoilage. A pitchfork or pickax will be the only winter-grazing tool you need to trim and straighten the pit face and knock down silage that extends above the reach of the cattle.

The livestock will clean up the silage right to the ground level, and if you monitor the wire height and refuse to give them more until they clean up what they can reach, they will keep the clamp face completely vertical and straight. If mud becomes a persistent problem around clamp faces due to insufficient ground freezing or extended periods of wet weather, you can locate small concrete, asphalt, or gravel pads around the farm adjacent to the various fields that you anticipate harvesting.

Transitioning to a Silage Diet

To make a smooth transition to the anaerobic silage diet (which requires a completely different set of microorganisms in the stomach to digest) and to train cattle unaccustomed to self-feeding under an electric wire, open the first clamp of the season while there is still at least two weeks' worth of grass left to graze in the same field. This enables the cattle to mix silage and grass in their diet for a short period while they become accustomed to grazing under the wire and allows their digestive systems to make the transition to anaerobic feed. To avoid frightening the cattle with loud electrical shorts that can also bleed into the silage face should any long silage material short out the wire, run the energizer at half power along the clamp face.

Planning for Drought

ONE THING WE CAN COUNT ON IN AGRICULTURE is the unpredictability of the weather. Although we use weather trends for our planning, abnormal weather happens more often than average weather, despite the fact that most management plans hinge on the average year. Always plan for the unexpected!

The simplest solution is to have a reserve of stored feed, ideally enough to see you through a full winter. A drought reserve, however, does not necessarily have to include expensive stored feeds if you plan your grazing rotation to continually carry over excess grass as a reserve.

The size of your drought reserve depends on how predictable the weather is in your area: The more arid the climate, the greater the weather fluctuations. In many arid and semiarid climates, it is wise to carry as much as a full year of grazing as a drought reserve at all times. This grass excess also helps to buffer the soil from flash flooding, excessive moisture loss, and sun baking.

When drought occurs, areas without a predetermined drought plan will become so seriously overgrazed that the first light rains will have no effect whatsoever on the grass and the soil. If the root reserves are decimated and the soil moisture completely depleted, the soil and grass will take much longer to recover after a drought. Usually, when farmers are caught without a drought plan, they fail to destock early, which leads to financial ruin as

their starved cattle become worthless in the forced destocking frenzy that sends market values falling through the basement. Only a well-thought-out drought plan will protect the soil, the grass, the herd's health and fertility, and the financial viability of the farmer.

A History Lesson

The typical response to a drought reads remarkably like a survival story from a nineteenth-century Arctic expedition. In the description of a typical drought below I have added parenthetical analogies to the typical disaster responses of those on such expeditions. It provides a remarkable insight into the predictability of human response to crises, in particular the optimism that we tend to harbor within ourselves that such terrible things won't happen to us, that we are smart enough and resourceful enough to successfully navigate through such situations when the time comes. Thus, it is a dire warning to us all that without preplanning our responses to a disaster/drought, all the best intentions and emergency measures invented after the onset of the disaster cannot prevent the wide range of terrible outcomes that result.

As the drought (disaster in the form of a shipwreck or of becoming icebound) begins, food supplies are rationed with a degree of optimism that the rains (supply ships or rescue ships) will soon arrive. As the drought (disaster) worsens, food rations are further reduced to becoming starvation rations. Still we hold on, hoping to pull through and save the herd (the crew).

When our luck holds out, the herd (crew) survives, rescued by expensive hay rations (expensive rescue attempts) brought in from afar, but the animals (crew) are emaciated and the land has been damaged by erosion and overgrazing. Herd (crew) productivity, health, and fertility have been severely compromised, sometimes beyond repair, and the farm (expedition) is financially destabilized.

So please, learn from history and don't get caught without a preplanned drought strategy.

In the worst-case scenario, the drought (disaster) does not end soon enough, and the herd (crew) starves or drastic measures are taken to sell emaciated animals at fire-sale prices (last-ditch survival strategies are attempted such as sending crew members on hopeless forays to reach help or resorting to cannibalism of deceased crew members). The financial, environmental, and emotional costs to the farm and family (expedition and crew) are irreparable. The business sinks and goes bankrupt (the expedition leader and the investors funding the expedition are bankrupt, the lives of those involved in the expedition are scarred, and the financial security of their families is severely or even permanently compromised).

We may shake our heads at the decisions early explorers made, but is our "buckle down and wait it out" response to drought really so different? I recommend reading about Sir John Franklin and his fatal Arctic expedition in search of the Northwest Passage and Sir Ernest Shackleton's two-year ordeal to rescue his crew after his ship, the famous *Endurance,* was crushed in Antarctic pack ice. They are remarkable studies of the optimistic denial and subsequent delayed reaction to the possibility of disaster

and the tragic suffering caused by failing to prepare a disaster contingency plan. There are countless farmers who have struggled just as valiantly in the jaws of a drought, with equally tragic results but without the international recognition afforded these polar explorers. So please, learn from history and don't get caught without a preplanned drought strategy.

Having said that, what should a drought plan look like beyond a predetermined grazing reserve carried over in the pastures at all times?

Preparing a Drought Plan

To survive drought, we must always take stock of our grass reserve and know how it is faring compared with our herd numbers; this is our point of reference. When our grass inventory starts looking inadequate — a warning sign — we need to be ready to start destocking. (To learn how to calculate cow-grazing days, or CGD, see chapter 25.)

A flexible herd size and knowing the size of your grazing reserve at all times are the keys to predicting and surviving a drought. Record your drought plan on paper or on your computer so that you can easily access it when you notice the first signs of trouble. In the stress and heat of the moment, it is easy to forget your plans and to be caught up in an instinctive wait-it-out attitude. Don't expect a level head to take you through the early warning signs of a drought. Arctic explorers were no tinhorns at their business either, yet time and time again they were caught unprepared. Our optimism will guarantee that our response to drought will be too little and much too late. *Write it down!*

Selling Early

When you notice the first warning signs of an inadequate grass inventory based on your cow-grazing-day calculations, begin selling off animals early so your grass reserve always exceeds what is needed to take you through the next year. Your CGD calculations will warn you of feed shortages that could lead to a drought long before the easily recognizable signs of a drought become apparent, allowing you to sell *before* the market rush starts, *before* you start rationing your animals, and *before* they begin losing weight. You want a premium for them, and you will get it only by offering fat animals for sale before the emaciated animals begin flooding the market.

By selling your animals early, you'll have money in the bank as a financial drought reserve, allowing you to restock your herd at the end of the drought, typically while prices are still low. Destocking early is the key to saving your land from overgrazing and abuse so it can recover quickly after the drought and so your financial situation can remain as stable as possible.

Culling the Herd

When your enterprise includes cow/calf pairs, stockers, and even grass-finishing animals, it is easy to start selling your nonbreeding classes without having to cut into your brood herd, especially because these animals will be seriously discounted later when the market rush begins, even if they are in good condition and at an ideal market weight. This is also the time to seriously cull from your herd the less-fertile, higher-maintenance animals; replacement stock will be cheap to buy at the end of the drought, particularly if your early intervention allows your land to recover quickly, before the drought-induced market collapse ends. By protecting your core brood herd and destocking primarily other livestock classes first, you ensure that you will continue to have an income at the end of the drought from the calves produced by your breeding animals. There is nothing more disheartening than to survive a drought but then suffer a cash-flow shortage afterward because the brood herd has been sold off or its conception rates have been affected by starvation. Businesses go bankrupt because of insufficient cash flow to pay the bills, not because of insufficient assets (see the discussion on page 130 about building a financial reserve against drought for more on this concept). Determine in your drought plan which animal classes you will cull first and *put it in writing,* so sentiment and stress do not cloud your decisions in the heat of the moment.

Avoiding Overgrazed Pastures

Never allow your herd to overgraze your land in a drought situation; a good grass residual is the only guarantee that your land will recover quickly after the rain returns. Land that can be restocked quickly is far more valuable to your survival than are emaciated cattle and overgrazed land that remains depressed in its production years after the rain returns.

Be sure to leave a good grass residual when the drought begins, so the root reserve stays deep and the soil remains covered with plant stubble and debris; both guard against wind and flash-flood erosion at the end of a dusty drought. The debris also shades the soil from sun baking and provides a protective barrier that slows evaporation and allows the water table to replenish itself more quickly when the rain returns. Do *not,* under any circumstances, allow your cattle to eat plants into the ground. Destock early.

Rotating Pastures Daily

Daily pasture moves will allow you to ration your grass supply most efficiently so that you can protect the grass from overgrazing and maximize grass recovery periods between grazes. It will also minimize grass waste due to trampling and fecal contamination.

You must rely on your cow-grazing-days calculations, discussed in chapters 7 and 25, to determine if you are allowing adequate regrowth of pastures before regrazing them. Only your CGD calculations will indicate whether your rotation is slow enough to account for grass shortages or your grazing rotation will catch up to itself, leading to depleted root reserves and overgrazing if you do not take preventive measures such as selling some cattle and finding additional pastures currently not included in the rotation. Thus "adequate regrowth" is not a quantitative measure of grass height, but rather a calculated indication (based on CGDs) of whether your herd's current progression through your grazing reserve is sustainable or happening too quickly to keep up with grass regrowth. By the time it's visually obvious that the grass is shorter than it ought to be when your herd moves to the next pasture, your grass reserve already will be seriously depleted and you will require considerably more-drastic measures to remedy the situation. During a drought it is tempting to regraze sooner than usual because the regrowth is not fast enough to maintain your rotation; doing so will seriously harm grass growth. In reality, you need to slow down your rotation to compensate for the slow growth.

There are three ways to slow down your rotation without risking overgrazing:

1. *Destock early.* Start by selling your non-breeding-class animals first, such as your empty and cull cows, stockers, and grass-finishing animals, and by culling from the herd the least-fertile, higher-maintenance individuals. By reducing your herd size as soon as you calculate that your grazing reserve will be inadequate, you will keep your remaining animals well fed, protect your land from overgrazing, and protect your breeding herd's fertility so it can provide you with a cash crop of calves as soon as the drought ends.

2. *Combine your herds* to increase grazing efficiency and lengthen the pasture recovery time (see chapter 9).

3. *Feed some hay.* This is done in a worst-case scenario if you still insist on maintaining a hay reserve, to allow you to lengthen the grass recovery time. It is far better to feed hay in the middle of a summer drought in order to give the grass more time to recover than it is to deplete your grass reserves first, overgraze, and then feed hay later.

For example, in southern British Columbia, summer can sometimes be so dry that without irrigation, almost all grass growth stops. If cattle farmers in such dry-summer areas begin to run short of grass, it is far wiser for them to feed hay during July or August and reserve lots of grass in the field so that grass growth can catch up with their rotation than it is for them to use up all the grass reserves through overgrazing.

By feeding early, the grass is capable of recovering in time to benefit from fall rains and cooler weather. A single month of feeding during the summer drought may well give you an extra three months or more of grazing in the winter, which you would not have had if you had overgrazed your summer rotation.

Supplementing Grass Quality

As grass quality further deteriorates during a drought, plan to supplement it so that you can still make use of poor-quality grazing rather than using stored feed (supplementation is covered in detail in chapter 7). Extra protein fed along with the mineral supplement can make a difference in getting cattle to successfully digest a lower-quality grass without having to abandon it.

If you decide to keep stored feed in reserve, hay will keep much longer than silage. Silage will remain edible in a clamp or bunker for only two years, but if you don't use your hay for ten years, it will still be perfectly edible. Furthermore, you can sell hay during years when prices are high, then replace it with fresh reserves when prices are low. Unused silage is not as easily transportable for sale and its shelf life is considerably shorter so it is much harder to resell and replace with fresh silage reserves.

Maintaining a Drought-Proof Water Site

Plan to have at least one drought-proof water site available in every corner of your grazing area so you do not have to abandon a valuable grazing reserve because you cannot provide your cattle with drinking water. A drought-proof water site is any water site that you can rely on to produce sufficient water to meet your cattle's needs during even the most severe drought, when many other water sources may fail. An ideal drought-proof water site could be a reliable deep well, access to a reliable river or lake that will not dry up during a drought, a connection to a municipal water supply that guarantees livestock water needs even if irrigation restrictions are enforced, or any other water site that can weather the effects of a severe drought. Many surface water sources such as dugouts, streams, ponds, and even shallow surface wells will dry up. If they do, either you will lose access to many parts of your grazing land or your cattle will have to traverse large expanses of pasture to reach water. These water routes will become overgrazed and compacted. With various drought-proof water sites spread across the farm, you can build makeshift alleys with portable fences and protect as much of the land as possible while still keeping your cattle watered.

Building a Financial Reserve

Build a financial reserve during the good years that you can draw upon during a drought. Even in a drought, bills will need to be paid whether or not you have an income. Part of drought survival is being financially buoyant through the drought period. Having enough grass but no cash flow to pay mortgages, rental fees, and other essentials will cause the bank to foreclose just as quickly as it will on the neighbor who has emaciated cattle and no grass reserve.

Remember that during a drought you can no longer use your cattle as a financial reserve because their market value will have collapsed. Any remaining equipment will also become temporarily worthless as everyone else begins selling his excess equipment to keep his bankers at bay.

Set aside a small part of your financial reserve as a vacation fund to use during the worst of the drought.

I remember phoning a friend during the middle of his normal grain harvest season in central Alberta during the drought of summer 2002. The drought on his farm was so severe that he had nothing whatsoever left to harvest, yet he had planned for it and had a financial reserve to survive the fiscal year. In fact, he was packing his bags to go on a fishing vacation. I was stunned! He told me that vacationing then was the best way to keep from becoming demoralized and to stay true to his drought plan. Sitting at home feeling powerless with absolutely nothing to do when he ought to be out harvesting his yearly income would have rubbed already raw nerves to the breaking point and would have sorely tested his ability to follow through on his drought plan.

Chapter

9

Managing Your Herd

GOOD HERD MANAGEMENT EXTENDS to the way we interact with cattle, how we organize their lives, and how we affect their sense of security in the world we create for them on our farms. Cattle are remarkably trainable if we understand how they perceive the world around them. Herd management is just cow psychology 101 with a little human psychology thrown in — after all, we have to overcome our own prejudices about how we expect cattle to react to us and overcome many of the bad habits we have developed in how we handle them.

Through integrated herd management, low-stress livestock-handling techniques, and learning how to build and troubleshoot handling facilities to reduce livestock stress, we can work with the way cattle naturally perceive the world. This allows us to treat our cattle more humanely and reduce livestock stress. We also benefit considerably ourselves through time and labor savings, increased cattle weight gains, higher-quality meat, reduced management costs, and having more enjoyable, stress-free interactions with our cattle.

Integrated Herd Management

One of the most laborsaving practices in a grass-based beef enterprise is to combine all the cattle on the farm into a single large herd. There is simply no place for multiple herds in a low-cost, high-profit natural production system. They are an unnecessary,

"make-work" practice that is justified by a number of excuses that are not valid and do not apply to a grass-based grazing scenario. Following are some of the reasons people cite for why they maintain several herds and my responses to them.

■ *To separate yearling heifers, first-calf heifers, steers, and weaker animals* from older, stronger animals for nutritional purposes. While this idea is valid in feedlots with limited bunk space, where feed access is limited and a source of great competition, it simply does not apply to a pasture-grazing situation where young, old, weak, and strong all have equal access to the grass beneath their feet.

■ *To preserve the social hierarchy of the herd.* This philosophy stems from the idea that cattle will not be able to remember the social hierarchy within the herd if it gets too large. But why should they? Pasture grazing is not the same as a feedlot system, in which feed and good bedding are limited resources that must be fought over until a hierarchy develops. Cattle on pasture don't have to fight for access to grass, a place to lie down, or access to water. (If they do fight over water, there is a problem with the water system, not with the cattle; see chapter 6 for more on this subject.) Cattle are natural herd animals that thrive on the security afforded by large groups. As is discussed below, as herd size increases, stress will decrease in the herd, a group identity will develop (cattle choose to graze as a group rather than spreading out), and their nutritional efficiency will increase.

■ *To separate cows that have calved from cows that have not calved* in order to address differing nutritional needs of the two groups, prevent calf stealing (discussed on page 82), and reduce disease pressure on newborn calves. These preventive measures simply do not apply to a grass-based summer-calving scenario because the herd has space, clean ground, and access to abundant high-quality summer grass.

■ *To keep a separate, registered purebred herd.* Registered purebred breeding programs are extremely expensive and time-consuming due to the paperwork volume, single-sire breeding programs required to establish paternity, extensive record keeping, and other hoopla called for by the registered-livestock business. They simply do not fit on the same farm with financially viable beef production, grass-based or otherwise; the production methods required by each are antagonistic to each other's success.

■ *To accommodate two calving seasons* in order to diversify market access and cut down on bull costs. Isn't the fiasco of one calving season enough? The slim benefits gained by two herds with separate calving seasons, at least one of which will be calving out of sync with nature's seasons, are far outweighed by the extra expense, labor, infrastructure, and decreased productivity (in terms of grass growth, extra management costs, and loss of valuable time that could otherwise be spent on planning, management, and marketing) that result from keeping and managing two herds.

■ *To separate steers from heifers* in order to prevent steers from trying to breed heifers (riding). As is discussed in chapter 3 and on page 135 of this chapter, although this is valid in the boring confines of a feedlot, it simply does not apply to steers and heifers grazing together in a daily pasture-grazing rotation. Even if steers and heifers that did not grow up together are lumped together on pasture, the riding will be minimal and short-lived and certainly not sufficient to warrant keeping them separated.

■ *To extend the breeding life of the bulls* by rotating them through separate herds. The added labor costs; decreased pasture productivity; increased complications in management; additional infrastructure; and subsequent time away from planning, management, and marketing simply cannot justify the comparatively slim financial savings of extending the breeding life of your bulls by keeping separate herds. You may save a few dollars on extra bull costs, but you will spend tens of thousands of additional dollars on extra labor, infrastructure, and management while your grass, soil, and grazing efficiency miss out on the significant benefits of grazing the herd as a single large unit.

Advantages of a Combined Herd

Once the herds are combined, you will need to manage only a single grazing rotation. You will require significantly fewer fences because the major pasture divisions can be much bigger and only a single rotation must be developed. This saves considerably on fencing infrastructure and water-site development costs. It becomes much more feasible to manage a daily grazing rotation that includes access to water from each daily pasture slice.

By the time you move your portable electric fences, portable mineral feeders, and water trough or water-access point, it takes almost the same amount of time to move a herd of twenty or a herd of a thousand animals to the next pasture. The difference: With a single herd you have to do it only once instead of having to repeat the process multiple times for various groups of animals, and you will have to refill only one set of mineral feeders, clean only one water trough, and maintain power in only one set of fences.

Managing your cattle as a single herd in a daily pasture rotation greatly simplifies pasture management. Grass recovery times are longer, rotations do not conflict with one another, rest intervals for your pastures are easier to plan, and it is much easier to achieve the ideal grazing interval to maximize grass growth, thereby increasing your farm's carrying capacity. A single large herd makes it easier to create an evenly distributed grazing impact on the land and helps to prevent overgrazing. It is also much easier to budget the grass reserve during winter and the drought season so your farm will be much more resistant to abnormal weather patterns.

Herd Behavior and Soil and Pasture Quality

The benefits of having a single large herd of cattle constitute more than just labor, infrastructure, and management savings. Even the herd's grazing behavior changes as the herd size increases. Competition for access to grass increases, and the cattle identify more strongly with the herd for security, causing them to graze as a group.

A minimum of one hundred head of cattle is required for the animals to begin grazing as a group rather than spreading out across the entire grazing area. As they move together, their hooves trample the weeds and fracture compacted soils much more uniformly, while their grazing becomes less selective and their manure is spread out evenly behind the herd. This is the *herd effect.* Multiple small herds simply cannot impact the soil and grass to the same degree. The bigger the herd, the stronger the effect; it is not an off/on phenomenon but rather an effect that increases almost exponentially with increased herd numbers. A quarter million bison grazing as independently migrating groups of five thousand per herd simply cannot achieve the same beneficial impact on the land as a quarter million bison migrating as a single herd.

The same is true for the cattle on your farm. The more cattle you combine into a single grazing herd, the more effective their impact will be on the land.

Social Benefits

The social effect of herd integration is also quite dramatic, because social ties are not severed as calves become replacement heifers, first-calf heifers, and cows. Family ties remain strong; it is not uncommon to see grandmother, mother, and daughter consistently spending time together within the main herd structure. This high degree of familiarity within a herd can only help decrease stress on the individual animal and increase the sense of group identity, stability, and security. Cattle introduced into an established herd take years to identify fully with the herd. It is not uncommon to find newcomers grazing together as separate subgroups to one side, even many years after joining the herd.

The added security and stability of the large herd and the resulting added resistance to stress increase not only disease immunity, but also the feed conversion efficiency of cattle, resulting in more efficient weight gains. Less stress equals more efficient digestion, something I think we can all relate to when we think of stress in our own lives.

One of the arguments for keeping separate herds is that it allows grass-finishing animals, stockers, and yearling heifers to get higher-quality nutrition (this represents the leader–follower grazing system, in which one class of animal, such as grass-finishing

The herd effect: A large herd of cattle graze together as a mob, increasing the positive impact on the soil and grass.

cattle in preparation for slaughter, is grazed one pasture ahead of the main herd in the pasture rotation so it can get access to better nutrition by skimming off the choice grass ahead of the main herd, which grazes the leftovers). The increased feed conversion efficiency of a large, integrated, single herd in a quick pasture rotation will achieve similar gains as that of a leader–follower grazing system, but with less complicated management and at a much lower cost per pound of beef produced.

Another argument for multiple herds is that different age groups have different nutritional requirements, but when you match the yearly calving cycle to the grass growth cycle, separating cattle by age becomes unnecessary. Admittedly, first-calf and second-calf heifers will require supplementation sooner than the adult cow herd during the winter months, but if the entire herd is managed to graze year-round without stored feed, the benefits and cost savings of single-herd management will greatly exceed the additional cost of supplementing the whole herd with a little extra protein or energy to address the lowest common denominator.

An integrated herd should include all age classes of cattle: adult cows, heifers, stockers, and even grass-finishing animals. Some people are concerned that the steers will ride the heifers when they start to cycle if the two groups are not separated (see page 133 for my response to this). It is true that riding will be more visible among cattle that are newly introduced to one another due to the stress of the unfamiliar. Nevertheless, it will fade quickly as the new animals get to know those in the group, as long as the herd is managed to reduce stress. Integrating the herd makes it much simpler to add stockers and grass-finishing animals to your beef enterprise, provided that you carefully inspect all the steers to ensure that none has slipped through with a testicle intact. Castrating with a knife instead of banding or ringing (see pages 50–55 for more on castration) helps provide an additional guarantee against unwanted or premature pregnancies among your heifers.

When to Separate the Herd

If you are concerned that marginal grass quality in your pastures will not result in the high weight gains required to produce a tender beef product, grass-finishing animals can be grazed separately on your best pastures in the last months before slaughter. Yet the additional stress of separating the grass-finishing animals from the herd will initially offset a portion of their gains and will increase your management effort and labor. This should be only an emergency measure if the grass quality becomes so deficient that you cannot continue grass-finishing animals along with the main herd.

Yearling heifers that you intend to grass-finish for slaughter should be separated from the main herd during the breeding season so their gains are not affected by pregnancy. This is especially important if any of the heifers are kept through the winter to be slaughtered the following spring.

Integrating the Bulls into the Herd

In the wild, there are two different behaviors that we see among the bulls of big herds. Some herd animals such as musk oxen, bison, and caribou remain together year-round as a large herd; the bulls either remain completely immersed within the herd or hang around as small subgroups along the herd's periphery. Despite this, the breeding season is still restricted to a very short period of time, controlled by a female's estrus period, which is determined by her nutrition and calving cycle. An alternative arrangement is seen among animals like deer and moose, in which the bulls completely lose interest in the females after breeding season. They are in each other's company only for the duration of the breeding season, when the combined stimuli of the weather and the photoperiod bring on the rut (the time when the bulls are physiologically and behaviorally capable of reproduction, sperm are produced, and the bulls are sexually excited) and cause the two groups to seek out each other.

Domestic cattle do not experience this same once-a-year rut. The high plane of nutrition that cattle enjoy in our domestic care enables the females to cycle year-round, which is why we separate the bulls from the cows for most of the year. Yet, presumably, this was not always so. On the Aleutian Islands in Alaska, a large number of cattle from a mixed ancestry were left unattended over many years and have become completely feral. Interestingly, these cattle have returned to the once-a-year calving season because of the severe nutritional constraints put on them by their environment and by the photoperiod effect on their fertility.

If we mimic the bison, musk oxen, and caribou by allowing the bulls to run with the herd year-round, and if we allow our animals to graze year-round so their body conditions can fluctuate naturally, in theory our own domestic cattle should also be capable of stabilizing their calving season around a single ideal season. The quick return of the estrus cycle after calving, however, would risk rebreeding too early and yearling heifers would be at risk of being bred too young.

Meat from relaxed animals is more tender and has a longer shelf life.

But we can combine the bulls with the herd from the onset of breeding season through winter until just before calving begins. The bulls have to be managed as a separate herd only from just before calving (when the heifer calves reach ten months of age and start to cycle) until breeding, a time period of approximately four months, to prevent cows or yearling heifers from being bred too early.

If you manage your grass reserve, body condition scores, supplements, and breeding turnout date carefully to ensure maximum fertility during the breeding season, and you time your calving season to maximize the benefits afforded by nature's seasonality (see chapter 3 for a detailed discussion about timing your calving season), your herd will have extraordinarily high conception rates during the first cycle and the remainder will be bred during the second cycle. Because the bulk of the herd will be bred by the end of the second cycle, the bulls can simply accompany the herd for the remainder of the winter.

Your culling program, however, must maintain the ideal calving season/breeding season length that is normally controlled by the rut season in nature.

You must immediately cull cows that are *bred* after the end of your ideal breeding season (forty-two days, or perhaps even twenty-one days). Those that are bred late but slip by without your notice will invariably still be pregnant when the end of the ideal calving season arrives and finally *calve* after calving season is meant to be over. Now you cannot fail to notice that they are out of sync; now they too can be marked/recorded for culling at the earliest possible convenience. Cows that breed/calve late can be culled as soon as the calf is at least four months old, leaving the calf behind to be raised by the herd.

Low-Stress Handling

When they think of stock handling, many people are overcome by the romance of the rodeo arena — human overpowering and outwitting beast — as if the roar of the crowd will follow them from the event into the pasture. The macho hype and bravado of the rodeo belong in an arena and to rodeo performers (cowboys), a mop-up crew to pick up the pieces (appropriately called *pickup men*), and rodeo clowns (to please crowds, not cows). Unfortunately, though, the rodeo often spills into the cattle world, where good stockmanship is the key.

The Costs of Stress

Stress causes more than frustration, injury, and lots of commotion. It exacts a tremendous economic cost and a significant toll on an animal's health and productivity. And stress from improper handling does not simply end when the handler disappears; it stays with the herd and affects its health and productivity for weeks and even months afterward. Decreasing stress significantly increases weight gains, calms cattle, makes processing and loading easier, decreases labor costs, and improves meat quality.

Stress raises the acidity of the meat. In the short term, this makes meat tough; in the long term, an animal under prolonged stress produces meat that becomes dark and sticky, like a bruise. These animals are called *dark cutters*. Meat from relaxed animals is more tender and has a longer shelf life. By contrast, rodeo steers barely make good hamburger.

I grew up around rodeo-style-cowboy handling practices. One day, however, during a high-speed calf pursuit, my dad and his horse ended up in a terrible accident. He suffered a massive head injury and has not farmed since.

You might think that I learned from his unfortunate example. But no, rodeo-style cattle handling came quite naturally to me; stockmanship did not. I was well on my way down a path of cattle prods, noise and hollering, intense frustration about "stupid" cows, rattling gates, bucking cattle squeezes, and an ironclad conviction about how cattle should be handled and how stressful cattle-processing days should be. Eventually I stumbled across the principles of low-stress stock handling — and then 200 year-ling heifers broke through a fence and wound up in the wrong lush pasture.

To return them to their pasture, the heifers had to be pushed across a huge field of boggy swamp and out through a fairly complicated gate. I had time, but I was alone and on foot, without fast horses and ATVs to force the situation into my favor. I had no choice; I had to work through this one gently and slowly. Whooping and hollering on my own would have turned the situation into a rodeo much bigger than I would have been able to cope with on my own. Those heifers taught me more about stockmanship in one long afternoon than I had learned in a lifetime. They taught me about flight zones (the invisible zones of personal space around animals, which initiate the fight-or-flight response if penetrated; for further information, see page 138), driving cattle from the front or the side (by manipulating the herd's flight zone and understanding livestock behavior) instead of through pressure applied at the rear of the herd, how to keep their momentum going, and how to give them time to figure out what direction I wanted them to take. More than anything, they taught me to shut my mouth and be patient.

It was one of those beautiful *ah-ha* moments when the heifers finally turned into a long winding line of cattle moving in the right direction through all the obstacles and out the gate, while I walked alongside the herd, feeling like a shepherd from a fairy tale.

The Audiovisual World of Cattle

Cattle don't perceive the world in the same way we do. Their evolutionary universe was a very quiet one: wind blowing, cattle chewing, and the occasional sounds of predators stalking them. Like many animals, their hearing is significantly more sensitive than ours, in both range and pitch. With those big ears, this is hardly surprising. As prey animals, their finely tuned sense of hearing is one of their first lines of defense in escaping predators. Yet because a sound in the normal range to us may be very loud, shrill, and painful to a cow, we should never raise our voices above a normal talking level around them if we expect them to keep calm. Ideally, we should work around cattle in silence: no whoops and hollers to push them through a squeeze chute or up into a truck.

Banging gates, squeaks, clatters, and metal-on-metal noises inside the processing corrals, in loading chutes, and in the surrounding area are sure to stress your cattle. Amazingly, processing cattle or loading them into trucks can actually be calm, stress-free, and efficient if we simply stop all the handler's noises, oil gates, close gates quietly, dampen the rattles of the squeeze chute, put away the cattle prods and whips, and quietly work the edges of flight zones (see the discussion about flight zones on page 138) instead of crowding cattle and collapsing their personal space.

Because their eyes are located on either side of the head, like most prey animals, cattle see the world in a two-dimensional way, with each eye having an independent field of vision. In only a very narrow zone directly in front of the cow's head do the two eyes' visual fields overlap; this is the only area in which they have binocular (three-dimensional) vision. In total, they have nearly 360 degrees of vision to watch for predators, but without our quality of depth perception. In addition, their visual perception is tuned to vertical rather than horizontal relief because predators appear as vertical relief above the horizon. Because of this, vertical rails make a more imposing corral fence than horizontal rails, shadows can be quite confusing or scary, and sharp alleys or poorly placed gates are easily mistaken for dead ends.

By contrast, predators' eyes evolved to be located close together on the front of the skull so that the visual fields of both eyes overlap, permitting depth perception. This three-dimensional vision is crucial to the hunting predator, who must judge distances, evaluate the contours of the land surface during high-speed hunts, and successfully reach the jugular of its chosen dinner entrée during a killing lunge.

To prey animals, keenly aware of eyes, two eyes visible at once are the mark of the predator. If a prey animal can see both eyes, then its fight-or-flight response will kick in. You can therefore use your direct eye contact to exert an enormous amount of pressure on cattle. Direct eye contact alone can be so powerful that for very fearful or nervous cows, it can tip the fight-or-flight response to provoke an attack. Be particularly cautious of how much eye contact you use with aggressive or unpredictable bulls.

Following this same principle, if you want animals to move past you, turn so only one of your eyes is visible to the animals. If you turn your head so that both eyes become visible again, the herd will either speed up to rush past you or balk and stop altogether in response to the increased pressure of your direct eye contact. If a herd refuses to move through a gate, chute, or alley, perhaps someone's direct eye contact is causing it to balk.

Another trick I learned from dealing with musk oxen and their frightening horns is to use a shield around aggressive or particularly stubborn animals. A three-by-five-foot piece of plywood with a handle bolted to the back works well because it increases your bulk and creates a barrier, which will boost your confidence. Typically, the shield's presence provides an added pressure to persuade animals not to challenge or crowd you. This can be a saving grace if you have to work on a newborn calf alongside its protective mother. Normally, when you squat down over a calf, your bulk disappears, giving the protective cow a great deal of confidence. A shield acts to maintain your bulk and protect you if she does try to physically throw you off her calf.

The Flight Zone

The key element of low-stress handling is to understand and use to your advantage the flight zones of your livestock. A flight zone is like an invisible force field that surrounds an animal, a kind of comfort zone of personal space that it seeks to maintain. Anything that penetrates this zone, including threatening sights or sounds, initiates a fight-or-flight response — the brain's response to panic, incorporating either "run away" (flight) or "attack and fend off the 'attacker'" (fight) if fleeing does not appear to be an option. The intensity of the response depends on how far someone or something penetrates into the flight zone and how threatening the presence is.

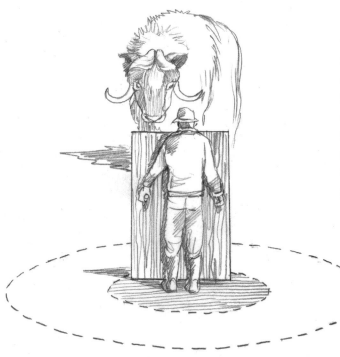

A 3-foot by 5-foot plywood shield is used for protection while herding a musk ox. The bulk created by holding the shield increases the amount of pressure exerted on the animal (the larger dotted circle) without the need to move closer to the creature.

The Approach

The size of the flight zone changes depending on the nature of the approach (sound, appearance, smell, or gesture), its speed, the size of the approaching object, and even the angle of approach. As livestock handlers, we must learn how to work gently along the outer edge of the flight zone to trigger a calm response that takes an animal in the direction we want it to go. Take the time to spend a few hours playing in a field with some cattle so that they teach you about their flight zones; don't wait until cattle-processing or sale day to learn.

The key to using the cattle's flight zone to calmly move the animal is to learn how to gently apply pressure to the zone from a direction and angle that will produce a calm and predictable response. Remember, cattle will move away with the sole objective of escaping/defusing the pressure that your presence is exerting on them. The more slowly you approach cattle, the more gradually you will elicit a response. The closer you approach, the faster they will move. If you move quickly, their flight zone will grow larger

and their response will be more dramatic. Slow and steady wins the race, giving you time to pull back out of the flight zone and slow the animal or stop it altogether by turning and walking away.

Pressure exerted on a cow's flight zone will also produce different responses depending on the direction and angle of the handler's approach. An angled approach toward the side of the cow allows you to pressure her flight zone gently, creating a directed, predictable, and calm response because there is only one direction the cow can move to escape your pressure. By changing the angle of approach and which part of the cow you target in your approach (the side of the head, the shoulder, the midsection, or closer to the rear), you can control the direction of the cow's reaction (see the illustration on page 140). Experience will quickly teach you that subtle changes in your approach angle can dramatically alter the cow's direction of travel, allowing you to steer the cow's movements from a distance.

When you approach a cow directly from the front, it is faced with the decision of which direction to

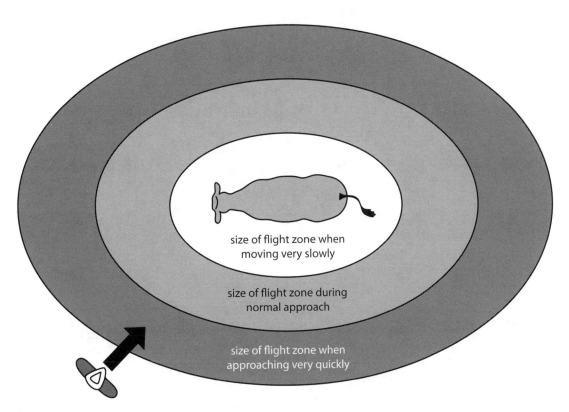

size of flight zone when
moving very slowly

size of flight zone during
normal approach

size of flight zone when
approaching very quickly

The size of an animal's flight zone changes depending on how threatening an approaching intruder (lower left) is perceived to be. The speed of the approach dramatically alters the perceived threat, and consequently the size of the flight zone.

move to best escape your pressure. Either direction, left or right, is an equally valid escape, but movement in either direction means that you will follow. By the time the cow reaches its decision, you will have penetrated much deeper into its flight zone, causing significantly more pressure, so the cow will react with a much stronger fight-or-flight response without a reliable indication as to what direction she will choose to escape your pressure.

If you approach from behind, the animal will turn at least partially to face and evaluate you. The animal thus faces the wrong way, and you have to figure out how to turn her again before you get her moving. This is why you should never approach from behind to gather a herd. If you circle around and approach at an angle from the front, the animals will already be facing the right direction when they begin to move in order to avoid you.

The area directly behind cattle is their only blind spot. If you try to push cattle from the rear, they will constantly turn around in an effort to see you, causing them to break the momentum of the herd movement or bolt outward to escape your pressure. If you stay along the side of the herd, where they can see you, they will continue to move ahead to escape your pressure.

You can use the blind spot behind cattle to steer them by gently threatening to move into it from the side. The cattle will turn slightly to keep you in sight. If you need to walk behind cattle to keep stragglers from stopping, move back and forth across the entire rear of the herd, and turn around only well beyond the edges of the herd so you are always visible when you change direction. If you change direction in their blind spot, they will expect to see you reemerge and will panic when you do not.

EFFECTS OF THE APPROACH

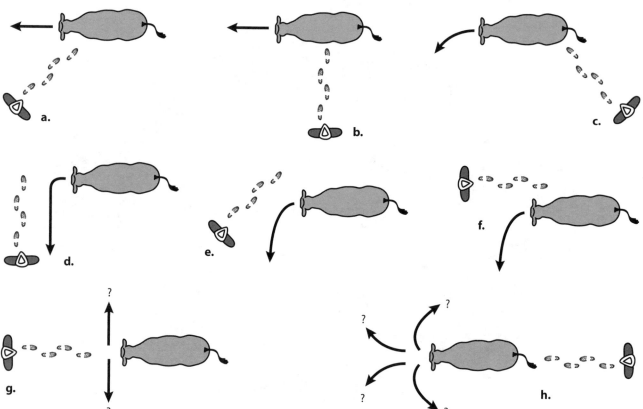

Note the effect of a handler's approach angle on a cow's direction of travel as she attempts to escape the pressure the handler exerts on her flight zone.

GATHERING AND MOVING A HERD

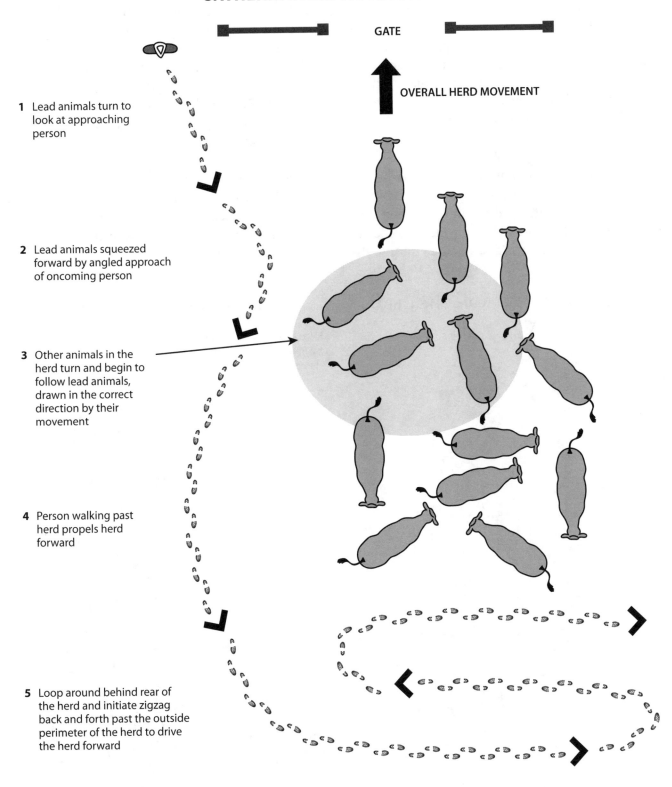

GATE

OVERALL HERD MOVEMENT

1 Lead animals turn to look at approaching person

2 Lead animals squeezed forward by angled approach of oncoming person

3 Other animals in the herd turn and begin to follow lead animals, drawn in the correct direction by their movement

4 Person walking past herd propels herd forward

5 Loop around behind rear of the herd and initiate zigzag back and forth past the outside perimeter of the herd to drive the herd forward

To sort cattle from a herd, use an angled approach from either the front or the rear to wedge the chosen animal out of the herd, as shown in the illustration. By altering the angle of your approach, you can determine the direction the animal takes and continue to steer it by modifying your angle. A direct approach would evoke a dramatic startle response, usually back to the safety of the herd.

Pressure and Herd Dynamics

Once cattle start moving, the leader pulls the rear and the rear pushes the leader. If you keep the leader moving, the rest of the herd remains in motion. This is why you want to work cattle from the sides and *not* from the rear. From the sides, you are visible to the leader and can pressure her. If you are behind her, and therefore not within her sight, the leader will stop or even turn to find you, thus holding up the entire herd.

Cattle will follow whatever direction the herd first takes. If the herd stops, the first animal that starts to move will be followed, regardless of the direction. If you keep the leader moving, you will be able to control the direction of the herd.

Cattle will speed up if you move in the opposite direction alongside them like an oncoming car, but they will slow down if you walk in the same direction alongside them like a passing car. You can use this to guide the direction of the lead animal and propel the herd forward.

SORTING CATTLE FROM THE HERD

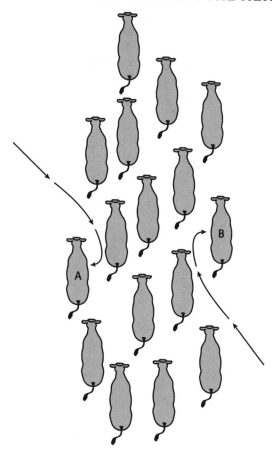

By manipulating your angle of approach, animals can be gently wedged out of the herd. Arrows approaching cow A show the angle of approach to successfully sort an animal when approaching from ahead; arrows approaching cow B show the angle of approach to successfully sort an animal when approaching from behind.

FOLLOWING THE LEADER

Herd follows motion initiated by herd leader; thus, the leader draws the herd forward

Animals at rear of herd provide momentum for cattle ahead of them: They "push" the animals ahead of them as they move forward.

When moving cattle, the most important animal in the herd is the leader. Steering the leader's direction of travel is your most powerful tool in accomplishing a successful herd move.

Start by moving toward the back of the herd along the edge its collective flight zone and then return to the front at a distance that keeps you outside its flight zone. Next, turn and sweep back in toward the lead animal and do it all over again. Your steps create a triangular motion. If you have an additional handler, that person should remain just behind the shoulder area of the lead animal, moving closer or farther away to apply pressure on her and steer her forward motion.

If you are trying to go through a gate, time your triangular motion to arrive at the gate at the same time as the leader does, and then repeat the triangle until the whole herd has been propelled through it. If an animal breaks from the herd, it is far more important to keep the herd moving than to retrieve the escapee. Nine times out of ten the escapee will be drawn back to the herd to avoid being left behind, but woe to you if the herd stops while you are out galloping after one escaped animal!

Size

You can also use your size to apply pressure to get animals to move: Lift your arms or raise a jacket to increase your pressure on the flight zone; drop your arms or squat to instantly remove pressure. Take the time to wait for animals to look toward or face the direction you want them to go, and give them time to see the gate before applying pressure to make them move. Waiting for the right movement is significantly easier (and faster) than trying to turn them after the movement starts.

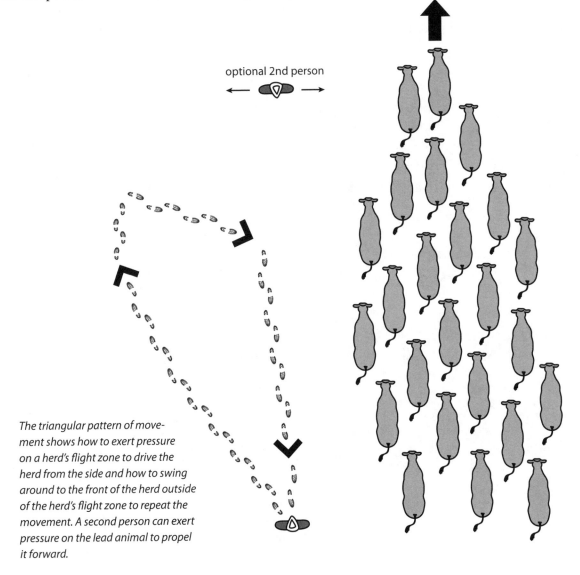

optional 2nd person

The triangular pattern of movement shows how to exert pressure on a herd's flight zone to drive the herd from the side and how to swing around to the front of the herd outside of the herd's flight zone to repeat the movement. A second person can exert pressure on the lead animal to propel it forward.

Speed

Never, ever run. The faster you move, the more you wind up the cattle and the faster everything happens from there. In a cow's world, fast movements are associated with alarm, so its reactions to you will become more erratic. Move the cattle at the same speed at which they normally walk in a pasture. This means you have to move slowly; you can't run and expect them to stay calm. Calm cattle naturally walk at about the same pace you would push a baby stroller or escort your grandmother along a sidewalk.

Horses

Cattle do not fear horses in the same way they fear humans, so they are less prone to react to a horse pressuring their flight zone, and will allow a horse to approach much closer than a human before they become uncomfortable with its presence.

The farther the stock handler is from the cow, the more time he or she has to correct an error or modify an approach before the cattle change direction or bolt. But because cattle generally allow horses (and their riders) to approach much closer than handlers on foot, a horseback rider must be far more delicate to trigger the desired response in the cattle and the horse must react more quickly to changes in the cattle's movements, though fast reactions tend to startle the cattle and wind up the stress levels of the herd.

Rare indeed is the rider who has that kind of fine-tuned control over a horse without succumbing to the temptation of using the horse's enormous speed to correct for mistakes or force compliance from the cattle. Once the running starts and the stress rises, it is hard to slow the cattle. It is far better to stop and take a moment to let the cattle calm down before gathering them than it is to get into a rodeo race, even if it means you have to start over and collect them from the far side of the paddock.

The amount of pressure exerted on an animal is directly proportional to the size (and therefore the perceived threat) of the approaching handler. Notice how much closer the cattle come to the crouching handler (bottom) than to the standing handler (top).

Dogs

Dogs and humans have long worked together to move stock, but most dogs are trained to work quite close to cattle. Their instincts come from separating and running down prey, not from gently guiding them. Dogs are trained to "push" the cattle in the right direction from close range, which causes undue stress to the cattle and requires constant pressure to keep the animals from turning to face the threat. Instead of allowing the cattle to be dragged along by the herd leader, dogs put constant pressure on the back of the herd, causing the cattle to bunch up as the animals look for ways to escape the relentless pressure on their flight zone. Although dogs can be trained to work the edges of flight zones, I have yet to see this. If you want stress-free cattle handling, get a family dog that you can take fishing on the weekend.

All-Terrain Vehicles

Nothing beats the delicate control you have on foot when you work the flight zones of cattle. This makes an ATV perfect as transportation for the farmer: You can use it to move quickly around the farm, but then you can abandon it without any fear that it will wander away in search of grass while you are working a flight zone on foot.

Keep your ATV's engine rpm down when you are around cattle. A revving engine, regardless of traveling speed, winds up cattle and people incredibly fast and is the equivalent of shouting. Around it, people are especially prone to getting excited and forgetting the delicacies of working a flight zone.

Teaching a Herd to Follow

When we work the flight zone of our cattle, we are assuming the role of predator. Yet we can also train cattle to follow our cues by using positive reinforcement (rewards) to motivate behaviors and then mold them to the desired response.

It is remarkably easy to train cattle in a rotational grazing system because you have a perfect reward for your cattle — fresh grass — and a predictable routine that the cattle can rely on. It doesn't take long to get them to associate the pasture move with whatever accompanying behavioral cues you use. Once these cues are established, you can use them to lead cattle across great distances. The cattle will continue to follow you until you give the cue that you have arrived at the best eating place. (See chapter 5 for more on this training process.)

Training your cattle to follow your cue (using a *come-cow* call and a moving mineral feeder, for example) has a number of benefits: It requires even

A dog can gently pressure the outer edge of the cattle's flight zones to move the animals (top). Notice how much distance the dog maintains between himself and the cattle, leading to their calm movement. More often, dogs aggressively invade the cattle's flight zones to push the stressed animals in the intended direction (bottom).

less labor than working a flight zone and it is completely stress-free for the cattle if you move them at their comfortable walking pace; narrow gates and confusing turns are more negotiable because the animals are calm and more likely to follow the animal directly in front of them and ignore the winding path that the front or rear of the herd is taking; and it becomes much easier to move a herd through delicious distractions that would otherwise cause the herd movement to stall.

Enlisting the Help of a Lead Animal

You can also train a lead steer or lead cow to help you with your pasture moves. A resident lead animal is particularly useful if you run new animals each year, such as in a stocker or grass-finishing operation, or if you choose to split your yearlings from the main herd. This lead animal knows the routine and is trained to follow a bucket of treats. It should be tame so you can approach from anywhere in the field, give the animal a good rubdown, and start luring it with your bucket of treats. Repetition makes perfection.

You will not pick the lead animal; the animal will pick you. As your herd calms down with the continuous use of low-stress handling techniques, you will become a curiosity in the field and the herd will surround you to check you out every time you wander through. Over time they will get braver; there may even be a few individuals that will be become brave enough to let you begin to touch them. Because your presence is not associated with trauma and you don't ever lock up the lead animal or force a halter onto it, it will begin to trust you. Once you manage to start scratching under its chin and behind its ears, the deal is done. Soon you can scratch its entire body. From there it's only a small step to get your lead animal to follow a bucket and finally to extend the distances that it will follow you.

---

*You will not pick the lead animal;
the animal will pick you.*

Be consistent with your lead animal; don't send mixed messages or forget to reward the behaviors you want. Ignore but never punish undesirable behaviors. If you react, even to punish, you are reinforcing negative behaviors. By contrast, immediately reward positive behaviors from your lead animal. The cow will quickly learn which behaviors are rewarded with a nice scratch behind the ears and a bucket of treats and which negative behaviors (such as being pushy) are ignored. Only positive reinforcement and consistent rewards will train lead steers to start the movement of a herd and follow you to whatever destination you choose.

My first lead steer chose my brother while he sat immobile in the pasture to stimulate the herd's curiosity. Once first contact was established, it was easy to build the steer's trust and begin using the animal to manipulate the herd's behavior. He would just about tip over with pleasure during his in-pasture belly scratches, his eyes closed and his whole body swaying to the vigorous rubdown. While the herd gathered around to sniff at me, I could do health and pinkeye checks as well as observe their eating habits for our forage analyses. This kind of contact becomes a good game that is much more fulfilling than the negative interactions associated with cattle-processing days. I guarantee you'll get a good laugh sitting in your field while your herd calms down and your future lead steer chooses you.

A trained herd attends the handler and follows him to its next pasture, happily ignoring the tall, lush grass underfoot.

Enlisting the help of a lead steer: The herd follows the movement of the lead steer, who falls in behind the handler carrying the bucket of treats.

Herding: Rotational Grazing without Fences

Low-stress handling techniques and training cues can be expanded even further by using shepherds instead of fences in a rotational-grazing program. Herding is becoming increasingly popular to manage rotational-grazing programs with large herds and in areas that have low stock densities such as community pastures, extensive rangelands, and large dryland farms. In the past, these areas were typically managed as continuous-grazing pastures, causing serious overgrazing and undergrazing problems as well as a great deal of trampling and manure damage along creeks, lakes, and other water sources.

In many cases, fences were not even considered because of their effect on wildlife movements and because having relatively few animals in such a sprawling space made fencing uneconomical. Often fences in such vast areas do very little to take the pressure off overused water sites and overgrazed land. Herding can eliminate these problems and significantly increase the productivity of these lands without requiring the expense of constructing and maintaining fences. Furthermore, a single shepherd using low-stress handling techniques and training the cattle to the herding routine can replace dozens of cowhands. One shepherd can usually manage 1,000 to 1,500 head of cattle once the cattle are trained to a routine.

The shepherd mentally divides the grazing range into a pasture rotation. Every morning he or she arrives in the pasture and grazes the animals on some fresh grass until they have had their fill. Then the shepherd gently guides the cattle to the water site. The cattle's thirst initiates the movement toward water, but the shepherd directs them toward the site of his or her choice. Cattle always drink two times, usually separated by a rumination session, before they are satiated and ready to return to the grazing area. If you try to herd them away before they have finished their second drink, they will balk and try to return to the water site.

After drinking, the herd is guided back to the grazing area, where it is rewarded with fresh grass. The animals essentially direct their own grazing pace across the pasture, but the herder is there to guide the direction so the rotation can be successful. Flies hatching from the manure piles act as the back fence to protect the regrowth after grazing.

This routine is very important in training the cattle to stick together and to wait for the shepherd's return each day to the area where they were left the previous night. Because there are no fences to direct the cattle to any specific water site, each site is used for only a few days to prevent it from getting muddy and fouled. If during cool weather the traveling distances between the daily grazing area and the water site become quite far, the cattle can even be taken to water every second day.

Calving

Calving season in a shepherding management program occurs much as it does in a daily pasture move. The cows tend to time their calving so they will give birth during the middle of the day, after the herd has returned from the water site to the day's fresh grass, and tend to give birth in the midst of the herd so the calves will associate the herd with their safety zone and follow it immediately. Calves born in the interior of the herd in a daily rotation or daily shepherding program will travel several miles back and forth to water without any problems on their very first day of life.

Training

When training cattle to a herding scenario, it is best to start with a small bunch of around two hundred head. The shepherd trains them to the routine of waiting for him or her to initiate the day's graze and direct the trip to water. Once the herd becomes a cohesive group, more cattle can be progressively added to the initial bunch to create any herd size you desire. In community pastures, for instance, a single shepherd can manage the collective herds of many farmers that share the grazing area.

Some individual cattle may resist the group identity and continually try to escape during herding. Although they can be trained to stick with the herd if you consistently return them to the group and remain vigilant to their attempts to escape, it is usually easier simply to sell these high-strung individuals.

Your Attitude

Your attitude will determine the success of your handling. Animals are extraordinarily perceptive about human feelings and emotions. The ways in which dogs and cats pick up on human feelings are well known. While working at a musk ox research farm in Alaska, I met a tame musk ox called Churchill that would seek me out for petting whenever I felt depressed, just like a six hundred-pound horned dog that comes to put its head on your lap. Cattle are no different; they sense your positive or negative energy with pinpoint accuracy. If you are having a bad day or if you feel frustrated or impatient, your cattle will know immediately and their stress level will skyrocket. In these cases, go home; you have no business being with the animals until your attitude again becomes calming and safe for the cattle.

THE BENEFIT OF THE DOUBT

Always have the attitude that if something goes wrong, it is your fault for not being clear, never the cattle's fault for not figuring out what you wanted them to do. Cattle can read your emotions, not your mind.

Turn every interaction with your cattle into a game. If something goes well, pat yourself on the back for being clear; if something goes wrong, be a detective and try to figure out what impeded the flow of the cattle, where your messages got confused, or what stress distracted the cattle from understanding what they needed to do. Every glitch is a clue that will help you determine a rough edge in need of smoothing, thereby improving your handling of the cattle.

Handling Practices versus Corral Layout

Although an ideal corral design is an asset to low-stress livestock handling, it certainly doesn't mean you have to tear down your old corral and start over. Reducing noise, working the animals using their flight zones, and allowing the animals time to figure their way through the system will make the biggest difference, regardless of the facility. I have worked cattle in corrals made of little more than thornbushes and some dry logs tied together with wire and baling twine. A loud sneeze would have knocked over most of it, but it still worked. Even an ideal, well-built, high-tech corral design will not work without low-stress handling techniques.

When the cattle's stress and fear begin to rise, their flight zone expands. If they have enough space to escape, they become flighty, pushy, and impatient, or even aggressive if they are crowded too much. If they are physically trapped so they cannot fight or flee, however, such as in a squeeze chute or a loading alley, their flight zone collapses inward until they surrender their bodies to fate. This is why cattle balk in a chute, refusing to move regardless of how much you yell, twist their tails, and cook their flanks with a cattle prod. Even in the most rudimentary handling facilities, if you keep the animals calm and quiet, give them space, allow them time to figure out where to go, and gently use the cattle's flight zone to move them, your handling will be successful without all the props and electric persuaders.

Corral Etiquette

When moving animals into a pen, crowd tub (a small, semicircular pen with a swing gate, used to funnel cattle from a pen or wide alley into a narrow chute or livestock squeeze), or chute (a narrow, single-file lane used to transfer cattle into squeezes or load cattle into cattle transports), never jam in too many. It is not an airline ticket sale; more is not better. The crowded conditions will stress the animals and make them flighty or cause them to balk. They may even refuse to move if their flight zone collapses too far. Distracting noise, pressure from the rear, and clanging gates will also cause the animals to mill around and turn in the wrong directions. If you give them silence and space, they will be able to focus on one another and follow each other through the corral or up a loading chute. In many cases, their progress will be so easy that you won't even need crowding gates. If the excitement level does rise, stop everything and let the herd settle down. A twenty-minute coffee break for you will give the cattle time to calm their nerves and start thinking again instead of acting out in panic.

You have to train your cattle to understand that the corral is a safe place. Give them a reward (such as a sprinkle of hay or alfalfa pellets) before letting them out of the corral so the reward is the last thing they remember of the experience. You should also have treats waiting for them as soon as they enter the corral as a reward for following your *come-cow* cue. A sprinkle of hay also helps hold the leader's attention until the whole herd has entered so animals don't turn and rush out against the incoming traffic.

Never just swing open the gate at the end of the processing day to let the herd rush out of the corral. Running will be associated with fleeing, and you don't want the animals to remember the corral as a scary place from which they must flee. Stay ahead of the herd to make them walk out. This also helps avoid cattle jamming into a gate or running into posts in the mad rush to get out of the corral.

There is also good etiquette to use in a squeeze chute (a specialized chute fitted with a head gate — a pair of locking swing gates used to restrain cattle and shaped to hold cattle behind the head — and collapsible sides that squeeze the animal to further immobilize it). Never allow animals to roar into a squeeze chute at full speed and then catch them by the shoulders to stop them. This causes massive bruising, makes a lot of loud scary noise, and leaves the animals with a painful panic memory of the

chute. To stop a headlong rush, close the head gate prematurely to trap a nervous, fast-moving animal inside the squeeze chute before catching it around the neck.

Make the environment around the squeeze as stress-free as possible. Cattle are not bothered by the sight of blood, but if an animal panics inside a squeeze, it will excrete pheromones in its urine and sweat (literally the smell of fear) that will stick to the squeeze. The other cattle will smell it, which will make them nervous and fearful when they enter. If this occurs, you will have to wash the squeeze thoroughly to keep animals that enter later on from panicking. The smell of fear sticks to the facility for a long time, and it will continue to frighten cattle until it is washed away.

If you have a gate or corral entrance that cattle have difficulty finding their way through, there is a trick to help prevent animals from escaping along the rear of the herd while the leader figures out where to go. Normally, the leader will turn away from the gate to face the distraction of your desperate attempts to hold the stalled herd. But if the cattle are trained to electric fences, you can use a long piece of rope or portable wire to apply constant pressure along the back of a herd while you remain quiet and immobile. In the calm atmosphere, the leader will find its way through the gate because it will not be distracted by the commotion you would otherwise be causing at the rear of the herd.

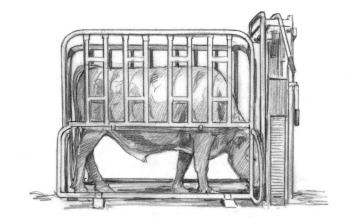

To avoid bruising of their shoulders, fast-moving, flighty, or nervous cattle should be trapped inside the squeeze by a closed head gate to stop their forward momentum.

An additional tactic is to have someone walk ahead of the herd or even drag a jacket or a piece of burlap through the gate ahead of the leader to show the animals the way. Movement will catch the animals' attention and draw the herd forward, just as a herd follows the movement of its leader. Movement draws movement.

Troubleshooting Problems in Your Corral

Inevitably, you will encounter situations in which the flow of your cattle through a corral is disrupted, the stress level increases noticeably during handling, or your cattle balk and refuse to move. Instead of blaming the cattle, these behaviors should be seen as clues that uncover problems in your handling technique or corral design. View these as opportunities to identify and fix the problems, rather than reasons to lose your sanity with your cows.

Doing a Walk-Through

You can also troubleshoot your corral ahead of time. Walk through the whole corral, chute, squeeze, and loading alley, aiming your eyes at the same level as those of your cattle, and try to imagine what might bother them. Metal-on-metal noises, clanking chains, and high-pitched sounds should all be eliminated. Pad with rubber loose metal parts such as removable squeeze sides; oil gates; tie up loose chains; and eliminate mechanical noises. Strange sounds from a hollow or textured floor can also provoke stress. Duct tape becomes a secret weapon for dampening unnecessary noises.

REMEDYING A PROBLEMATIC CORRAL ENTRY

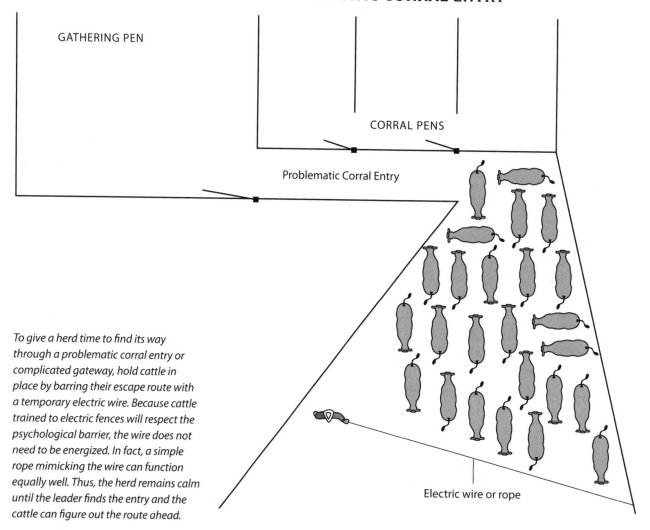

GATHERING PEN

CORRAL PENS

Problematic Corral Entry

To give a herd time to find its way through a problematic corral entry or complicated gateway, hold cattle in place by barring their escape route with a temporary electric wire. Because cattle trained to electric fences will respect the psychological barrier, the wire does not need to be energized. In fact, a simple rope mimicking the wire can function equally well. Thus, the herd remains calm until the leader finds the entry and the cattle can figure out the route ahead.

Electric wire or rope

Visual "noise" may also affect cattle. Flapping objects can startle them. An unfamiliar item such as a dropped hat or a jacket hung over a fence can cause a serious traffic jam. Bright objects, brightly colored clothing, and shadows can be frightening. With their two-dimensional vision, shadows can look like deep holes, which is why cattle guards (bridges over a ditch that consist of parallel metal bars to allow pedestrians and vehicles, but not cattle, to pass) and even white stripes painted on a black pavement to imitate a cattle guard will deter cattle from stepping across them. The harmless yellow divider lines on a highway appear frightening to cattle. They will balk and mill around before each animal finally hurls itself across the paint marks to escape falling into the bottomless pit between the two lines. Similarly, dark alleys can look like dead ends and deep mud or slippery surfaces are sure recipes for balking.

Bright lights are very scary to cattle. Reflections off metal, glass, water, or snow will cause balking. Even strong drafts can create a disturbance. I recall a very quiet day of working cattle in the most rudimentary of corrals. Toward evening, the flow suddenly stopped and within a very short period of time the stress level of the cattle went through the roof. The cause: The sun had migrated across the sky so the cattle now had to make a crucial turn directly into the setting sun. Furthermore, the shadows from the handlers operating the squeeze had also turned and were now moving back and forth across the alley. Once we realized the cause of the cattle's sudden "uncooperative behavior," we parked a livestock trailer to block the worst of the sun's glare, switched the handlers to the opposite side of the chute so their shadows would cease to frighten the cattle, and by being particularly gentle, patient, and conscientious of any remaining shadow movements, we were able to coax the remainder of the herd calmly through the rudimentary chute to finish the processing day.

Using Curved Alleys

Ideally, alleys should be curved gently to take advantage of cattle's natural circling behavior. If an alley is curved too sharply, however, or if a gate is positioned in the wrong place, the cattle's two-dimensional vision may perceive it as a dead end. A short, straight stretch at the entrance before the

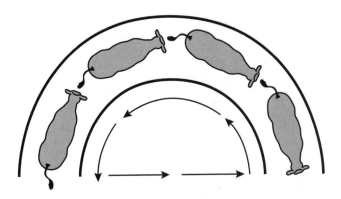

Curved alleys take advantage of cattle's natural circling behavior. The arrows inside the curve indicate how handlers should move to propel cattle forward inside the alley.

alley begins to curve will help show the cattle that the alley is not a dead end.

Gently curving alleys also make it easier to work the flight zone of the cattle. Handlers should walk along the insides of the curve, against the flow of traffic, to propel the animals forward (see the illustration above). They should never return to the front of the alley by walking parallel to the flow of the cattle; doing so will stop the cattle and cause them to back up. In a curved alley, it is much easier to cross through the center of the curve, outside the animals' flight zone, to return to the front of the alley. If the alley is straight, the triangular motion to propel a herd forward during a herd move (see page 143) will work equally well inside a corral.

Shielding Animals from Fear

The movements of people working in the corral can also be a problem. Shield key work areas such as the squeeze with plywood so they are not visible to the cattle from inside the chute and alleys. Board up the sides of loading chutes to help acclimatize the cattle to the dimmer lighting inside the truck and to hide the movements of handlers along the outside of the truck. Position backup gates so people don't frighten cattle when they approach to shut the gates. Set up the backup gates with ropes so you can operate them from a distance. Boarding up the sides of the back half of your squeeze can also help hide the squeeze operator from incoming cattle. And make sure the cattle can see daylight through the front of the squeeze before they enter it so they know it's not a dead end.

Safeguard Animals from Harm

You also should identify and fix sources of bruising in the corral, both to reduce pain-induced stress and to safeguard the meat quality of your animals. Cushion with rubber any sharp edges at gate openings or at the ends of rails that protrude into alleys. Padding the head gate will reduce the impact if your cattle are nervous about moving through a squeeze. If your animals are horned, you will need to give them significantly more space in the handling facility to avoid bruising, and you can expect animals crowded in the vicinity of a horned animal to become much more agitated if they are not given enough space to stay out of reach of the horns.

CORRAL DESIGN

For anyone designing a corral or looking for further information on how to improve existing corrals, I recommend the Web site of Dr. Temple Grandin, associate professor of animal science at the University of Colorado in Fort Collins; see resources page 359. In my opinion, she is the handling-facility design guru in North America. Most new slaughter facilities and stockyards are based on her designs, which you will see throughout the catalog pages of many livestock-handling-equipment suppliers.

The illustration below, from her Web site, is an example of a very efficient livestock corral layout that is designed to utilize the natural circling behavior of cattle and allow handlers to work the cattle's flight zones despite being in the confined space of a handling facility.

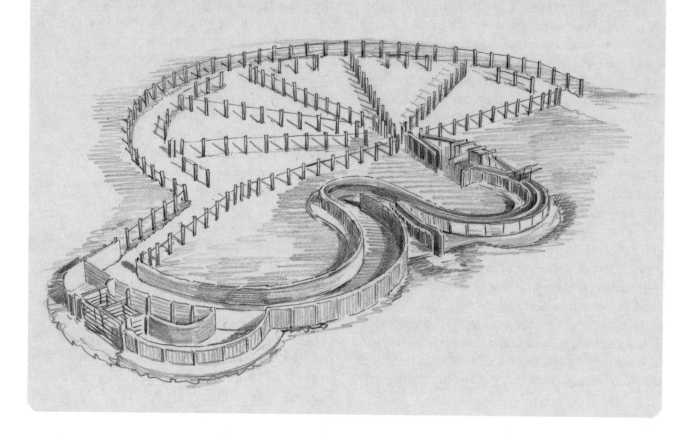

Chapter 10

Pests, Parasites, and Diseases

PESTS, PARASITES, AND DISEASES AFFECT FARM profitability in more ways than just the direct economic cost of providing veterinary care and pharmaceuticals. Consider the enormous labor investment it takes to sort and treat sick animals and how much these animals complicate your grazing management when they must be quarantined from the rest of the herd. Then consider how little time is invested in managing the healthy animals in the herd. Also consider the impact on your bottom line from pest-, parasite-, and disease-related weight loss or poor weight gain, not to mention the devastating effects that pest-, parasite-, and disease-related stress and the pharmaceuticals used to prevent or eradicate them have on meat quality if you are trying to market natural or organic beef. The only way to minimize these situations is to understand how pests, parasites, and diseases work so you can avoid creating conditions that allow them to flourish. An ounce of prevention through good management is worth a pound of cure.

Good preventive management begins by understanding the nature of the "beast" that you are trying to prevent or eradicate, be it pest, parasite, or disease. Of the three, disease represents by far the broadest category. A disease is any impairment of health or any condition of abnormal functioning that has a specific cause and specific symptoms, resulting from infections, nutritional deficiencies, toxicity, unfavorable environmental factors, and genetic or developmental problems. The first category — disease that

results from infections — can be caused by a number of agents such as bacteria (those that cause pneumonia, pinkeye, *E. coli* infection, salmonella, and the clostridial diseases of blackleg, tetanus, botulism, and so forth), viruses (such as infectious bovine rhinotracheitis, or IBR; bovine viral diarrhea, or BVD; parainfluenza-3, or PI3; and foot-and-mouth disease), fungi (such as ringworm), and protozoa (any of various single-celled organisms that are often parasitic in nature: These include coccidian, giardia, cryptosporidium, and the protozoa that cause African sleeping sickness in cattle).

An ounce of prevention through good management is worth a pound of cure.

Second are pests: any unwanted or destructive insects or other small animals that harm or destroy crops or livestock. Common livestock pests include blackflies, horn flies, stable flies, face flies, horseflies, deerflies, tsetse flies, midges, and mosquitoes. In addition to the physical harm and irritation they inflict on cattle, they can transmit a number of infectious diseases. For example, the tsetse fly is known for its role in transmitting African sleeping sickness through its bite.

Finally, parasites are essentially specialized pests that have a vested interest in keeping their victims alive to facilitate an ongoing relationship from which they benefit. Thus, these pests exist in or on the living tissue of a single host animal of another species for a portion of or the duration of the pest's life cycle. While living on its host, it obtains all its nutritional requirements from the host, generally to the disadvantage of the host's health, productivity, and well-being. Notable internal parasites include warble fly larvae, pinworms, roundworms, tapeworms, flatworms (such as liver flukes), and lungworms. Notable external parasites include ticks, fleas, biting and sucking lice, mites, and even vampire bats, which rely on livestock and wildlife blood to feed. Parasites can also transmit a wide range of serious or fatal diseases to cattle.

Although the microbe or parasite may be a disease carrier, it is stress, nutrient deficiencies, and management shortfalls that depress an animal's immune system and cause diseases and parasitic infestations.

Pests, parasites, and diseases are predators in the natural world. Like wolves and lions, pneumonia, scours, coccidiosis, flies, and fleas all prey on the same individuals: the young, the weak, the sickly, and the old. Livestock become significantly more vulnerable to these predators if the animals' living conditions or yearly life cycle weakens them or exposes them to unnecessary challenges. For example, heavy predation by wolves, coyotes, and lions occurs only when calves are born out of sync seasonally from the rest of the animal world. This exposes them to harsh weather, weakening them and making them easy targets because other prey animals are not as readily available. Another challenge occurs if seasonal forage availability does not match cattle's changing nutritional demands due to breeding out of season, or when calving out of season exposes the livestock to unnecessary environmental pressures.

This lesson applies equally to the world of tiny predators. All pest, parasite, and disease occurrences on your farm are red flags that point out problems with your management. If a cow is not stressed, its immune system will provide necessary resistance to illness and parasites as long as its life cycle is in sync with the seasons. Although the microbe or parasite may be a disease carrier, it is stress, nutrient deficiencies, and management shortfalls that depress an animal's immune system and cause diseases and parasitic infestations.

The Mechanisms of Disease

The life cycles of most pests, parasites, bacteria, protozoa, and viruses are extremely short: a few hours to a few weeks at most. This quick turnover of generations allows them to mutate extremely rapidly in response to changes in their environment. That's why they quickly become resistant to new pharmaceuticals and chemicals designed to defend against them and why there will never be a single, effective long-term cure for any specific disease or one chemical capable of killing or preventing pest and parasite infestations for all time. As quickly as pharmaceutical companies invent new antibiotics, pesticides, and parasiticides and biogenetic engineering companies create disease-resistant plants and animals, the parasites, pests, and microbes mutate to circumvent or adapt to these countermeasures. In doing so, the microbes become "superbugs," stronger and increasingly resistant to our efforts to control them.

Unfortunately, livestock can adapt to mutated diseases, pests, and parasites only as quickly as the generations of animals renew themselves over years rather than hours and weeks. The more we resort to laboratory cures for these diseases, therefore, the more resistant the microbes become and the less familiar they are to the immune systems of our livestock.

Complicating matters is that while antibiotics attack the invading disease, they also depress the animal's immune functions, making it dependent on the drug and more vulnerable to disease pressure. Pharmaceutical poisons to control internal and external parasites and pests fare little better under close scrutiny. Again, when the strength of the poisons is increased, the pests and parasites quickly adapt. This is why laboratory cures or treatments can create a spiral of long-term addiction to their use, increasingly dangerous diseases, and rising costs to fight them.

Another factor is that despite the slaughter withdrawal times for these chemicals, the possibility of residues in the meat makes their use highly unattractive to natural-meat consumers. Indeed, their use is prohibited by the organic certifying associations. In addition, the chemicals take a toll on the health and immune systems of the livestock as their body tissues and organs — especially their liver, responsible for filtering toxins — absorb, metabolize, or otherwise process them.

Stress

A stressed animal or plant subtly advertises its weak condition to predators. To the human eye, the signs of stress may not be visible, but detecting such weaknesses is crucial to the survival of disease microbes, pests, and parasites. For example, if a plant is nutrient-deficient, its sugar content is lower than that of a nutrient-balanced plant. Thus, its ability to refract light is slightly altered. While we may not see this, the eyes of a grasshopper have uniquely evolved to distinguish specific light wavelengths and a plague of hungry grasshoppers is able to detect subtle differences in sugar levels through smell and sight. Sometimes the visible signs of stress are obvious to us. A cow that is nutrient-deficient will often not shed her winter coat as early in spring, or her coat may become noticeably dull or matted. It follows that this animal is preyed upon far more heavily by flies and disease organisms than are its healthy companions.

Stress to an animal can come from many sources: environmental challenges, handling, social factors, nutritional deficiencies, digestive challenges (such as a grain diet), mineral deficiencies, and unsanitary conditions, among others. Although each individual stress may seem negligible, added together they eventually reach a point when the animal's capacity to cope with them is overtaxed and its ability to live in balance with its surroundings is disrupted. Its immune system becomes compromised and its ability to fend off diseases, pests, and parasites is diminished. Even if this point is never reached, continual stress is taxing; the animal may remain healthy, but its weight gain or fertility will be compromised. Our job is to manage and minimize our livestock's stress loads. If we reduce or eliminate the stress factors we can influence (such as environmental, nutritional, and handling stressors; seasonal timing; oral-fecal contamination; and poor water), we give the animals' immune systems the greatest chance of taking care of disease, pest, and parasite stress on their own and maximize the herd's productivity and fertility. If we keep their significant stressors to a minimum, a small addition of stress related to daily challenges will not disrupt an animal's balance with its environment or make it vulnerable to disease and parasites.

Instead of stocking up on medicines, take stock of your production and management practices. Become expert at identifying the factors that allowed disease through the door. A keen eye for troubleshooting and preventing disease will be far more valuable and will save more cattle's lives in the long term than being an expert in animal emergency first-aid.

Preventing Disease

Disease is a symptom of an underlying problem in our management system, yet often we treat the disease without ever addressing the cause. We don't ask what caused an animal to become susceptible to a pneumonia outbreak or a worm infestation. Once we control the immediate problem, we tend to feel satisfied, when in reality we have addressed only the symptoms of an underlying cause. This is not the time to be smug about the effectiveness of your antibiotic or pharmaceutical treatment program. Disease signals that you've missed something important in your management strategy; when you put away the antibiotics, start investigating what caused the disease or the parasite to get hold of your cattle and what overtaxed their immune systems.

Antibiotics and other pharmaceuticals should be used only as a last resort to control diseases, pests, and parasites, just as rifles and traps are extreme ways to deal with predator problems, and serve to point out an urgent management or production issue that needs to be addressed. If you are excelling at crisis management, it means you haven't been doing your homework in prevention. It is the farmer whose herd doesn't have any diseases who should get a pat on the back, not the one who prides himself on being a cattle rescuer.

Effective Culling

The first practice in successful disease prevention is effective culling. Any animal that shows signs of disease, becomes ill, or requires treatment must be culled from the herd *without exception* at your earliest possible convenience. Animals that show a weakness that might make them vulnerable to disease or parasites, such as excessive weight loss during the winter, poor hair shedding, or even giving birth to a sickly calf, should all be sent to auction. These are high-maintenance animals and their genetics are not suited to thrive in your management program.

Scrutinizing Your Operation

Disease prevention is about examining your operation to see how well it matches nature's conditions.

- Does your calving season match nature's seasons and grass cycle?

- How does your climate affect the vulnerable stages in the development of your calves and the vulnerable periods in a cow's life?

- Is your rotation preventing oral-fecal transfer?

- Is your rotation disruptive to the life cycle of pests and parasites? For example, do your cows migrate away from manure piles before fly larvae hatch?

- Is your grass providing adequate nutrition and mineral content? Is your rotation maximizing the nutritional value of your herd's daily grass intake and of the grass reserve throughout the year?

- Is your grass healthy? Does it require any soil amendments?

If a disease occurs, try to figure out how it gained an advantage. For example, I remember a terrible outbreak of coccidiosis among the calves one year during August. This was a disease I remembered well from the days of winter calving — it attacked calves during the peak of the spring mud season — but it had never occurred on grass and I had not seen it since the family farm switched to summer calving.

After my initial bewilderment, I realized that a number of conditions had occurred to give the dis-ease its advantage: August had been abnormally wet, so the young calves were struggling to keep warm and dry. On its own, the wet weather should not have been enough to cause the disease, but inadequate pasture subdivisions were still restricting pasture moves to every ten days to two weeks. The slow pasture rotation was exposing the calves to oral-fecal contamination. While rigorous and labor-intensive antibiotic treatment had failed to stop the outbreak and had only marginally succeeded in saving some of the calves' lives, as soon as I moved the animals onto fresh grass and increased the pasture rotation speed, the disease seemed miraculously to disappear.

Don't Harrow; Rotate

Even some of our well-intended management practices can dramatically increase disease pressure on cattle. One such practice is harrowing pastures. Many farmers harrow in spring to break up manure and distribute it evenly throughout the pastures. Some even harrow each pasture every time it is grazed. Although this makes sense from a fertilizing standpoint, it does not make sense for cows. Cattle avoid grazing old manure and urine patches. In fact, they will avoid manure patches for an entire grazing season, long after the manure has disintegrated. It seems contrary, as the dark lush grass around the old patch looks particularly delicious to our eyes, but the cow's sensitive nose continues to locate and avoid the old excrement.

This avoidance behavior evolved to combat the life cycles of internal parasites that are excreted in the cattle's manure. To hatch, the eggs of many of these parasites must come into contact with the soil, and the larvae must spend part of their life cycle in the external world before being ingested again by the cattle. When the larvae are ready to reenter the cattle, they migrate up the grass and are consumed by the host cows during grazing. In detecting and avoiding old manure patches, cattle avoid ingesting the larvae for a season. Harrowing spreads the manure (and its parasites) over the entire pasture, therefore increasing the parasite load the cattle are exposed to — and increasing the need to combat these parasites with pharmaceuticals or expensive natural remedies.

Another problem from harrowing: Because the manure is spread across the entire pasture, all the grass smells less appetizing to the cattle, so their daily grass intake (and weight gains) will be slightly reduced.

Rather than harrowing, you need to rely on an intensive rotational grazing pattern to keep the grass and soil healthy and to distribute the manure evenly. Healthy soil contains more healthy organisms to help to break down the manure more quickly. By grazing year-round, manure will not accumulate around hay or silage feeding areas, but will instead be spread evenly over pastures; there will be no manure concentrations to break up by harrowing in spring, and you will save on equipment, fuel, and time.

Specific Diseases, Pests, and Parasites

In the section below I have selected several common health issues faced by cattle producers to demonstrate how a good preventive management strategy can replace the symptomatic treatment protocols typically used against disease, pests, and parasites. Each example illustrates how understanding a disease, identifying its root causes, and subsequently re-evaluating the management practices that facilitated the disease can significantly reduce or eliminate the occurrence of the disease in the herd. Included are diseases, pests, and parasites of particular concern to natural and organic beef producers because of the high labor and veterinary costs associated with their typical treatment and prevention protocol, and because the antibiotics and pharmaceutical chemicals typically used to treat and/or protect against them cannot be used on animals marketed as natural or organic beef. Following the sections on various diseases is a discussion about the treatment and prevention of various pests and parasites.

Pinkeye

Pinkeye, or infectious bovine keratoconjunctivitis, a highly contagious infectious disease of the cattle eye, is most commonly caused by the bacterium *Moraxella bovis*. The first sign of pinkeye is excessive tearing of the affected eye(s). As the disease progresses, discomfort will cause the animal to squint or even shut the affected eye. If untreated, the cornea (clear surface of the eyeball) becomes inflamed and turns white, followed by possible ulceration, which can cause a permanent white scar in the center of the eye. Affected calves often lose weight and will be discounted at auction. Damage to the eye may lead to permanent blindness. The disease is spread through direct or indirect contact between cattle. Irritation to the eye caused by sunlight, dust, pollen, and grass seeds can bring about primary inflammation to the eye, which allows the *M. bovis* bacteria to invade the eye. Flies are also an important factor in the spread of the disease, as an eye irritant, by directly damaging the eye through their bites as they feed on the eye, and by transmitting the disease from infected to uninfected animals.

Cattle become vulnerable to pinkeye when their eyes are stressed. Lack of pigment around the eyes increases the light that reaches the eyes, which is especially problematic in areas exposed to high ultraviolet radiation or intense sunlight. Full pigment around the eyes reduces a cow's susceptibility to pinkeye.

Dust, Pollen, and Seeds

Dust can also be irritating and very stressful to the eyes. This problem is most common when pastures are overgrazed and when cattle must travel along dusty alleys to reach water. Dust containing very fine, dry manure particles (such as occurs around water sites, in alleys, and in feed pens) is particularly irritating; the manure makes the dust slightly acidic and also contains a high bacterium load.

When you don't manage the pasture rotation to prevent grass from going to seed, dry grass can poke the cattle's eyes as they graze, and pollens and grass seeds can be especially irritating if they get into an animal's eyes.

Flies

Flies are not only irritating to the eyes; they also carry diseases that can be introduced into stressed eyes. Problems with flies can be controlled by a quick pasture rotation so fly larvae in the manure do not hatch while the cattle are still grazing near it. In addition, an adequate water supply will prevent cattle from lingering around water sites, which will in turn prevent the buildup of fly-incubating manure in these areas.

Nutrient Deficiencies

Nutrient deficiencies also increase an animal's susceptibility to pinkeye. Iodine is particularly important to the eye's immunity against disease. Although synthetic iodine is lethal to cattle in high doses, natural iodine is selectively absorbed by cattle in only the amount they need. A mineral program that includes kelp, which is very high in natural iodine, tends to significantly reduce the occurrence of pinkeye and lowers the instance of pinkeye infection from other infected animals, even if they are in direct physical contact with each other.

Cancer Eye

Cancer eye, or bovine ocular neoplasmia (BON), includes a variety of benign and malignant skin tumors of the eyeball and eyelids, often characterized by a pink, fleshy growth on or around the eyeball. The most important factors affecting the susceptibility to cancerous growths in and around the eyes are intense sunlight or intense ultraviolet radiation and lack of skin and/or hair pigmentation around the eyes.

Like pinkeye, cancerous growths in the eye can be caused by a persistent irritation to the eye or eyelid from, for example, seeds, thorns, and burs that lead to scarring. The most significant irritation related to cancer eye, however, is ultraviolet radiation, which is why cattle without dark pigment around their eyes (which reduces solar glare) are most suscep-

tible. Farmers in areas prone to intense ultraviolet radiation should select cattle that have full pigment around the eyes. Perhaps the genetic predisposition to cancer eye is a characteristic of being more sensitive to intense sunlight due to a lack of pigment around the eye.

Foot Rot

Foot rot, also known as interdigital necrobacillosis or interdigital phlegmon, is an acute infection of the soft tissue between the claws of the feet caused by two anaerobic bacteria, *Fusobacterium necrophorum* and *Bacteroides melaninogenicus,* both of which are common in the environment. *F. necrophorum* is also present in the rumen and feces of normal cattle. Because these bacteria cannot penetrate intact, healthy skin, they enter the foot through abrasions, cuts, bruises, or otherwise damaged tissue. The disease is characterized by swelling and lameness of the affected foot accompanied by the soft tissue between the claws becoming necrotic (dying) and cracked, a foul odor but little pus discharge, a slight fever, loss of appetite, and weight loss.

Wet, muddy, or manure-laden conditions make it particularly easy for the anaerobic microbes to flourish, and because cattle's feet cannot dry and wounds and scratches are kept dirty and full of bacteria, these microbes find easy entry. Cattle in stockyards and feed pens are more prone to foot rot than cattle on pasture, although those on pasture are also candidates for the disease if they are exposed to muddy pasture conditions, muddy water sites, or an insufficient water supply that results in cattle bunching around water sites in mud and manure for long periods of time.

An injury through which the bacteria enter the foot can be caused by, for instance, stepping on barbed wire that has been left on the ground, stepping on the sharp edges of a mineral feeder, or scratching the feet on the exposed edges of concrete pads around water troughs or feed bunks. Extended soggy conditions can cause the soft tissue between the hooves to weaken and break down due to moisture saturation (much as our bodies turn into prunes if we sit too long in the bathtub), which is exacerbated still further by the manure's acidity, thereby allowing the disease to enter the foot.

EYE PATCH

Once the eye becomes infected, all further irritation will promote the disease and weaken the eye's recovery. A popular alternative to antibiotic injections is to cover the eye with an eye patch (available commercially), which seals the eye from sunlight, wind, dust, and debris. A small amount of antibiotic ointment can be sprayed or smeared on the eyeball before gluing the patch around the eye. To prevent further irritation, be sure that the patch does not touch the eye. Usually, within a week the glue weakens, the eye patch falls off, and the eye has healed without any antibiotic injections.

My first portable mineral feeder designs were built on metal skids, but the upturned skid ends were sharp enough to injure the cattle's feet. By curling the skid ends to eliminate the sharp edges, I greatly reduced the incidence of foot rot in the herd. Although accidents happen, most of time the conditions that promote foot rot microbes in the soil and the injuries that invite the microbes into the body are preventable human errors and management shortcomings.

Scours and Coccidiosis

Scours and coccidiosis are epidemics of a winter- and spring-calving herd. *Scours* is the common term used to describe cattle diarrhea, particularly among calves, typically resulting from a contagious intestinal infection and leading to potentially life-threatening dehydration and electrolyte imbalances. The most common infectious agents that cause scours are *Rotavirus, Coronavirus, Cryptosporidium parvum* protozoa, salmonella bacteria, and *Escherichia coli (E. coli)* bacteria.

Coccidiosis, another type of cattle diarrhea, is caused by parasitic protozoa called coccidia, which destroy the lining of the small intestine, causing diarrhea, intestinal hemorrhage, emaciation, and sometimes fatal dysentery. The diarrhea may contain blood, mucus, and stringy masses of tissue from the sloughing off of the epithelium of the intestine.

Both diseases spread when uninfected animals come into contact with the infectious bacteria, viruses, or protozoa that are shed by infected animals in large amounts through their feces, urine, saliva, and nasal secretions. The infectious organisms that cause both scours and coccidiosis can survive for months in the soil in the right conditions and some can even be transmitted by dust in the air. They favor wet, muddy environments, though most are killed by direct sunlight and dry weather.

Young calves are particularly prone to scours and coccidiosis because their guts are still immature and their immune systems are more vulnerable than those of older animals. In addition, calves born in wet, snowy, or muddy conditions are weakened, thereby significantly increasing their susceptibility just when weather conditions favor the infectious organisms that cause the diseases. Switching to summer calving on grass in a pasture-grazing rotation all but eliminates these diseases. The calves are born onto clean grass, the weather is warm and comparatively dry, and a rapid pasture rotation prevents oral-fecal transfer.

Pneumonia

Pneumonia is an acute or chronic disease marked by an inflammation of the lungs caused by a contagious infection by any one of a host of viruses, bacteria (or both at once), other microorganisms, or even by physical or chemical irritants that can lead to secondary bacterial infections of the lungs. Pneumonia is characterized by panting, high fever, coughing, loss of appetite, and, in some cases, nasal discharge. The disease, which is highly unpredictable, ranges from mild to rapidly fatal. Susceptibility to pneumonia is increased by stress, including cold stress, heat stress, dust, dampness, injury, nutritional stress or deficiencies, a compromised immune system, exposure to high loads of infectious bacteria, shipping stress (marked by conditions of exposure and/or exhaustion), and dehydration.

Although many of the infectious agents that cause pneumonia favor similar weather and environmental conditions as those that cause scours and coccidiosis, pneumonia is more prevalent in all age groups. For example, my family farm had an annual outbreak of pneumonia among the three-month-old calves that were born in late July. It arrived every year in late October or early November, just as the cattle were moved from their summer pastures to their small, winter feed pens and as they were switched from grass to a silage diet. The weather in late fall and early winter always turned cold and wet at exactly this time, rain alternating with snow for a period of several weeks. To make matters worse, all these conditions coincided with the time when the calves' immune systems had to take over from the effects of the mothers' colostrum.

I eliminated this yearly pneumonia outbreak from the herd without resorting to vaccinations through a few simple management alterations: I moved the calving season to begin on June 10, allowing the calves time for their immune systems to make the crucial transition to independence during the warm, dry September weather. Extending the grazing season meant that the calves did not need to make the transition from an aerobic grass diet to an anaerobic silage feed ration during the stressful wet fall weather and also ensured that the calves had a fresh, manure-free pasture to sleep in and eat from every single day. Additionally, their mothers' udders were kept clean by the pasture grass during damp, muddy fall weather.

Flies, Fleas, Worms, and Warbles

Pests and parasites can have an enormous range of negative effects on cattle, including irritation, physical discomfort, blood loss, poor feed utilization, secondary infection in wounds caused by biting insects, and transmission of infectious disease carried by a pest or parasite. Their impact on cattle can result in decreased livestock productivity, poor fertility, weight loss, and reduced milk production (see the discussion of pests and parasites at the beginning of this chapter).

Flies, for example, can significantly hamper cattle performance by feeding on them or by interrupting the animals' feeding and resting through their constant irritation. This causes reductions in weight gain and milk production due to poor feed utilization and blood loss. Biting and bloodsucking lice cause a disease characterized by dry, scaly skin; hair loss; and itching. The afflicted animal may also become anemic as a result of the louse infestation. Fleas produce similar symptoms, but a flea infestation can be identified by the black specks of dried blood or flea feces on the host's skin or hair.

Cattle can be affected by an enormous variety of worms that can live in the lungs, body cavity, gastrointestinal tract, liver, beneath the skin, or even in the lacrimal ducts (which connect the nose with the tear ducts in each eye). Though the life cycle for each worm species will vary, many worm infestations begin when adult female worms lay microscopic eggs that pass out to the pasture in the cattle's manure. These eggs hatch and the larvae (or cysts, as flatworm larvae are called) develop. Within a few weeks the larvae/cysts are capable of infecting cattle by climbing up the grass stalks and being ingested as the cattle graze or by being consumed through water contaminated by manure containing

the larvae. After being ingested by cattle, the larvae complete their final stage of development inside the cattle's intestines, where they subsequently mate as adults. Fresh worm eggs are shed by cattle within two to four weeks of ingesting the larvae and the cycle begins again.

Gastrointestinal worms most commonly affect cattle in one of two ways: They can puncture holes in the cattle's tissues and feed on their host's blood, potentially causing anemia, poor growth, and occasionally death; others can penetrate the glandular tissue, causing diarrhea, malnutrition, and poor growth. Lungworms cause a lung disease in cattle that is similar to viral or bacterial pneumonia. Liver fluke, a type of flatworm that burrows through the liver, causes its cattle host to experience symptoms of acute or chronic liver disease, such as digestive inefficiency and malnutrition, which are also the symptoms of gastrointestinal worm infestations.

Warbles are lumpy abscesses that form on the back of cattle, under the hide. They are caused by the larvae of any of several species of large, parasitic hairy flies known as warble flies, heel flies, botflies, gadflies, or bomb flies, found on all continents of the northern hemisphere, primarily between 25 and 60 degrees latitude. The warble flies lay their eggs on cattle near the hocks or on the body. The hatching larvae then burrow through the body of the cow, including through the meat and stomach lining, to the cow's back, where they remain under the skin, breathing through a hole cut in the hide, before the inch-long mature larvae finally emerge and drop to the ground to complete on open pasture their development into flies. This process causes considerable pain to the cattle, can lead to serious infection, makes the meat unfit for human consumption, renders the hide valueless, and can cause milk yields to suffer.

Pest and parasite infestations can be exceedingly difficult to eliminate without the help of pharmaceutical chemicals; prevention is the best offense against them. The first step is a healthy rotational-grazing program. By moving ever onward, the cattle stay ahead of flies hatching from the four- to six-day-old manure patches.

But if the cattle are healthy, their bodies have defenses. The immune system of a healthy cow exposed to a very low-level presence of these insects and worms is stimulated to build defenses against them and prevent further infestation. Because pharmaceuticals kill everything, the immune system does not learn to defend itself and, without repeated treatments, remains vulnerable to future infestations. If your management system strengthens your animals' immune systems as much as possible through good nutrition, effective mineral supplementation, effective pasture rotations, stress reduction, and healthy soils, the cattle will be equipped to fend off many parasite invaders. A rigorous culling program is also crucial; high-maintenance animals will be most heavily targeted by parasites.

Fire Ants

In areas where fire ant mounds are numerous, these ants can cause baby calves serious injury and even death. If a calf cannot get on its feet quickly enough to escape the ants' stings — often the case if the birth was difficult, the calf is tired, or the calf was dropped into or immediately adjacent to a fire ant mound — the ants can overwhelm it in a matter of seconds. The frequency of injury and death is most prevalent during the hot summer months, when the voracious ants are starved for food and moisture.

Again, prevention is the best offense. Calve during the cooler months, when other sources of food and moisture are readily available to the fire ants. Because fire ants prefer open landscape that is exposed to the sun, such as short grass or bare ground, and are most at home in sandy soil, during the calving season maintain a rigorous daily pasture rotation through tall, lush pasture grasses and avoid pastures containing a large number of fire ant mounds (see chapter 5 for a discussion of how to manage a daily pasture rotation during calving season).

Because fire ants require surface or subsurface moisture to survive, they can be deterred from colonizing pastures by eliminating standing water and preventing access to moisture sources within or near calving areas. If fire ants persist, colonies can be broken up by frequent harrowing before calving or by watering pastures sufficiently to flood the colonies.

In worst-case scenarios, calving can be delayed until the cool, fall weather, when the ants are dormant, though this will lead to the additional financial, nutritional, fertility, and health costs associated with calving out of season. Meticulous calving and grazing planning and a good understanding of the behavior of fire ants are by far the most cost-effective strategies against fire ant attacks.

Nonpharmaceutical Solutions

If a parasite infestation does occur, there are a number of nonpharmaceutical options to help get rid of them.

■ *Garlic.* Garlic is a natural wormer and immune-system booster. With some success, I have fed fresh minced garlic in alfalfa pellets and molasses to individual cows with acute health or parasite issues.

■ *Herbs.* I have heard of quite a number of other herbs with worming properties, although I will leave prescriptions of these to a veterinarian, naturopath, or herbal practitioner. They include wormwood, wild ginger, conifer needles or pitch, goosefoot, neem, plants from the mustard family (mustard, turnips, radishes, horseradish, and nasturtiums), seeds from vine plants such as squash and pumpkin, ferns, lupines, some nuts or nutshells, tobacco, pyrethrum (chrysanthemum family), plants in the Apiaceae (Umbelliferae) family (carrots, anise, cumin, juniper, fennel, and parsnips), tansy, yarrow, blackberry, raspberry, elder shoots, calendula, lady's slipper, blue cohosh, pokeweed, savory, skullcap, skunk cabbage, nettle, valerian, and common knotgrass. I find it surprising how many of these plants are familiar weeds that we tend to disregard when we spot them in our pastures.

■ *Effective grazing.* Part of a holistic approach to farming is to keep the grass competitive through effective grazing principles while allowing a wide diversity of wild herbs to be part of the smorgasbord. If livestock have access to many of these so-called weeds in their grazing rotation, they can supplement their diet with some of them to take advantage of their herbal properties. Just as our own bodies experience cravings in response to nutrient deficiencies, cattle will seek out some of these herbs by smell to satisfy their cravings and address some of their health issues.

■ *Diatomaceous earth.* Diatomaceous earth has been used on internal and external parasites with varying success. It consists of the microscopic fossilized remains of various marine algae called *diatoms*. In a powdered form, the fossilized remains serve as sharp little objects that rip the parasite's exterior membrane and innards. For internal parasites, small amounts of diatomaceous earth can be mixed into the mineral supplement; it will also provide additional calcium to the animals' diet. To treat external parasites, you can dust it over the hides of animals to kill fleas, lice, mites, and so forth. Be cautious of either you or your cattle inhaling the dust; although diatomaceous earth seems to be safe for large mammals to digest, the sharp, fossilized fragments may be irritating to the lungs.

Surfactants

Surfactants are liquids that break down the surface tension between water droplets (think of them as fancy soaps), making water "wetter" (it soaks in many times faster and makes ingested water more efficient). Natural surfactants can also be used as worming agents, but only if they are deemed safe for animal consumption. The theory behind their effectiveness is that when the surface tension of water droplets is reduced, the surface membranes of the parasites are distorted, thereby killing them. It is not entirely clear which worm species are affected, though according to the theory, worms must come into direct contact with the surfactant, making it suitable primarily for gastrointestinal worms.

At the time of writing, Shaklee's Basic H soap is the only commercially available natural surfactant that I am aware of that has been consistently used as a natural wormer, though this is not a practice endorsed by Shaklee Corporation. I have used it successfully at a rate of 4 ounces of soap per 150 gallons of water. Basic H is made of soybean enzymes, so it is advertised as natural, biodegradable, and environmentally friendly. Unfortunately, because Shaklee Corporation does not want to divulge its ingredients to protect its trade secrets, not all organic certification associations are willing to certify it.

I mix it into the water trough or into a water-supply tank and let the cattle drink it for two days to ensure that all the animals get a good drink. During this time, I prevent the cattle from using any untreated water sources. Because the herd does not need to be run through a corral, it's remarkably easy to apply. I use it every spring and fall and a couple of times during summer, or every time I notice the cattle starting to get dirty rear ends, a sure clue that worms are present.

You are what you eat; your health can only be as good as the food you eat, and the food you eat can only be as healthy as the soil in which it is grown.

Making Healthy Soil for Healthy Animals

Pasture soils play an important role in disease resistance. Healthy soil will produce more nutritious, healthier grass to keep the animals healthier and more disease-resistant. You are what you eat; your health can only be as good as the food you eat, and the food you eat can only be as healthy as the soil in which it is grown.

But soil health also plays a much more direct role in reducing disease pressure on your animals. If the soil nutrients are balanced and agrochemicals (chemicals such as pesticides, herbicides, and fungicides that are used for agricultural purposes) and synthetic fertilizers have not depressed the vitality of the soil, a healthy soil full of oxygen and an aggressive host of soil microbes becomes a hostile environment for parasites, pests, and disease organisms.

For example, clostridial diseases thrive in an anaerobic environment such as a calcium-deficient soil that is waterlogged and lacks oxygen. An effective grazing rotation to loosen the soil and speed nutrient cycling and an application of lime fertilizer to correct calcium deficiencies will help aerate the soil and make it a less suitable place for clostridial bacteria to hide. Soil amendments, organic fertilizers, and grazing efficiency are therefore crucial to pest and disease management. In some cases, soil amendments can directly reduce the disease and parasite presence in the soil. For example, lime and copper sulfate fertilizers are harmful to the snails that are part of the liver fluke's life cycle.

Complementary Grazing Species

When we look at the wildebeest herds on the Serengeti, we see small numbers of additional species accompanying the great herds on their migrations. The presence of multiple species in the migration is beneficial to everyone because they reduce one another's parasite loads.

Many parasites are host-specific. For example, cattle parasites cannot survive inside sheep, or vice versa. When multiple species with different grazing habits graze together, they can break up the life cycles of each other's parasites by ingesting significant quantities of each other's parasite larvae. When the host-specific parasites are ingested by the wrong species, the parasites cannot adjust to the unfamiliar species and simply die in their digestive systems.

Some host-specific parasites also rely on the distinct grazing habits of their specific host species to help them complete their life cycle. The cattle-specific

Wildebeest on the Serengeti are accompanied by many complementary grazing species, including zebra, kudu, and ostriches, that follow or even merge with the wildebeest migration. This multiple migration is beneficial to the animals and helps improve the soil and grass along the route.

Cattle and sheep can graze together as a herd, improving soil fertility and grass productivity, disrupting each other's parasite life cycles, and providing one another with protection against predators.

Pests, Parasites, and Diseases | 163

gastrointestinal worm larvae and other cysts discussed earlier, for example, climb perhaps eight to ten inches on grass stems so that cattle will consume them. By contrast, because sheep crop grass much closer to the ground during grazing, the sheep-specific worm larvae/cysts climb only two to three inches up grass stalks to guarantee consumption by sheep. If cattle and sheep alternate grazing pastures, they will reduce one another's parasite loads still further, effectively cleaning pastures of each other's parasites. Though sheep will always ingest cattle-specific parasite larvae as they graze, if cattle are to be used to clean up pastures for sheep, they would have to graze the grass down to two to three inches.

Another factor to consider in mingling cattle and sheep is that just as the wildebeest herd protects zebras and other migrating herd members, the cattle herd provides the sheep with a feeling of security. Sheep numbering fewer than a hundred in a cow herd will mingle with the herd for security, but if the sheep number more than a hundred, they will feel safe without the protection of their bovine cousins and will separate to form their own herd.

Birds kept in the vicinity of the cattle herd will also reduce pest and parasite pressure on the cattle by eating insects and their larvae. Mobile containment pens, laying barns, and electric mesh fences can be used to graze poultry, ducks, turkeys, and other birds alongside cattle. Even goats, horses, and pigs have a place in parasite control and plant species management (by debrushing or rooting). The behavior of these animals helps loosen soils, and their manure content and grazing habits beneficially affect the nutrient cycle.

Domestic birds graze alongside cattle in movable containment pens in another beneficial multi-species grazing arrangement.

Soil Fertility

SOIL FERTILITY AFFECTS FAR MORE THAN the productivity of the grasses growing in the pasture. The consequences of good or poor soil fertility reach into every single aspect of your farm's production, management, and marketing success.

■ Effective soil-fertility management improves the mineral and nutrient contents of your forage supply, thereby promoting the health, well-being, and productivity of your cattle. Consequently, the quality and flavor of your beef will improve and the animals will become more resistant to diseases, pests, and parasites.

■ Good soil fertility increases the nutrient and mineral content of your winter-grazing reserve, improving its palatability through the long dormant season and reducing your mineral and feed supplement costs.

■ Good soil fertility increases the competitiveness of your desirable grass species, reduces weed pressure, and facilitates effective natural weed-management strategies.

■ Good soil fertility improves soil moisture retention and water filtration into the soil and helps prevent soil erosion and nutrient leaching. As a result, pasture productivity increases, your land becomes more drought-resistant, and your irrigation costs decrease.

■ Good soil fertility decreases the prevalence of disease organisms and pest and parasite populations by reducing and/or eliminating environmental conditions that allow them to flourish. They are all interconnected, and it all comes back to the soil.

We have at our disposal an enormous variety of tools with which to manage our soil fertility. These include the herd impact of our grazing animals, composts, soluble and insoluble fertilizer amendments (from natural and/or synthetic sources), tillage equipment, green manure crops, bacterial stimulants, and even certain plant species such as nitrogen-fixing legumes and specialized plants that can survive in nutrient-deficient soils or can help reestablish soil nutrient balance. But to use these effectively and to avoid unintended negative side effects, we must understand how plants, animals, soil organisms, and nutrients work together and affect one another in soil fertility. A solid understanding of soil fertility also enables us to understand how other seemingly unrelated management decisions (such as the use of irrigation, our choice of weed-control methods, and winter-grazing strategies) might indirectly affect soil fertility, either positively or negatively, and promote or inhibit the vital soil organisms that we rely on to create and maintain a healthy, fertile soil.

The most important soil-fertility management tool on a farm is the grazing rotation of the cattle herd. Your cattle are the domestic equivalent of the wild bison herds that created the thick, rich soils of the northern prairie. Dollars spent on herd infrastructure (fences and water sites) to manage grass rotations, herd impact, nutrient transfer, and manure buildup are *self-maintaining fertilizer expenditures* — that is, one-time expenditures that will continue to improve soil fertility year after year without repeated investment. By contrast, regularly adding fertilizers to maintain fertility requires annual financial expenditure.

Self-maintaining fertilizer expenditures should always be made before purchasing off-farm soil amendments. In this way, your natural resources (the grass and the herd) will be working for you to their maximum potential before you add an annual fertilizer bill to your expenses. Although you may still apply fertilizer amendments in the future, the herd will remain the driving force behind grass productivity and soil fertility, leading to significantly fewer fertilizer additions to maintain the same soil fertility levels.

Soil amendments (fertilizers, compost, bacterial stimulants) are a supplemental tool to correct gross soil deficiencies and to boost production beyond what your grazing rotation can produce. Fertilizers can also play a role if specific soil imbalances are blocking your grazing rotation's ability to improve soil fertility through herd impact alone. Fertilizers, however, are only a tool and must be well understood if you want to avoid becoming dependent on their yearly application.

The Importance of Soil Analyses

The herd's grazing impact provides a balanced approach to improving the entire spectrum of our soil's fertility. To a lesser extent this is also true of green manure crops such as ryegrass that are grown and incorporated (tilled) into the soil before they flower to increase soil fertility and nutrient availability. Even in the case of green manures, however, can we be truly certain that the plant will achieve its intended effect on our soil's fertility if we do not know *precisely* what the nutrient shortage or nutrient imbalance is within the soil? This lack of certainty increases exponentially with other soil fertility management tools, in particular fertilizer amendments,

including compost and other natural/organic fertilizer amendments, because in using them, we are adding individual nutrient elements to the soil that will dramatically alter the ratio of nutrients to one another (the importance of nutrient ratios is discussed at length below).

Fertilizers do not simply stack up in the soil, independent of one another, like preserves in a pantry; they are intimately interconnected. Without a precise understanding of which nutrients are deficient in the soil, how the deficiency is affecting the availability of other nutrients, and how the addition of a fertilizer or compost will affect the overall balance of nutrients, a fertilizer amendment can, and most likely will, do more harm than good to your pasture productivity, even if the long-term consequences are not immediately obvious or are overshadowed by short-term growth increases in the grass supply.

A case in point is the frequent use of composted manure in organic agriculture. Because this amendment comes from a natural source and because composting is a natural process, it is frequently assumed that compost is going to be good for the soil — the more you add, the better it will work. This assumption is dead wrong. Compost is good for the soil only if it contains the specific nutrients and nutrient ratios that are required to address specific nutrient deficiencies and imbalances on the specific patch of soil to which it is applied. But how do you know if it will correct your soil's problems? It might contain the wrong nutrients and nutrient ratios for your soil, or it might be beneficial only if some other missing nutrient is added to the compost mix.

The only way to know for sure what will benefit your soil is to carefully sample and analyze both the soil *and the compost* for their nutrient contents and their nutrient ratios. Without doing this, nitrogen in the compost will likely produce a short-term increase in grass growth, but in the long term you could be severely interfering with your soil's fertility. For example, because composted cattle manure is very high in potassium, repeated applications of compost made with cattle manure can cause potassium levels in the soil to rise excessively, thereby seriously affecting the nutrient availability of other minerals and impairing healthy plant growth. Over time, pasture productivity will decrease further, causing the erroneous application of still more compost, which will drive potassium levels still higher until you have a vicious cycle of decreasing soil fertility, unproductive pastures, and decreasing profitability.

You will not realize consistent, reliable results if you apply fertilizer amendments based on what you did last year, what the neighbor is using, what worked in the adjacent field, Uncle Bob's "surefire recipe for fixing that unproductive pasture," or by "eyeballing it" according to how the soil and plants look (a great management tool, but not for designing fertilizer applications). You will not maximize your pastures' potential and you can cause serious detrimental effects to your soil fertility and profitability,

PREPARING SOIL SAMPLES

If you do your own soil sampling (many reputable soil consultants and fertilizer companies offer this service free of charge as part of their fertilizer recommendation package, or may even insist on collecting the samples themselves to provide consistency for their soil analysis laboratory), there are a number of things to keep in mind in order to obtain representative, uncontaminated samples.

- The soil probe must be clean and rust-free. Do not use galvanized probes (because of zinc contamination) or chipped probes (chips will also contaminate the samples). Nickel-plated probes are ideal because nickel is not one of the nutrients for which soil laboratories test.

- Don't touch the sample with your hands; there is salt in your perspiration. Instead, use a clean screwdriver dedicated to soil sampling to push or knock soil from the probe into your collection bag.

- Collect samples directly into a *new* plastic zip-closure bag; debris in used bags can dramatically affect results.

- *Do not* use paper bags for your samples; the glues in the paper may contain elements such as boron that can leach into your

samples. Remember that laboratories are dealing with elements in parts per million, so even the slightest contamination will affect a sample.

- *Do not* collect samples in a bucket. It is not necessary to mix each sample before you bag it (the laboratory will thoroughly mix them). Buckets can contain contamination from feed, salt, and so forth that soaks into the plastic or rubber and cannot be completely removed through washing. Metal buckets risk introducing metals into the sample.

- Label every bag so you know which field each sample was taken from. Keep a record (a map is best) of all the fields and their corresponding sample bag names or numbers.

- One probe sample for every 1 to 2 acres is ideal; a minimum of five samples is needed to be representative of a single area.

- No more than 20 to 30 acres should be grouped together as a sample unit.

- Avoid sampling manure sites (including those old dark green spots and the tall mature grass left on them), bird droppings, fertilizer storage sites, and field ends where

the fertilizer truck tends to turn around and cause some fertilizer overlap.

- Any differences in the pasture (soil color, slope, growth, soil texture) should be sampled separately if they represent an area large enough to be fertilized separately. If they are not this large, they should be discarded. You want your sample to be representative of the bulk of the area, not a random sample of everything in the area.

- Pastures should be sampled to a depth of 4 inches (this will be the depth of the broadcast fertilizer's influence).

- Soils disturbed by tillage should be sampled to a depth of 6¾ inches or to the depth of the soil disturbance, whichever is deeper.

- When you send the samples to the lab, include a complete history of each field on which the lab can base its recommendations, including previous crops, current crop (or predominant pasture species), date of planting if it is not an established pasture, lime applications (list brand names used over the past three years), which fertilizers you have available, and which fertilizers you have access to.

which may require additional thousands of dollars of fertilizer amendments to repair.

Thus, before you add any fertilizer amendment (natural, organic, or synthetic), compost, or bacterial stimulant, you must take representative soil samples of your land (see the box on page 167 to learn how to prepare effective soil samples) and have them analyzed by a soil laboratory. You must also know the exact nutrient content of each fertilizer you will apply (so get your compost tested before you use it; no two composts are likely to be the same), and if you plan to use other means such as bacterial stimulants to alter your soil fertility, you need to know exactly how they will affect your soil before you use them. This means you must work closely with a soil laboratory, soil consultant, or reputable fertilizer company that is well versed in the delicate science of making fertilizer recommendations.

The remainder of this chapter will give you a solid understanding of soil fertility. You will also gain the ability to interpret your soil analyses — one of the single most powerful tools at your disposal. Soil analyses provide vital feedback about the effectiveness of your soil, pasture, and herd-management practices and serve to indicate what steps you need to take to maintain and improve your soil fertility, herd health, and other areas in which soil fertility affects your farm's production, management, and marketing. Finally, your understanding of soil analyses will give you unprecedented control over your nutrient-management program so that you are not forced to blindly follow fertilizer recommendations designed by companies, laboratories, and consultants who do not know your farm or goals.

TIP
You will do more to increase overall production with fertilizer amendments by fixing large problem areas than by trying to boost the yield of your better-quality soils.

What Does the Cow Think of Soil Fertility?

We may have a difficult time detecting in grass the subtle differences in soil nutrients, yet to cows, these differences are enormous. They can taste the differences in sugar levels or nitrate content and can see the differences in light refraction between healthy and unhealthy plants. Even the varying smells of grass with different nutrient levels are apparent to their discerning noses. They will go to great lengths to seek out the best stuff; even if we can't detect these differences, they seem to know what is best for them.

One of the clearest examples of livestock selectiveness for high-quality nutrition comes from the Albrecht papers (Dr. William A. Albrecht, *The Albrecht Papers,* vol. 2: *Soil Fertility and Animal Health.* Kansas City, MO: Acres U.S.A., 1975, 88–89), in which the late Dr. William A. Albrecht, the legendary professor of soils at the University of Missouri College of Agriculture, who contributed so substantially to eco-agriculture, describes a group of hogs that were let into a cornfield to graze. The hogs traversed the forty-acre field to the farthest corner, the one most distant from water and supplements, and proceeded to knock down and eat the corn there while ignoring the rest of the field. It turned out that several years before, the corner had been limed and given phosphate to grow alfalfa. To the farmer, the corn in the far corner was indistinguishable from that in the rest of the field, but to the hogs it was clearly different.

When a cow leans through a fence to reach some indiscriminate weed in the fence row while ignoring a lush field full of grass, she is likely seeking out some nutrient that she can distinguish by smell. Her delicate nose is working as hard as any chemical laboratory to appease her cravings, much as I might go to great lengths to satisfy a craving for Belgian chocolates or salted pistachios.

How Healthy Soil and Grass Work

Soil fertility is controlled not by the individual minerals in the soil, but by the overall balance or ratio of minerals. A *colloid* is the smallest soil particle, made of either clay minerals or organic matter. It is like a little wafer that attracts and holds mineral ions through electrical charges. (Mineral ions, also called cations, are nonorganic soil elements — calcium, magnesium, nitrogen, phosphorus, potassium, sodium, sulfur, copper, zinc, and so forth — that have

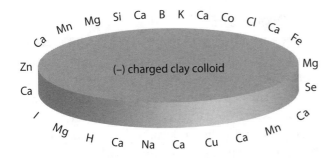

The clay colloid carries a negative electrical charge, allowing it to attract positively charged mineral ions (cations). The base saturation of a mineral element refers to the percentage of sites occupied by each cation around the colloid.

acquired an electrical charge.) This is essentially how soil nutrients are stored in the soil. Each mineral element represents a certain percentage of the available sites (called *base saturation*) on each colloid (see the illustration above).

The plant draws nutrients from the colloid through its own electrical charges in the roots and root acids. When the balance of elements is correct, the plant is able to access each nutrient according to its needs. If the ratio of mineral ions on the colloid is incorrect, the interaction of electrical charges among the various mineral ions will cause some elements to be bound more tightly to the colloid while others are held more loosely, making it difficult for the plant to access the correct proportion of minerals.

Ratio of Calcium to Magnesium

The ratio of calcium to magnesium profoundly affects a plant's ability to exchange nutrients with the soil (see Cation Exchange Capacity at right). It is the base saturation (percentage of available sites on the colloid) occupied by calcium or magnesium, not the actual amount of each nutrient measured in the soil, that determines when calcium or magnesium fertilizer needs to be added. The base saturation of these two elements should make up approximately 80 percent of all available sites around the colloid, with calcium taking 60 to 70 percent of the sites and magnesium occupying 10 to 20 percent. The fertilizer amendment is meant to correct the base saturation of each element on the colloid, not the actual amount of each nutrient in the soil. Although the nutrient-holding capacity (number of colloids available in the soil) varies from one soil type to the next,

it is the percentage of colloid sites occupied by each nutrient — the base saturation — that actually determines soil fertility. Thus, by looking at the base saturation of each mineral element in your soil test, you can determine if the *ratio* of mineral elements in the soil is correct or needs adjusting.

Soil Acidity/pH

A measure of the soil's acidity or alkalinity is its pH, which ranges on a scale from 0 to 14, with 7 being neutral. A figure of less than 7 represents increasing soil acidity and more than 7 represents increasing soil alkalinity. The level of acidity/alkalinity of the soil is also important to the growing environment of plants. As soil pH changes, so does the solubility of each mineral element in the soil, making some elements more accessible to plants and other elements more difficult for the plant to absorb. The ideal soil pH that allows plants to absorb the correct balance of minerals is between 6.0 and 6.5, meaning that the soil is slightly acidic. This is also the pH at which the correct balance of soil fungi and soil bacteria exist and in which the majority of plants and soil organisms are most active. Soil pH on its own, however, does not control nutrient availability.

Cation Exchange Capacity

How easily available soil nutrients are to plants is referred to as the soil's *cation exchange capacity*. It is affected by a combination of the percentage of sites occupied by each mineral on the colloid (base saturation), the soil pH, and the number of total colloids available in the soil (determined by the soil type — clay, loam, silt, sand, peat, gravel or a combination — because finer soils and those with high organic matter have more ultra-small colloids capable of storing and exchanging mineral ions). The higher the cation exchange capacity, the greater the number of available nutrients. Yet because there are more colloids to hold nutrients in a soil with a high cation exchange capacity, larger amounts of fertilizer amendments are required to correct deficiencies, while in a soil with a lower exchange capacity (fewer nutrients are available to the plant and fewer colloid particles are in the soil), smaller amounts of fertilizer amendments are needed to correct deficiencies and rebalance the ratio of mineral ions.

Insoluble yet Available

Much of the synthetic fertilizer industry is based on the idea that water-soluble soil nutrients are most available to plants. Unfortunately, soluble fertilizer elements are highly susceptible to leaching; unused minerals are simply flushed into the groundwater before the next growing season. This, however, is only part of the story.

In truth, soil nutrients do not need to be soluble in water to be available to a plant. There are many forms of insoluble minerals that are protected from leaching so they can be stored for another year, yet the root acids of the plant can dissolve them as the plant needs them, hence the term *root acid solubility*. In the case of such minerals, the soil nutrients can either be locked in an insoluble inorganic form or they can be held as complex organic compounds by the soil organisms and humus (organic component of the soil) until they are released (dissolved) by the root acids. They are insoluble yet available. Herein lies the strength of good pasture rotations and manure management: Soil organisms are promoted to guard the nutrients for future availability and the nutrients are recycled back into the soil to be stored in an insoluble form for future use.

Soil Organisms

The role of soil organisms in soil fertility is often overlooked. On the same piece of ground there is more biologic mass supported below the soil surface, in the form of roots, insects, worms, rodents, and a host of microorganisms such as bacteria, fungi, microbes, and other microscopic soil organisms, than there are living organisms above the soil's surface. There is an army of living beings down there!

Soil organisms are responsible for breaking down organic material and converting it into humus, which the plant roots can access at a later time, and are responsible for the transport and redistribution of nutrients: They dissolve nutrients from rocks, fertilizer pellets, manure patches, and plant remains and redistribute them over a broader area. Soil organisms also transport insoluble nutrients to the plant from beyond the roots' reach.

Mycorrhizae fungi (one of the many microscopic soil organisms) will actually join the root zone of a plant to form a secondary rootlike network that transports nutrients from the soil to the plant. This network extends the root zone more than ten times beyond the plant roots, outward, downward, and through all the minute spaces among the root hairs.

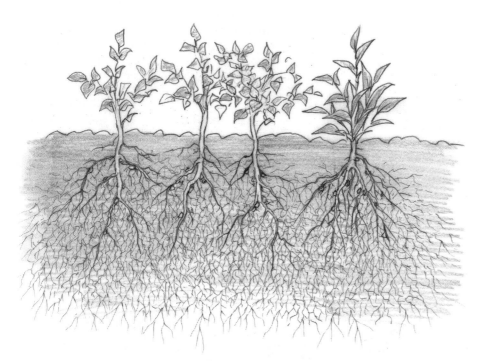

Mycorrhizae *fungi form a microscopic, secondary, rootlike network that links to the roots of plants to transport nutrients from the soil to the plant. Larger, thicker plant roots are interconnected by a complex, fine, hairlike web of beneficial fungi that extends well beyond the actual root zones of the plants.*

Risks of Conventional Fertilizer Management

Unfortunately, much of our conventional fertilizer management does nothing to keep soil organisms healthy and working for us. In many cases, soluble synthetic fertilizer amendments, high concentrations of fertilizer applications, pesticides, and herbicides harm these delicate soil organisms, reducing their vitality and slowing their effectiveness. It comes as no surprise that this harm, combined with groundwater leaching of excess soluble synthetic fertilizers, creates an annual cycle of dependency on soluble synthetic fertilizer use.

Misconceptions about Soil Preference

There is a misconception that every plant prefers a specific type of soil. For example, blueberries are said to grow best in very acidic soils, but this is not altogether true. More accurately, they are capable of growing competitively in acidic soils while many other plants are not, giving them a competitive advantage in such an environment. The best blueberries grow in the same balanced soils that other plant species prefer, but in normal soil conditions many other plants are more competitive than blueberry plants.

Once the soil is balanced, the management conditions determine which species flourish. Weeds and grasses (and blueberries) all have the same nutrient advantages, but herd impact, constant grazing, and trampling hooves harm many of the weed species, while the grass species thrive under a periodic herd impact and therefore are able to out-compete and outperform other plants.

If soil fertility declines, however, nutrient imbalances take on a greater importance in determining which species can cope and continue to be competitive. Soil amendments can be used to give desirable grazing species an opportunity to compete for nutrients. After this, grazing management determines which species survive and which ones are choked out.

Calcium and Magnesium Revisited: Proper Ratios Are Critical

Calcium and magnesium should be the priority of your long-term soil-fertility management, although other nutrients may be more important in addressing immediate deficiencies within a growing season. Soil pH is often thought to be controlled exclusively by calcium, and the strong effect that magnesium has on pH is overlooked. Therefore, some farmers lime their soils to address pH levels but fail to address calcium deficiencies and magnesium excesses.

Earlier, we learned that calcium should represent 60 to 70 percent of available sites on each soil colloid, while magnesium should represent 10 to 20 percent, with the remaining sites being held by the other soil nutrients. If calcium is in excess (or magnesium is deficient), the soil becomes too loose and prone to erosion. If magnesium is in excess (or calcium is deficient), the soil becomes hard and compacted and lacks air and water-holding capacity.

The two nutrients have a special relationship: When calcium is depleted, the empty calcium sites on the colloid are filled by atomically compatible magnesium. Yet once magnesium occupies these sites, it does not easily let go of them, even if calcium is added to the soil; its bond to the colloid is stronger. Magnesium ions raise soil acidity 1.67 times more than calcium ions do. Soil pH will be temporarily corrected by lime fertilizer long before sufficient calcium is added to the soil to fully correct the calcium deficiency.

When calcium becomes deficient in the soil, the high ratio of aggressive magnesium ions at calcium sites on the colloid interferes with the normal exchange between the plant roots and the other minerals on the colloid. In order to release magnesium's grip on the other minerals and to correct the calcium–magnesium ratio, calcium needs to be added above and beyond the amount required to correct soil pH, despite the effect that this additional calcium will have on soil pH. Once the calcium–magnesium ratio is corrected, the pH will decrease to normal levels (rather than increasing further), because the weaker calcium ions do not boost pH as vigorously as the magnesium ions do. In effect, then, the correct calcium–magnesium ratio makes all other minerals available to plants.

Considerations When Liming

Not all soil laboratories make liming recommendations based on calcium–magnesium ratios. Some recommend liming based only on correcting soil acidity. If only pH is corrected, the calcium–magnesium ratio will progressively decrease after each lime application — calcium becomes increasingly deficient while magnesium artificially maintains the correct soil pH. Correcting the pH should be just a side effect, not the primary goal of liming.

---

*Without humus,
the soil becomes sterile.*

It is also important to know what kind of lime to use. Although limestone is typically thought of as a source of calcium, it can also be very rich in magnesium (sometimes known as *dolomite*), or range anywhere between calcium-rich and magnesium-rich extremes. Unfortunately, without a chemical analysis of the lime, you cannot tell them apart. Do not automatically assume that the lime recommended on your soil analysis by the soil laboratory will be the lime that is supplied to you by the fertilizer company. In this area mistakes happen more often than you would think. If you use the wrong form, your calcium-magnesium problem will be compounded, other nutrient imbalances will deteriorate instead of improve, grass productivity will decrease, and you could be buying truckloads of "correct" lime for years to come to fix the problem.

When liming, it takes three years for the full effect of the calcium to become available in the soil, while magnesium does not become available until the second year. If a magnesium-rich lime is used when magnesium is not needed, you will see an improvement in soil fertility the first year when the first third of the calcium is released, but then soil fertility will collapse again in the second year, when the magnesium in the lime is released, further unbalancing the calcium–magnesium ratio.

Organic Matter, Humus, and Nitrogen

The presence of only soluble nutrients can be measured in a soil analysis. Unfortunately, when the soil sample is removed from the protection of the ground, the soluble nitrogen component begins evaporating from the sample very quickly because nitrogen is such a volatile element. If it is not measured by a laboratory within hours after sampling, the soluble nitrogen will disappear from the sample.

As soil organisms break down organic matter, their digestive processes release nitrogen. The more humus the soil has, the more organisms the soil will support, the larger the volume of organic matter the organisms will be able to digest, and the more nitrogen they will release. As discussed above, although nitrogen may not show up in your soil analyses due to its evaporation, if your soil has a 5 percent humus content, you can expect approximately one hundred pounds per acre of available nitrogen, less if your calcium–magnesium balance is incorrect. At 2½ percent humus, your soil biology will release approximately seventy pounds of nitrogen per acre into the soil, and so on.

Organic matter and humus are not the same thing. *Organic matter* is anything organic in the soil (such as roots and dead grass); *humus* is the portion of the organic matter that has been broken down by soil organisms into a stable form. Humus does not include undigested plant remains, woody bits, and so on. The digestive processes of the soil microbes create humus as they break down fibrous plant remains. Green vegetative material (green plant remains, or green manure) that has not become mature and fibrous will not become humus; it will be digested entirely by the microbes so all its nutrients become available to the plants in the next growing season. By contrast, plant remains build humus over time.

Soil organisms live in the humus component of the soil; without humus, the soil becomes sterile because the organisms in it cannot survive. The insoluble forms of many plant nutrients are stored as complex organic compounds in the humus component, where they can be accessed by the root acids as the plants require them. Humus gives the soil its texture, allowing air and water to flow through it, and it provides the glue that makes the soil resistant to erosion. Without humus, all the nutrients required by the plants would have to be added yearly as soluble fertilizer amendments and the soil would become hard, compacted, and prone to erosion.

Grazing to Build Humus

The most effective way to build humus in the soil is by having livestock periodically graze the grass. After a herd migrates through a pasture, the roots that die back in response to grazing and the material that is trampled into the soil by the animals' hooves provide organic matter for the soil organisms to digest and turn into humus. This herd impact created the rich humus soils of the prairies when the migrating bison herds roamed them. If, however, the sod skin protecting the soil is removed or buried by tillage and the soil is exposed to the atmosphere, the organic carbon compounds and nitrogen in the humus become vulnerable to the oxidizing air

and are lost as carbon oxidizes to carbon dioxide gas (CO_2) and nitrogen degases as nitrogen gas (N_2).

Humus loss through tillage (and grazing practices that expose the soil directly to the air) is dramatically measurable. We have only to consider how much the rich prairie soils in the American Midwest have been depleted of their high humus content since pioneers first broke the ground with plows. In fact, worldwide there has been a dramatic reduction in the organic matter of agricultural soils in response to one hundred years of mechanized, large-scale agriculture. Where did all this organic matter go? Because the reduction is so widespread, the best guess is that the bulk of it simply degased out of the soil and into the atmosphere. The amount of carbon dioxide needed to account for the reduction in organic matter in soils worldwide corresponds with the increase in global carbon dioxide levels in the atmosphere, which are thought to be a fundamental cause of global warming (CO_2 gas is by far the most abundant greenhouse gas). Industrial greenhouse gas emissions pale in comparison to the vast volumes of carbon dioxide gas released from soil by tillage agriculture.

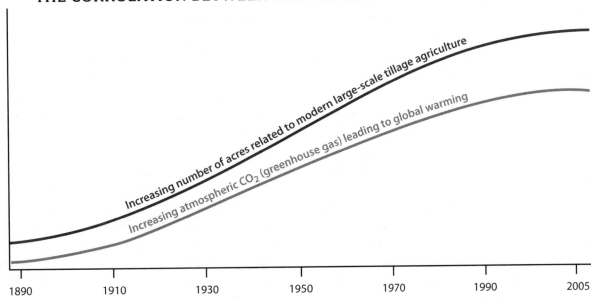

THE CORROLATION BETWEEN TILLAGE AND ATMOSPHERIC CO_2 GAS

Increasing number of acres related to modern large-scale tillage agriculture

Increasing atmospheric CO_2 (greenhouse gas) leading to global warming

1890   1910   1930   1950   1970   1990   2005

The decrease in organic matter in soils caused by tillage agriculture and the exposure of bare soil to the oxidizing effects of the atmosphere correlate all too well with the increase in carbon dioxide gas in the atmosphere. Is modern agriculture at least partially to blame for global warming?

Even on my parents' farm, I remember a permanent grass pasture with 6 percent humus content in the sandy soil, but the main crop field adjacent to it had approximately 3½ percent humus content after only twenty-five years of tillage agriculture since my father first cleared the timber from these fields. Thus we can presume that both fields must have started with at least 6 percent humus in the soil before the land was cleared.

Nitrogen

Legumes are able to draw nitrogen from the air to supply a significant portion of their nitrogen requirement. While the plants do not contribute a large amount of nitrogen directly to the soil if they are harvested and removed, when livestock graze legumes, the nitrogen the plants capture will be returned to the soil via manure and urine so that soil organisms can make it available for future plant use.

Locating Sources of Organic Nitrogen Fertilizer

Finding organic nitrogen fertilizer sources is likely the most difficult challenge for the organic producer. Compost certainly works well, but its use must be accompanied by comprehensive soil analyses and an analysis of the specific compost being used, as discussed on page 166.

Before spreading compost on your fields, you should know the exact content and ratios of the other nutrients in the compost so that its application does not lead to further excesses, deficiencies, or mineral imbalances, and so that it does not affect the appropriateness of your other fertilizer amendments. Further complicating matters is that composting is not cheap, requiring that you use expensive winter feed to create concentrations of manure (thus depriving the fields of this resource) and operate costly equipment to compost and spread it, or you need to find off-farm compost, which is costly to transport and spread and may contain diseases or chemicals.

Large concentrations of raw, uncomposted manure should never be used. Manure patches left behind by grazing cattle are small enough to be exposed to air and for soil organisms to quickly digest their nutrients, but unless they are mechani-cally composted, large concentrations of raw manure create a substantial anaerobic environment that is prone to breeding many dangerous bacterial diseases (for example, clostridial diseases such as tetanus and blackleg) that cannot survive when oxygen is present. Furthermore, because the nitrogen content in uncomposted manure has not been processed into a stable form by aerobic bacteria, it evaporates into the atmosphere when the raw manure is spread in the fields. The terrible stink from the spread of uncomposted manure is the smell of valuable nitrogen escaping your farm forever. Conversely, composted manure smells earthy.

Overfertilizing with synthetic nitrogen is the leading cause of humus loss from soils.

Nitrogen can also be added in the form of fish emulsion, seaweed, and protein meal. Though these sources are organic, they can be cost prohibitive when nitrogen is required over large areas. Read on for more cost-effective options for adding nitrogen to your pastures.

Biological Fertilizers

Nitrogen and a host of other nutrients can be increased in the soil through the use of bacterial stimulants, otherwise known as *biological fertilizers* and sold under names such as Vitazyme, Agrispon, and Agralife. These stimulate the digestive process of soil organisms to release more nutrients (including nitrogen) from inorganic soil particles (sand, silt, clay) and from *undigested* organic plant remains, *but they should not digest the humus component of the soil.* The discussion below about microbial fertilizers explains the danger of using overly aggressive stimulants that cause the soil microbes to begin digesting valuable humus reserves, a practice that has devastating long-term consequences to soil fertility.

Over the course of the growing season, unlike their synthetic counterparts, biological fertilizers aid soil microorganisms to process more raw undigested organic matter in the soil to create more humus and build up additional available nutrients in the soil in a stable form that will not leach out or evaporate. This makes additional nitrogen available to the plants as needed at a relatively low purchase and application cost. Biological fertilizers should be applied once in spring and usually again in fall to allow the bacterial action to occur, but they must be used according to the manufacturer's directions to avoid damaging your humus layer or creating undesirable bacterial growth. The more balanced your soils are, especially the calcium–magnesium ratio, the better the biological fertilizers will work.

Beware of Microbial Fertilizers

Microbial fertilizers are biological stimulants designed to release nitrogen from the humus component of the soil by specifically increasing microbial digestion of humus — different from the biological fertilizers describe above, which stimulate the microorganisms to increase non-humus organic digestion as a way of building up the humus component while the existing humus component remains untouched. Microbial fertilizers are extremely dangerous because, although they will release a tremendous amount of nitrogen from the humen in the short term, within a few growing seasons, they will quickly and effectively deplete your soil of its humus content, thereby destroying your soil's current nutrient reserve and future nutrient storing ability, as well as destroying the habitat of your valuable soil organisms.

Be aware of what a biological nitrogen fertilizer does before you use it on your fields. If the humus content is destroyed, you cannot fix the problem with next year's fertilizer budget and you cannot simply replace humus by adding organic matter. It takes decades for soil organisms to process sufficient raw organic matter to rebuild the humus in soil, and it is possible to do so only through a rigorous, well-managed herd impact.

Dangers of Synthetic Nitrogen Fertilizers

Conventional synthetic nitrogen fertilizers come with a host of problems. Unless you use it with a great deal of caution, excess nitrogen will bond with

PROHIBITIONS ON SYNTHETICS

Organic certifying associations consider many synthetic fertilizers harmful because their soluble nature makes them prone to leaching, which produces algal blooms and water contamination. Furthermore, their highly acid nature (a by-product of the refining process) makes them harmful to soil organisms.

Sometimes organic associations prohibit their use for far less obvious reasons; for example, the agent used to granulate the fertilizer may not be allowed even though the fertilizer itself is perfectly safe. In the case of a potassium-magnesium fertilizer I once wanted to use, the corn-gluten-granulating agent was not certified organic because it might have been made from genetically modified corn.

other soil nutrients such as calcium to form water-soluble compounds that will be leached from the soil through groundwater movement, or other minerals to form compounds that are inaccessible to plants, thereby creating copper, phosphorus, and many other deficiencies that may not be apparent in soil tests.

Costs of Excessive Nitrogen Fertilization

Overfertilizing with synthetic nitrogen is the leading cause of humus loss from soils: The excess nitrogen is used by soil microbes to speed up their digestive process, much like the microbial fertilizers described on page 175, allowing them to break down humus to form soluble nitrogen and CO_2 gas, which escape from the soil with disastrous long-term results. Liquid or gaseous nitrogen fertilizers injected directly into the soil are particularly dangerous. These fertilizers, such as anhydrous ammonia (one of the most common, made by a reaction between natural gas and nitrogen) and raw liquid manure, are applied by specially designed fertilizer injectors mounted on cultivators, seeders, or other specially designed equipment capable of cutting beneath the soil surface to deliver the liquid to its chosen depth. This creates a series of injection bands (the fertilizer is applied only within each strip of tilled soil created by the cultivator shank fitted with an injection applicator) around which the high concentration of injected nitrogen fertilizer (95 percent of it remains within three inches of the injection band) causes nutrient leaching, nutrient tie-up, and humus burning through increased microbial digestion. Calcium is the hardest hit by leaching because the calcium-nitrogen compounds that are formed are exceedingly soluble in groundwater. The leached calcium is replaced by magnesium on the colloids, which causes the soil to harden. Tellingly, temporary military airplane runways are sometimes made by adding excessive amounts of nitrogen and magnesium to the soil to leach calcium and create a hard surface.

To be effective, injected nitrogen typically requires even higher application rates than broadcast nitrogen because of the spacing of the banded injections. Anhydrous ammonia is also extremely toxic to the soil biology, further contributing to fertilizer dependency.

One of the most common and most volatile forms of broadcast nitrogen fertilizer is urea. Unless it is immediately tilled or washed into the soil by at least ½ inch of rain or irrigation, as much as 70 percent of the urea will evaporate within a week. The safest synthetic nitrogen fertilizer is ammonium sulfate; it is released slowly (over sixty days) and is not as vulnerable to leaching.

Invest in Infrastructure Instead

Nitrogen fertilization is no picnic: Organic nitrogen is often costly and cumbersome to apply in the quantities needed by grasses, and conventional nitrogen is toxic to soil biology, hard on soil nutrients, and creates a fertilizer dependency that is difficult to break. Therefore, investing money in water sites and electric fencing is truly the most cost-effective and sustainable fertilizer program, allowing you to recycle and spread out the manure and urine through daily cattle moves and benefiting your entire production system without the annual costs, pitfalls, and risks of other fertilizer amendments.

The Scoop on Macronutrients

Macronutrients are any of the chemical elements required by plants (and animals) in relatively large amounts: calcium (Ca), magnesium (Mg), nitrogen (N), phosphorus (P), potassium (K), sodium (Na), and sulfur (S).

Until the calcium–magnesium balance in your soil is right, many of your other fertilization amendments will be hindered. Correcting the ratio between calcium and magnesium will actually help access previously inaccessible or deficient minerals as the soil pH and the base saturation of these two major minerals change. All other soil minerals are most available when the calcium–magnesium ratio is balanced (see Ratio of Calcium to Magnesium on page 169) and the soil pH is between 6.0 and 6.5. At a higher or lower soil pH, some minerals may become even more soluble while others become less so or are entirely unavailable to the plant. (I will not list ideal base saturation values of the various macronutrients, or the ideal numerical quantities of the micronutrients, because each laboratory uses different ideal values specific to its testing methods. But it is important to know these values when interpreting your soil analyses. The ideal base saturation values or ideal measured quantities related to your laboratory's specific testing method will be supplied on your soil analyses or, if not, they can be requested directly from the laboratory, soil consultant, or fertilizer company making the fertilizer recommendations.)

Sodium

Although sodium is rarely deficient in the soil, the base saturation of sodium shows up on soil tests not only to alert you to excesses, but also because it takes up a certain number of sites on the colloid, thus affecting the base saturation of all the other macronutrients. In this way, excess sodium ties up other nutrients in the soil. The base saturation of sodium must always be less than the base saturation of potassium. If sodium exceeds potassium on the colloid, a plant will absorb it instead of potassium into its cell walls; in hot weather the sodium will expand, killing the plant when the cell walls begin to rupture. Excessive sodium must be removed or buffered by sulfur and other fertilizer amendments to allow other macronutrients to take their place on the soil colloid.

Potassium

When soil gets too much potassium via manure around feed areas or bedding sites or an overapplication of compost, base saturation of potassium increases too much, which knocks calcium and magnesium from their respective sites on the colloid and results in the same mineral imbalance in the bloodstreams of your grazing animals. If the potassium–sodium level increases above the calcium–magnesium level in the soil, the animals will die without warning and with no visible signs of the cause of death, although a blood sample will reveal the severe mineral imbalance. Because calcium plays a role in making plants taste sweet, calcium deficiency caused by excess potassium in the soil makes the grass taste bitter, which makes it less palatable to grazing cattle.

Phosphorus

Like potassium, excess phosphorus can also knock calcium from the colloid. The resulting mineral imbalance in the soil can cause serious nutritional deficiencies in cattle because the plants they eat will reflect these soil imbalances. Even human health is affected when we eat mineral-deficient plants and meats because these affect the mineral balance in our bodies. Thus, phosphorus excesses in the soil and the resulting deficiencies and imbalances created in the grass supply affect cattle health and productivity and reduce weight gains.

Rock Phosphate, Soft Rock Phosphate

Rock phosphate and soft rock phosphate are two organically approved sources of phosphorus that are not well understood and are thus frequently frowned upon by the synthetic fertilizer industry. As is the case with other organic fertilizers, their soluble phosphorus content is quite low, which makes them unattractive when compared to their soluble synthetic fertilizer cousins, yet the phosphorus in these compounds is readily available to the root acids. Like other insoluble organic fertilizers, rock phosphate and soft rock phosphate release phosphorus into the soil for years after they are applied, whereas the unused portion of soluble fertilizers is leached from the soil before the next growing season.

Sulfur

Although the roles of potassium and phosphorus in plant growth are well known, the role of sulfur, just as essential, is often overlooked. Yet it is certainly as important as potassium and phosphorus to crop yields, plant health, and the nitrogen-fixing ability of legumes. Among other benefits, it has the ability to regulate excesses of other elements in the soil, such as sodium, magnesium, potassium, and even calcium, because it bonds with these to form soluble compounds that are easily leached from the soil.

Sulfur will remove whatever element is most in excess. Two pounds are required to remove each pound of excess magnesium or calcium. Yet because it has been overlooked for so long by conventional nitrogen-phosphorus-potassium (N-P-K) fertilization, it is often deficient. Making matters worse, the sulfur content of many N-P-K fertilizers has been reduced as fertilizers became more refined and yields increased.

ORGANIC MATERIALS REVIEW INSTITUTE

The best available lists of organic fertilizer sources are provided by the Organic Materials Review Institute (OMRI; see resources, page 360). The institute lists fertilizer sources by generic material, supplier, or brand name, and rates each fertilizer as either allowed, restricted, or prohibited for organic farming. A subscription to the Web site is not required to view the Web editions of the OMRI lists, though you must subscribe to view the full list of fertilizer materials.

And a Pinch of Micronutrients

Micronutrients, or trace elements, are the chemical elements required by plants (and animals) in relatively small concentrations. Among them are boron (B), chlorine (Cl), cobalt (Co), copper (Cu), iodine (I), iron (Fe), manganese (Mn), molybdenum (Mo), nickel (Ni), selenium (Se), silica (Si), vanadium (V), and zinc (Zn).

Micronutrients are the most neglected elements in many fertilizer programs. Many conventional N-P-K fertilizer recommendations barely address calcium and magnesium, and, much like sulfur, micronutrients and their sometimes minute deficiencies are often ignored altogether. Although the amounts needed by the soil and grass are tiny compared to the large volumes of macronutrients required, the role of micronutrients in regulating soil and plant functions is vitally important. Minute deficiencies can reduce grass productivity, cause weed and parasite infestations, and seriously compromise the fertility and health of your herd.

If a soil deficiency in your pasture is obvious enough for you to see, it is extremely serious.

An imbalance in micronutrients illustrates a concept known as *hidden hunger*, when there are no easily recognizable visual signs of nutrient deficiency. Instead of manifesting as stunted or discolored leaf growth, for instance, micronutrient imbalances and deficiencies are expressed as drops in yield, plant health issues, and pest infestations that are further magnified by health issues in the livestock that eat these plants. Only soil analyses, forage sampling, or blood samples of your livestock can detect micronutrient deficiencies before hidden hunger affects the health, fertility, and productivity of your plants and livestock, and the profitability of your business.

By the time you can actually see the signs of a micronutrient deficiency, the imbalance has become so serious that it will already have severely impacted your plant and animal productivity and health. For example, a copper deficiency in the soil may seem to be unimportant as long as the grass looks green and is growing vigorously, but the hidden hunger caused by the deficiency may lead to abortion in your most vulnerable breeding animals. Rough hair coats and coats that refuse to shed may be signs of a copper deficiency in the soil. Once copper levels in the soil rise above 2 parts per million, cattle's coats become shiny and sleek again, even in winter.

Only a soil test can detect micronutrient deficiencies early enough to prevent the devastating effects of hidden hunger. The most common micronutrients analyzed in parts per million (ppm) by tests are boron, copper, iron, manganese, molybdenum, and zinc. There are many more micronutrients than this, but these are sufficiently understood to be part of fertilizer programs. Two micronutrients to add to this list for livestock crops and pastures, though they are not usually considered for field crop fertilization, are selenium and cobalt, which, though vitally important to animal health, are difficult to supplement through natural sources of livestock mineral supplements. It is usually easier to correct selenium and cobalt imbalances by fertilizing the soil than it is to address them through a livestock mineral supplement program. Note that their deficiency is difficult to detect except through careful soil sampling and laboratory analyses.

Recognizing Deficiencies and Imbalances

Nine times out of ten, if a soil deficiency in your pasture is obvious enough for you to see, it is extremely serious and has already caused reduced gains, reduced fertility, increased susceptibility to disease and parasites, and reduced pasture yields.

Unfortunately, many of the recognizable signs of nutrient deficiencies are difficult to tell apart, particularly because they manifest differently depending on the plant species they are affecting. No matter what the signs look like, they are red flags. Immediately take a soil test to determine the problem and correct it with the proper fertilizer amendment before the deficiency cuts further into your profits. One way of identifying subtle deficiencies and

Soil Test Repor

UK COOPERATIVE EXTENSION SERVICE
University of Kentucky – College of Agriculture

Lexington 859-257-2785
Princeton 270-365-7541
www.rs.uky.edu/soils

COUNTY SAMPLE NO.: 2101	To: Van Meter, Ike 224 Barrow Rd Lexington, KY 40502
REPORT FORM: A	

Kimberly Poe   859-987-1895
Agricultural Agent

Date	Owner Sample ID	Owner ID	County Code	UK Lab NO.
4/16/2015	Low Tide Big Field	69	17	11196

Nutrient	Lab Results	Level of Adequacy					Calculated CEC Data	
		Very Low	Low	Medium	High	Very High		
Phosphorus (P)	17	>>>>>>>>>>>>>>>>					CEC (meg/100g):	16
Potassium (K)	181	>>>>>>>>>>>>>>>>					%Base Sat.:	64
Soil pH	6.2						%K:	1
Buffer pH	6.8						%Ca:	57
Calcium (Ca)	3609						%Mg:	6
Magnesium (Mg)	223	>>>>>>>>>>>>>>>>>>>>>>>>>>>>>>>>>>>					%H:	35
Zinc (Zn)	1.8							
Other Test								

Acres	Primary Crop	Primary Management	Primary Use	Previous Crop	Previous Management	Previous Use	Crop 2 Years Ago Tobacco	Soil Draina
80	Cool Season Grass	Annual Top Dressing	Pasture	Cool Season Grass	Hay or Pasture < 4yrs.	Pasture		Modera Wel

RECOMMENDATIONS:	N	P2O5	K2O	LIME	Mg	Zn
	see comments below	100 lb/ac	50 lb/ac	None	None	None

Apply 100-125 lbs 46% urea per acre

Apply 220 lbs 46% phosphate per acre

Apply 85 lbs 60% potash per acre

COMMENTS:
Mehlich III used for P, K, Ca, Mg, and Zn (lbs/acre). Crop response is highly probable with Very Low or Low soil levels, slight with Medium, and not likely with High or Very High. N, P2O5, K2O, Mg, and Zn recommendations are based on lbs of the nutrient. Fertilizer needed will depend on nutrient content in the fertilizer. Soil pH is calculated from 1 M KCl soil pH using: 0.91 x 1 M KCl soil pH + 1.34. 1 M KCl soil pH and Sikora II buffer pH are used for determining lime needs based on 100% effective lime. Lime quality in KY is defined by relative neutralizing value (RNV). RNVs for ag lime are determined by the KY Dept of Ag and are on the internet (publications at soils.rs.uky.edu).
TETANY PROBLEMS with cattle are sometimes encountered on straight grass pastures, particularly with nursing cows where grass pasture is the only source of feed. SUPPLEMENTAL FEEDING OF MAGNESIUM to nursing cows on such fields is recommended as a means of lowering tetany risks.
SCHEDULE FOR N: Feb 15 to Mar 15 apply up to 100 lb/ac N; May 1 to 15 apply up to 50 lb/ac N; Aug 1 to 15 apply up to 80 lb/ac N. When P and K are adequate (>30 - P and >200 - K) the use of nitrogen stimulates growth during peak production. 200 lbs N/acre may produce up to 4 tons of dry matter.
All the commonly used N sources are acceptable for topdressing grasses during late winter-early spring. After early May urea may be less effective. Efficiency of liquid nitrogen is between that of urea and ammonium nitrate.
For additional information, ask your local Extension Office for the publication AGR-103 "Fertilization of Cool-Season Grasses" and AGR-169 "Problems in Diagnosing Nutrient Deficiencies of Cool-Season Forage Grasses."
UK Forage publications can be found at the website http://www.uky.edu/Ag/Forage/ForagePublications.htm and variety trial

UK COOPERATIVE EXTENSION SERVICE
University of Kentucky – College of Agriculture

Soil Test Repor

Lexington 859-257-2785
Princeton 270-365-7541
www.rs.uky.edu/soils

COUNTY SAMPLE NO.: 2099	To: Van Meter, Ike 224 Barrow Rd Lexington, KY 40502	*Kimberly D. Poe*
REPORT FORM: A		Kimberly Poe 859-987-1895 *Agricultural Agent*

Date	Owner Sample ID	Owner ID	County Code	UK Lab NO.
4/16/2015	Bad Pasture	69	17	11194

Nutrient	Lab Results	Very Low	Low	Medium	High	Very High	Calculated CEC Data
				Level of Adequacy			
Phosphorus (P)	34	>>>>>>>>>>>>>>>>>>>>>>>>					CEC (meg/100g): 15
Potassium (K)	208	>>>>>>>>>>>>>>>>>>>>>					%Base Sat.: 73
Soil pH	6.4						%K: 2
Buffer pH	7.0						%Ca: 66
Calcium (Ca)	3811						%Mg: 5
Magnesium (Mg)	179	>>>>>>>>>>>>>>>>>>>>>>>>>>>>>>>>					%H: 28
Zinc (Zn)	1.4						
Other Test							

Acres	Primary Crop	Primary Management	Primary Use	Previous Crop	Previous Management	Previous Use	Crop 2 Years Ago Tobacco	Soil Draina
15	Cool Season Grass	Annual Top Dressing	Pasture	Cool Season Grass	Hay or Pasture < 4yrs.	Pasture		Modera Wel

RECOMMENDATIONS:	N	P2O5	K2O	LIME	Mg	Zn
	see comments below	60 lb/ac	40 lb/ac	None	None	None

Apply 100-125 lbs 46% urea per acre

Apply 135 lbs 46% phosphate per acre

Apply 70 lbs 60% potash per acre

COMMENTS:
Mehlich III used for P, K, Ca, Mg, and Zn (lbs/acre). Crop response is highly probable with Very Low or Low soil levels, slight with Medium, and not likely with High or Very High. N, P2O5, K2O, Mg, and Zn recommendations are based on lbs of the nutrient. Fertilizer needed will depend on nutrient content in the fertilizer. Soil pH is calculated from 1 M KCl soil pH using: 0.91 x 1 M KCl soil pH + 1.34. 1 M KCl soil pH and Sikora II buffer pH are used for determining lime needs based on 100% effective lime. Lime quality in KY is defined by relative neutralizing value (RNV). RNVs for ag lime are determined by the KY Dept of Ag and are on the internet (publications at soils.rs.uky.edu).

TETANY PROBLEMS with cattle are sometimes encountered on straight grass pastures, particularly with nursing cows where grass pasture is the only source of feed. SUPPLEMENTAL FEEDING OF MAGNESIUM to nursing cows on such fields is recommended as a means of lowering tetany risks.

SCHEDULE FOR N: Feb 15 to Mar 15 apply up to 100 lb/ac N; May 1 to 15 apply up to 50 lb/ac N; Aug 1 to 15 apply up to 80 lb/ac N. When P and K are adequate (>30 - P and >200 - K) the use of nitrogen stimulates growth during peak production. 200 lbs N/acre may produce up to 4 tons of dry matter.

All the commonly used N sources are acceptable for topdressing grasses during late winter-early spring. After early May urea may be less effective. Efficiency of liquid nitrogen is between that of urea and ammonium nitrate.

Educational programs of the Kentucky Cooperative Extension Service serve all people regardless of race, color, age, sex, religion, disability, or national origin. UNIVERSITY OF KENTUCKY, KENTUCKY STATE UNIVERSITY, U.S. DEPARTMENT OF AGRICULTURE AND KENTUCKY COUNTIES COOPERATING.

From:

Bourbon County Extension Office
603 Millersburg Rd
Paris, KY 40361

U.K **COOPERATIVE EXTENSION SERVICE**
University of Kentucky – College of Agriculture

Lexington 859-257-2785
Princeton 270-365-7541
www.rs.uky.edu/soils

COUNTY SAMPLE NO.: 2100	To:	
	Van Meter, Ike	*Kimberly D. Poe*
REPORT FORM: A	224 Barrow Rd	
	Lexington, KY 40502	Kimberly Poe 859-987-1895
		Agricultural Agent

Date	Owner Sample ID	Owner ID	County Code	UK Lab NO.
4/16/2015	Alfalfa	69	17	11195

Nutrient	Lab Results	Level of Adequacy					Calculated CEC Data	
		Very Low	Low	Medium	High	Very High		
Phosphorus (P)	207	>>					CEC (meg/100g):	14
Potassium (K)	344	>>>>>>>>>>>>>>>>>>>>>>>>>>>>>>>>>>>					%Base Sat.:	71
Soil pH	6.5						%K:	3
Buffer pH	7.0						%Ca:	63
Calcium (Ca)	3554						%Mg:	5
Magnesium (Mg)	166	>>>>>>>>>>>>>>>>>>>>>>>>>>>>>>>>>					%H:	29
Zinc (Zn)	1.9							
Other Test								

Acres	Primary Crop	Primary Management	Primary Use	Previous Crop	Previous Management	Previous Use	Crop 2 Years Ago Tobacco	Soil Draina
15	Alfalfa	Annual Top Dressing	Pasture	Alfalfa	Hay or Pasture < 4yrs.	Pasture		Modera Wet

RECOMMENDATIONS:	N	P2O5	K2O	LIME	Mg	Zn
	None	None	100 lb/ac	see below	None	None

0.5 T/A of 100% effective lime is required. This can be supplied with 1 T/A from Bourbon Limestone (51% RNV), OR 1 T/A from Spartan Products (RNV from spr '14) (52% RNV), OR 1 T/A from Walker Company-Montgomery Stone (45% RNV)

Apply 170 lbs 60% potash per acre

Apply required lime listed above

COMMENTS:

Mehlich III used for P, K, Ca, Mg, and Zn (lbs/acre). Crop response is highly probable with Very Low or Low soil levels, slight with Medium, and not likely with High or Very High. N, P2O5, K2O, Mg, and Zn recommendations are based on lbs of the nutrient. Fertilizer needed will depend on nutrient content in the fertilizer. Soil pH is calculated from 1 M KCl soil pH using: 0.91 x 1 M KCl soil pH + 1.34. 1 M KCl soil pH and Sikora II buffer pH are used for determining lime needs based on 100% effective lime. Lime quality in KY is defined by relative neutralizing value (RNV). RNVs for ag lime are determined by the KY Dept of Ag and are on the internet (publications at soils.rs.uky.edu). Lime is calculated based on quarries in your area.

For alfalfa production, apply 1.5 - 2.0 lb of elemental boron (B) per acre every two years. If fertilizer B or B-containing materials such as solid waste from coal fired power plants has been routinely applied in successive years to established stands of alfalfa, a soil test for B should be run. If hot water extractable levels of B exceed 2 lb/acre, B should not be applied.

Topdress nitrogen applications are not recommended for legume and grass-legume mixtures.

For alfalfa yields above 5 T/acre and red clover yields above 3 T/acre, fields should be sampled every year.

For long-term production of alfalfa and alfalfa-grass mixture, the pH should be about 6.5 - 7.0.

For additional information, ask your local Extension Office for the publicatons, AGR-64 "Establishing Forage Crops"; AGR-76 "Alfalfa, The Queen of Forage Crops"; and AGR-90 "Inoculation of Forage Legumes".

UK Forage publications can be found at the website http://www.uky.edu/Ag/Forage/ForagePublications.htm and variety trial information can be found at http://www.uky.edu/Ag/Forage/ForageVarietyTrials2.htm

excesses in soils is by observing which weeds grow in your pastures; certain weeds cope better with certain fertilizer deficiencies (see chapter 12 for more information). Following is a list of nutrients and the most obvious signs of their deficiency.

Nitrogen

A nitrogen deficiency appears as yellowing or browning of the plant leaves starting at the tips of the oldest leaves. The discoloration moves inward from the tip to form a yellowing V up the leaf (see illustration at right).

Calcium

A calcium deficiency in plants is usually not obvious because the high acidity of the soil limits plant growth before the gross deficiency is apparent. If, however, you find buried plant residues that are two or more years old, then you know you have a calcium deficiency, which reduces the rate of decomposition in the soil. A calcium shortage is noticeable when the soil becomes hard and tight. An excess makes the soil loose and prone to erosion. Remember that calcium and magnesium usually work together, so a deficiency in one often creates an excess in the other.

Magnesium

A magnesium excess shows itself by causing the soil to become hard and the soil to crack when dry. The wider and deeper the cracks, the greater the magnesium excess (or calcium deficiency). A magnesium excess also makes the soil sticky; large clumps of wet dirt accumulate under your boots and kick off only with effort. The thicker the sticky layer, the worse the magnesium excess.

Aerators and other tillage equipment are often advertised as the tools best designed to address compacted soils, but if calcium levels are rebalanced by liming to drive out the magnesium excess, the soil compaction will remedy itself without mechanical intervention.

A magnesium deficiency typically creates loose soil and most often occurs in coarse-textured, sandy soil. Magnesium-deficient plants show whitening between leaf veins and a purplish tint on the underside of leaves. Be cautious, however: A magnesium

A nitrogen deficiency in plants causes a V-shaped yellowing or browning at the tips of the oldest leaves.

deficiency in grass can actually be due to a magnesium *excess* in the soil; insufficient calcium is present to help the plant absorb the magnesium and other soil nutrients. Remember that the absorption of all other nutrients by plants is determined by the base saturation of calcium in the soil. Too much or too little magnesium reduces microbial activity in the soil and prevents magnesium uptake by the plants.

Low magnesium concentrations in the blood resulting from a magnesium deficiency in grass can cause grass tetany, also known as hypomagnesemia, in cattle. Symptoms include low appetite; dull demeanor; staggering, stiff gait; irritability; excitability; nervousness; muscular tremors; falling over; spasms; and death. It is most common in lactating cows grazing lush early-spring pastures because lactating cattle require high magnesium and calcium and early-spring grass growth is low in magnesium. (Cool, spring temperatures cause the plants to preferentially absorb potassium, rather than magnesium, from the soil.) Calving in late spring or early summer greatly reduces the risk of grass tetany because the high demands of lactation begin long after magnesium uptake into plants is corrected by a return to warm weather.

Phosphorus

Phosphorus is the workhorse responsible for cell division and cell enlargement. A deficiency, then, is seen as slow seedling growth, slow maturity, and poor seed formation. It will also appear as a slight stunting and darkening of the foliage or as a purple discoloration on the leaves of young plants.

Potassium

As in a nitrogen deficiency, a potassium deficiency first affects the oldest leaves of a plant, causing yellowing or browning, but in this case the discoloration shows up along the outside edges of the leaves and moves inward (see the illustration below). A potassium shortage also causes blockages or thickening at the nodes of grass stalks (the slightly enlarged portions of the grass stalk that bear new leaves).

Sulfur

A sulfur deficiency manifests much like a nitrogen deficiency, only the light green or yellow discolorations appear first on the newest leaves, as opposed to the oldest ones.

Boron

The main functions of boron relate to cell wall strength and development, cell division, seed development, and sugar transport. Consequently, a boron deficiency affects young leaves first and results in

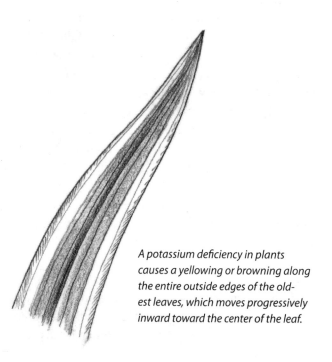

A potassium deficiency in plants causes a yellowing or browning along the entire outside edges of the oldest leaves, which moves progressively inward toward the center of the leaf.

stunted growth, slow seed development, and poor quality seeds. Young leaves may become misshapen, thick, brittle, and small, while older leaves continue to appear healthy. Dark brown, irregular discolorations may form on a new leaf, followed by a yellowing of the entire leaf.

Copper

Copper is responsible for stalk strength and plays an essential role in photosynthesis and seed production. When copper is deficient, leaf tips whiten, become twisted, and die. Young leaves are affected first: They may not unfold properly, curling into a tube, or they may become limp and discolored. A general lightening of the color of the grass crop, particularly of emerging leaves, may be visible, while the oldest leaves remain dark in color.

Iron

Iron is required for chlorophyll formation and electron transport during photosynthesis and is essential to many enzymes involved in metabolism and transpiration. Insufficient iron causes a yellowing among the veins of the youngest leaves; the veins themselves remain green. There is usually a sharp distinction between the yellowing and the dark green veins, which sets it apart from zinc or manganese deficiency, in which color change is gradual. Leaf tips may become dry and brittle.

Manganese

Manganese is involved in the production of chlorophyll and activates enzymes to assist photosynthesis. A manganese deficiency reveals itself as yellowing among the veins of the youngest leaves while the veins themselves remain dark green, but with a gradual distinction between the vein boundaries rather than the sharp distinction seen in an iron deficiency. White or gray specks may appear on older leaves partway up the stalk, spread to the oldest leaves, and then finally affect the youngest leaves.

Molybdenum

Molybdenum is required by certain enzymes involved in plant growth. It is also required for nitrogen conversion in plants and especially for nitrogen fixation in legumes; thus, a molybdenum deficiency is easily confused with a nitrogen deficiency. A molybdenum deficiency causes leaves to become narrow and twisted (like a whip), affecting the oldest leaves first. The oldest leaves also turn yellow among the veins and the leaf edges look scorched. Plant growth is usually severely stunted.

Zinc

Zinc is involved in photosynthesis, sugar formation, protein synthesis, fertility, growth regulation, and disease resistance. A zinc deficiency causes white striping along the youngest leaves of plants, a bleaching of the leaf tips, and a yellowing of the entire base of the leaves nearest the stalk. Leaf margins look puckered. Leaves may appear to be growing too close together due to a stunted stalk and leaves themselves may look stunted and narrow.

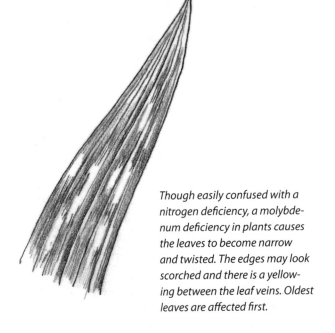

Though easily confused with a nitrogen deficiency, a molybdenum deficiency in plants causes the leaves to become narrow and twisted. The edges may look scorched and there is a yellowing between the leaf veins. Oldest leaves are affected first.

Chapter

12

Weeds

Specific plant species thrive in specific environments not because they perform best, but rather because other plants are not as competitive in those environments. This fact allows us to control undesirable plant species in our pastures by influencing their conditions so that only the plant species we want to keep remain competitive. Healthy grass, weeds, trees, shrubs, and herbs can be individually cultivated in the same balanced soil, but they won't all succeed if they are grown together to compete for the same space. The dominant plant species will be the plant that is most competitive in the specific combination of soil nutrient challenges, climatic restrictions, moisture conditions, land management, herd impact, and grazing pressure.

What Is a Weed?

What exactly is a weed? Many books seem to divide all plants into weed and nonweed species, which is terribly misleading. A weed is simply a plant that is not useful to a specific human activity. The first time I opened a book on medicinal herbs, many of the plants considered "weeds" from an agriculture standpoint shone like brilliant stars above many of the "valuable" plants. And the definition becomes even more confused: Many livestock production plants (for example, grasses and alfalfa) are considered terrible weeds if they invade a field crop. Likewise, when desirable crop plants such as berry bushes, asparagus, and medicinal St.-John's-wort invade our pasture

grasses, we classify them as weeds. Weeds, therefore, are simply plants that are undesirable given our production goals for at particular piece of land.

A cow does not distinguish between weeds and nonweeds. According to her palate, plants are simply ranked from most to least desirable on any given day. In fact, what may be virtually inedible to her on one day may be the first thing she eats the next day in response to a different craving or nutrient deficiency. This, of course, leads to a contradiction of our crop-oriented perception of what an ideal pasture should look like.

Variety Is All

We have become accustomed to wanting a very limited plant community in our fields in order to maximize hay production volumes and meet crop harvest dates. But when we enter the grazing world of the cow, we need to set aside this prejudice. The greater the variety of plants in our pasture, the more likely the cow is to meet all of her nutritional needs throughout the year. In addition, greater plant variety makes the soil healthier and the plants more resistant to pests, parasites, and disease. A variety of plants with different root zones will draw nutrients from a greater distribution of depths within the soil, which helps to spread out the nutrient drain on the soil and increases the likelihood that all of the plants will be able to supply themselves with adequate nutrients without the addition of fertilizer.

*The greater the variety of plants
we can encourage, the more likely the cow
is to meet all of her nutritional
needs throughout the year.*

If you look closely at the seed bank in the soil, there is an astounding variety and abundance of seeds represented, yet most lie dormant, waiting to germinate at some future time. The amount of seed transported by wind and animals is also impressive, despite the fact that most of the newly introduced seeds do not germinate. Each of these dormant seeds just lies in wait until the dominant plant spe-cies becomes less competitive and it can burst forth. The presence of the seeds does not guarantee a weed outbreak; certain conditions must exist for them to take hold, and those conditions are different for each plant species.

When and Why Certain Plants Emerge

The presence of some plants in a pasture indi-cates management problems such as too much soil compaction, a slow herd rotation that makes the pasture grass varieties less competitive, and not enough rest for the dominant grass spe-cies between grazing intervals. Other plants are barometers of specific soil fertility issues that limit

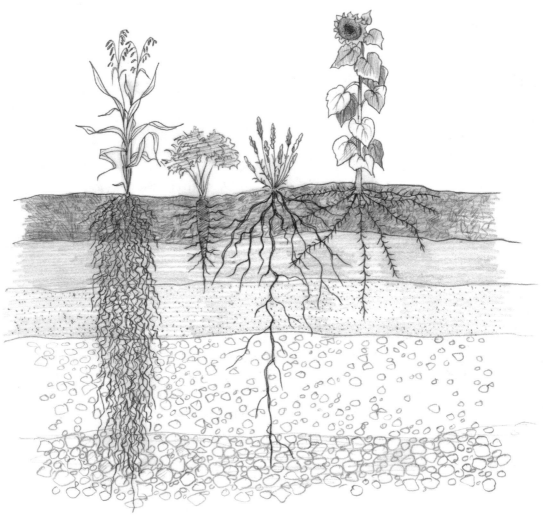

A variety of plants in a pasture, each with different root depths, will draw upon nutrients from a wide range of soil depths, resulting in improved plant and animal health and a reduced dependency on fertilizers to support healthy pasture productivity.

the competitiveness of desirable plants and allow weeds to find a niche in which to thrive. The drop in competitiveness of the grasses can be caused by soil nutrient imbalances or deficiencies, water shortages or excesses, weather, drought, and other conditions. Each plant species has evolved to withstand a different environmental, soil, or climatic challenge so it can take over when that challenge reduces the competitiveness of species that cannot cope with those conditions.

But nature's designs do not stop there. When a weed emerges in response to changes in soil fertility, its presence will, over time, correct the nutrient deficiency — which will cause it to lose its competitive edge. The species that takes its place is determined by the management practices used after the deficiency is corrected.

The roots of weeds can loosen compacted soils for other plants. Weeds are nature's mechanism for correcting nutrient imbalances and soil problems.

For example, if your soil is deficient in calcium, only plants that are capable of overcoming the deficiency will survive, by continuing to meet their calcium needs through excreting specialized root acids that dissolve calcium from larger soil particles. Or the plants may cope by extending their roots deeper into the soil, below the root zones of other plants, to access minerals the other plants cannot reach. After the specialized plants die and are reincorporated in the soil, the new nutrients released from inaccessible parts of the soil are added to the nutrient bank near the soil surface, where all plants can access them.

Beyond coping with deficiency, weeds can also remove nutrient excesses from the soil, allowing other plants to return after the excess is gone. In addition, the roots of weeds can loosen compacted soils for other plants. Weeds are nature's mechanism for correcting nutrient imbalances and soil problems. Nature's security lies in the enormous variety of dormant weed seeds waiting for an opportunity to gain a competitive edge, an army of specialized

weeds ready to combat any environmental, nutrient, or land-management challenge as soon as it occurs.

The plants you see in your pastures merely reflect existing pasture conditions. When weeds move in, they are telling you that something in your system — your management, soil fertility, or something else — is making the grass less competitive. The changing plant community points out that the plant species currently in the pasture are not in balance with the current nutrients, land management, climate, or herd impact. In many cases, the weeds reveal the problem if you are able to identify the conditions they have evolved to withstand or the purpose they have evolved to fulfill.

For example, if sage, knapweed, and woody plants such as willow and poplar invade your pastures, it's a sign that the grasses are not able to compete with them. These plants do not handle repeated trampling and their seedlings will not be competitive in a healthy, fast-growing grass sod. Their presence, therefore, indicates that your pasture management is not keeping the pasture grasses competitive. Livestock may be overgrazing the grass, it may not be getting sufficient rest between grazes, or undergrazing may be making it less competitive. All are signs of a poorly designed herd rotation.

Alternatively, the livestock density may be too low to trample evenly all the land, allowing the weed seedlings to survive. Poor herd impact may result from insufficient animals in the herd, not enough bunching of livestock, and too-large pasture divisions. The invading weed or woody species are telling you that the livestock migrations are not being managed efficiently enough to prevent the invasion. In recognizing this, the solution to the weed problem is revealed: a simple adjustment of your livestock rotation and herd-management practices.

Quick Fixes Versus Long-Term Solutions

I could cite myriad examples, yet each would indicate the same problem: The plants you want in your pasture are not supported by your current herd management. The solution, too, would be the same for each: Fix your rotation. Even soil nutrient deficiencies will slowly correct themselves over time as dead root and plant material is recycled by the herd impact. If, however, you determine that a weed problem is caused by a soil imbalance or nutrient deficiency, you can help speed the soil's recovery by using fertilizer amendments. Fertilizer is just a tool to speed up the natural soil recovery process. A correctly applied herd impact would achieve the same end, only more slowly. Remember that the fertilizer is only a tool; once the problem is corrected, the soil should be maintained by herd and land management.

Other tools to speed up the natural process of regaining a grass prairie in your pastures include mowing, spraying herbicides, and burning woody and dead plant material that chokes the grass. But note that unless these methods are followed by a management system that keeps your desirable plant species competitive, you will have to repeat them at considerable financial and environmental costs. Most of these methods, though they may be suitable short-term fixes, will be very damaging to your pastures if used repeatedly. Synthetic fertilizer will eventually leach nutrients and harm soil organisms. Burning may add calcium (from the ash), but it also reduces the organic matter and humus in the soil and makes the soil susceptible to erosion, carbon dioxide loss, and nitrogen degasing. Mowing eventually compacts soil in the tractor ruts, requires expensive fuel, and does not recycle nutrients with the same efficiency of manure, which decomposes after grazing. Herbicides are hard on soil organisms as well as on the health of animals and desirable pasture plants. These methods should be used only to get the plant community back on track so you can maintain it by properly managing your herd.

Educate Yourself, Then Put Your Cattle to Work

When you battle weeds, it is important to know something about the plants you are trying to eliminate: In what soil conditions do they thrive? What grazing habitat do they prefer? How will grazing affect the weeds during their various growth stages? This knowledge is vitally important; it can allow you to use grazing pressure to target weeds when they are most vulnerable and avoid grazing when it might further promote their competitiveness.

Annual weeds are most vulnerable when grazed *at first flower,* when all their nutrients and energy reserves are redirected from the roots and stalk to seed formation. Perennial weeds have the lowest reserves when they are grazed *after full leaf, but just before first flower formation.*

Use your cattle to graze weed reservoirs that you would otherwise have to mow, spray, burn, or turn a blind eye to: areas alongside driveways, in machine yards, and other out-of-the-way places that are prone to weed buildup. The beauty of working with electric fences is that these weed reservoirs can be grazed with minimal infrastructure and time investment.

Managing Inedible Weeds

Inedible weed species can be controlled by trampling and by managing the grass species so that it out-competes the weeds. If the grass is kept competitive, eventually the advancing grass sod will choke out weeds and the tall-grass residual will shade them, robbing them of important sunlight. It's important that the grass residual is never grazed lower than 6 to 8 inches during the growing season so the grass can continue growing quickly and the grass sod can out-compete surrounding weeds. If grasses are grazed shorter than weeds, then the weeds will rob the grasses of sunlight and maintain the competitive advantage. In cases of severe weed infestations, mowing weeds after grazing will give the grass a competitive advantage; it will grow over the tops of the weeds and choke them out.

Weeds to Remedy Nutrient Deficiencies

Pastures are not grain fields, hay acreages, or soybean plantations, where purity is half the measure of market price. You don't need to eliminate every last weed. Only if they become significant enough in number to lower pasture production do weeds become a problem. In many cases, small numbers of weeds can actually be beneficial: They allow cattle, using their taste or smell, to single out specific plants that address certain nutrient deficiencies in their diet. Although these weeds will not be the mainstay of their diet, they can provide important mineral supplements.

THISTLES: A CASE STUDY

Here is a wonderful example of how to use rotation to control a wicked weed infestation by thistles such as bull thistle (*Cirsium vulgare*) and Canada thistle (*Cirsium arvense*). Very young thistles do not stand up to trampling, but once they begin to grow, their height enables them to stand well above the surrounding grass to make use of the sunlight. Large patches of thistles develop as they rob sunshine from the shorter grasses. If they are mowed or clipped during their growth stage, they will regrow quickly.

If the season determines that it is time to begin seed formation, the thistles are even able to create seed heads close to the ground before they have regrown to their full size. This makes them particularly resistant and allows them to persist in poorly managed pastures, especially those that are consistently overgrazed, where the grasses are weak and short, and the thistles — avoided by animals because they are prickly — quickly overshadow the surrounding grass and take over the pasture.

On the other hand, if you use a well-timed, healthy pasture rotation, trampling will keep the thistles short while you manage the grass to remain tall and overshadow the short thistles. The thistles will be choked for sunlight while the competitive grass sod takes away their ground.

Over time the problem will resolve itself. If you want to enlist mechanical help, thistles are most vulnerable to mowing during their flowering stage (the prettier the flower, the more vulnerable the thistle), but only if they are mowed so high that just the flowers are clipped while the rest of the thistle remains intact. Once they begin to flower, all their energy reserves are invested in seed formation. When only the flowering tops are gone, the plant must continue to support the remaining plant mass after losing all the valuable energy reserves that it stored in the seed head. This proves to be too much, and the thistle will be severely weakened in its ability to regrow a seed head, or it may even die.

By contrast, removing the whole plant allows the thistle to regrow because only a small remaining plant mass must be supported by the limited energy reserves remaining in the roots. Thistles are also a sign of either too much or too little phosphate, so fertilizer amendments based on a soil test will help eliminate the thistle's competitive advantage over the grass.

Before trampling, bull thistle towers over the poorly managed, uncompetitive surrounding sod, blocking sunlight from reaching the grass.

After trampling, tall grass overshadows and out-competes short bull thistles.

Similarly, don't forget that many weeds have medicinal value. Cattle have neither our prejudices nor our knowledge about plant phytotherapy, but if presented with a smorgasbord, they are remarkably adept at selecting the right plants to eat to address their health needs.

Dandelions

If dandelions *(Taraxacum officinale)* make a pasture look like a yellow carpet when they start to bloom, this is a sure sign of a weed infestation. They are *allelopathic,* meaning that they emit their own herbicide to stunt the growth of other plants around them, eliminating the possibility that these plants will rob them of sunlight. Unless the dandelions take over the entire field, however, their impact on grass productivity is minimal.

On the positive side, dandelions have long taproots that allow them to bring nutrients from deep in the subsoil to the surface, where shallow-rooted plants can access them after the dandelions die. Interestingly, cattle seem to love them; I have seen cattle consume every dandelion in a newly seeded pasture before touching many of the more "desirable" grasses in the field. This may be because they are extremely high in nutrients and vitamins and have liver-cleansing properties, to name just a few of their positive attributes. With repeated grazing timed to occur just before the flowers go to seed, even the worst dandelion infestations can be virtually eliminated.

Leaves from Young Poplar Trees

In the Yukon, for several weeks every spring the cattle ignore the new flush of grass and go absolutely crazy over the leaves of young poplar trees *(Populus* spp.). They continue to eat them periodically for a good portion of the summer, systematically stripping the branches and even knocking down the 4- and 5-inch-diameter trees so they can completely defoliate them. Certainly it is a convenient additional food source (and provides free logging), but the poplar leaves are also a detoxifier (a substance used to remove toxins from the body), so it is no coincidence that cattle are particularly interested in eating them after a long winter of consuming old freeze-dried grass.

DANDELION IN BLOOM

LEAVES OF YOUNG POPLAR TREE

Other Beneficial Weeds

Plantain *(Plantago lanceolata),* also known as ribwort, is another common weed whose herbal benefits include antiseptic properties (aiding lung, stomach, and even urinary infections) and soothing irritated lungs and upset stomachs. German chamomile *(Matricaria recutita)* has relaxing, healing, and antiallergenic properties and yarrow *(Achillea millefolium)* has a capacity for healing wounds (including internal bleeding) and stimulates digestion and circulation. Burdock *(Arctium lappa)* has antibiotic and detoxifying qualities, and curled dock *(Rumex crispus)* has the ability to extract high amounts of iron from the soil, thereby enhancing red blood cell production and providing a natural source of iron. (See page 161 for a list of plants with natural worming properties.)

There are far more "weeds" with medicinal properties than those mentioned here and many of them concentrate specific minerals and vitamins that are beneficial to our livestock. In fact, it is often the weed's natural ability to compensate for specific soil nutrient deficiencies that enables it to concentrate certain minerals, vitamins, and medicinal properties that are not available from grass alone. In small quantities, weeds provide cattle with access to a natural pharmacy. A healthy rotation will ensure that while they don't take over the cattle's dinner table to serve as the main course, weeds can be an optional hors d'oeuvre.

RIBWORT/PLANTAIN

SIGNS OF SOIL DEFICIENCIES

Following are a few of the soil deficiencies that your weeds may be telling you about. Familiarize yourself with them to get a sense of the advantages of using weeds as soil fertility indicators. The soil deficiency or imbalance a weed indicates provides a clue to which fertilizer amendment may be needed to resolve it.

Have fun exploring the fascinating world of weeds. They open an intriguing window into the soil and are a wonderful source of information to guide your pasture management strategy. Observing and then manipulating what cattle do with them can give you rapid feedback on the relative success of your grazing management strategy.

■ Redroot weeds (plants in the Amaranthaceae family) such as redroot pigweed *(Amaranthus retroflexus)* are a sign that the iron–manganese ratio is out of balance: There's either far too much iron or far too little manganese. Redroot weeds also indicate a soil that is excessively high in potassium and magnesium and low in phosphorus and calcium.

REDROOT PIGWEED

- Quackgrass *(Elytrigia repens)* is also a sign of an improper iron–manganese ratio. It makes an excellent pasture grass, so grazing and manure recycling will help correct this fertilizer imbalance over time.

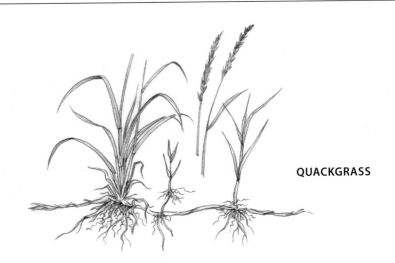

QUACKGRASS

BITTERWEED

- A whole host of plants indicate a calcium deficiency in the soil. They include bitterweed *(Helenium tenuifolium)*, stinging nettle *(Urtica dioica)*, broom sedge *(Andropogon virginicus)*, horsetail *(Equisetum arvense)*, trumpet vine *(Campsis radicans)*, and wild buckwheat *(Polygonum convolvulus)*. Because many of these weeds concentrate calcium, they are excellent sources of calcium for the cattle if they choose to eat them. Wild buckwheat also indicates low phosphorus and an excess of potassium.

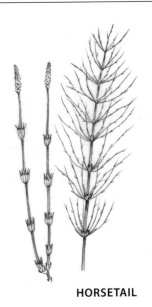

HORSETAIL

STINGING NETTLE

BROOM SEDGE

TRUMPET VINE

WILD BUCKWHEAT

- Burdock indicates low-calcium, high-potassium soils.

BURDOCK

- Yarrow likes low-calcium, low-phosphorus, low-humus soils.

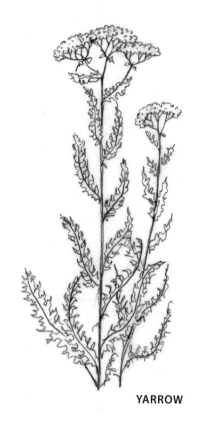

YARROW

KNAPWEED

- Knapweed *(Centaurea maculosa)* grows in soils that are low in calcium, low in humus, and very low in phosphorus.

- Oxeye daisy *(Chrysanthemum leucanthemum)* grows in low-phosphorus, high-potassium, high-magnesium soils.

OXEYE DAISY

- Foxtail barley *(Hordeum jubatum)* is characteristic of low-calcium, high-magnesium, compacted, poorly drained soils.

■ Broadleaf plantain *(Plantago major)* grows in compacted, poorly aerated soils.

BROADLEAF PLANTAIN

■ Lamb's-quarter *(Chenopodium album)* grows in low-phosphorus, high-potassium soils. Because cattle manure is quite high in potassium, these plants are often found in areas where manure concentrations have accumulated.

LAMB'S-QUARTER

■ Curled or sour dock *(Rumex crispus)* is another plant that loves compacted soils and is characteristic of soils that are low in calcium and extremely high in magnesium, phosphorus, and potassium.

CURLED DOCK

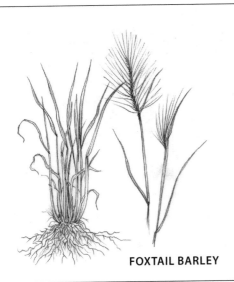

FOXTAIL BARLEY

ST.-JOHN'S-WORT

■ St.-John's-wort *(Hypericum perforatum)* favors soils that are dry, calcium deficient, phosphorus deficient, magnesium rich, potassium rich, and salty. Also be aware that the plant can cause extreme photosensitivity in cattle, leading to irritability; severe sunburn to nose, skin, and udder; and even death.

Soil Moisture and Irrigation

IN THE WORLD OF AGRICULTURE, WATER IS never far from our minds. As we wonder when the next rains will come or when they will finally end, one eye is always on the weather. We tend to feel that moisture is something over which we have no control. Is irrigation our only defense against water shortages in the soil, and can water excesses be controlled only with drainage ditches, runoff gullies, and sandbags? Are we really that helpless against the weather?

Utilizing cattle to manage stream- and riverbanks helps to prevent erosion.

When we look at what happens to water that falls on our land in the form of rain, snow, snowmelt, and irrigation water, the story is discouraging. On average, a full 66 percent of this moisture evaporates into the atmosphere and 25 percent of it disappears as runoff. That leaves only 9 percent, a remarkably low amount, to filter into the soil. If we can reduce evaporation and runoff losses and increase water infiltration by just a small amount, we should be able to dramatically reduce water shortages and water excesses.

Before resorting to artificial means to compensate for water shortages, droughts, declining water tables, and depleted groundwater levels or to slow erosion and flooding, we must examine how we can improve the water cycle in our soils. Only after we

have done all we can to increase water infiltration and reduce runoff and evaporation should we consider using technological methods to make up the difference. If we do our job right, these other methods will play only a minimal, peripheral role in our agricultural production, if they are needed at all.

If moisture from rainfall (topmost dark band) fails to soak through the dry layer of dust at the soil's surface (light band) to connect with the soil moisture layer below (lower dark band), the moisture will disappear through evaporation (arrows) without benefiting plants or the soil.

Decreasing Evaporation

Here's a familiar scenario: After it rains, you go out into the field to see what effect it had. The humidity is stifling; evaporation is clearly occurring full force. Everything looks thoroughly soaked, yet when you kick at the dirt, you can still find dust under the thin, wet, muddy top layer. When you dig down into the soil, you see that the rainwater has penetrated only about two inches, under which is a bone-dry layer of dust, then the soil moisture zone (see illustration on page 192). As the day progresses, the soil surface dries out until the two-inch layer on top dries completely without its moisture ever reaching the lower soil moisture. The rainstorm accomplished nothing.

This fast rate of evaporation can be slowed in a number of ways. When a healthy herd rotation leaves a tall-grass residual and a good soil cover of live and dead plant residues, the soil is shaded from the sun's direct rays, the tall plants help to block the wind, and the plant and residue cover helps to cool the soil and air just above the soil surface. The most significant role in decreasing evaporation is played by increasing moisture infiltration into the protective environment of the soil. This begins by reducing runoff. Learning how to slow the movement of water across the soil's surface after a rain gives the raindrops more time to soak into the soil and contribute to the soil moisture zone instead of being lost to lakes, streams, and rivers.

Reducing Runoff

Imagine exposed, dry soil that is sloped just enough to avoid standing water. With nothing to hold the water droplets in place as they accumulate, the raindrops turn to runoff, moving across the soil surface faster than they can be absorbed. Before long they have joined streams and rivers along with all the soil debris that is carried along by the fast-moving water. The goals are to give the water time to soak into the ground and to prepare the soil to quickly absorb moisture before it becomes runoff, flowing away over the parched soil surface.

Again, a healthy herd rotation leaves behind a tall-grass residual and other plant residues that cover the soil and slow the flow of water, giving it more

time to soak in and hold soil particles in place and protect them with sod, thus preventing soil erosion. Utilizing cattle to manage stream- and riverbanks helps to prevent erosion by covering the banks with sod and plant debris, which slows the stream's flow so more water soaks into the ground, contributing to the groundwater table. But along with establishing a healthy herd rotation, we must prepare the soil to increase water infiltration.

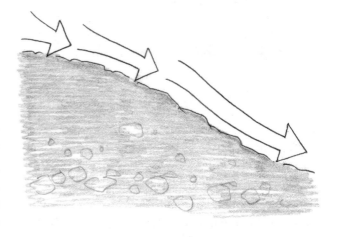

Vegetation and plant debris on a slope (top) slow runoff and protect the soil from erosion so that rainfall can be absorbed. Without vegetation (bottom), runoff heads down the slope and collects in puddles, ruts, or pools without benefiting the sloped land and carries away fine soil particles, leading to erosion.

Increasing Water Infiltration

Even if we slow evaporation of rainfall and decrease runoff so that moisture has more time to soak in after a rain, unless the ground is prepared to absorb moisture, water absorption will be negligible. Again, livestock play an important role in this ground preparation.

Soil Moisture

An exposed soil surface will dry out completely, whereas a soil surface covered with tall grass and debris will be shaded and will remain moist far longer. In dry, parched soil raindrops cling together on the surface, unable to soak through the air-filled dust layer below. A slightly moist soil, by contrast, absorbs water much more quickly. It follows that preventing soil from drying out helps to improve water absorption. When the surface moisture of soil is protected by a covering, the ground is primed to absorb a greater volume of rainwater through capillary action. The soil does not have to wait to be moistened first before it can absorb moisture. *Capillary action* is the movement of water within the spaces of a porous material such as soil by means of cohesive intermolecular forces between water molecules and adhesive intermolecular forces between water and other substances. In other words, water molecules stick to each other and to other substances, such as glass, cloth, organic tissue, and soil.

Preventing soil from drying out helps to improve water absorption.

Referring to the illustration on page 192, if the soil moisture reached to the soil surface, to the base of a thick grass sod, a rainstorm's water that would completely evaporate after a brief time in the two top inches of parched, bare soil would instead provide its long-term benefits by soaking in to the earth. The moisture already existing in the covered soil functions as a wick that draws added moisture down into the soil.

Spongelike Soil

As the fertility of the soil increases and particularly when any imbalance in the soil's calcium–magnesium ratio is corrected (see chapter 12), the structure of the soil becomes more spongelike instead of hard and firm like the soils of many overgrazed and mismanaged pastures. The roots of an active grass sod play a role in creating this texture by helping to break apart the soil structure.

In addition, healthy pasture management promotes an abundance of soil organisms that further break up the soil, creating pathways through which water can percolate. Earthworm trails, tracks of burrowing insects, spongelike soil bacterial growths, and even mole tunnels are all beneficial.

The dieback of roots in response to grazing also creates pathways for water, while the residual dead organic material wicks the moisture down from the surface of the soil. This healthy, spongelike soil created by careful grazing management has a great capacity for drawing in and storing moisture and shielding it from the atmosphere's drying conditions.

Herd Impact

In addition to slowing down evaporation and runoff through grazing, the herd tramples the soil, breaking up its crust so that rainwater can be absorbed. A crusted soil is resistant to moisture absorption and promotes further runoff, even if the soil underneath is loose and moist. This crust forms in response to rapid wetting and drying, nutrient leaching, sun baking, pounding of water droplets on the surface of the soil, and wind polishing, during which small soil and humus particles are blown away while remaining soil particles are realigned, thereby

ALL IN MODERATION

Cattle have a positive effect as long as they don't overstay their welcome and begin to compact the soil instead of loosening it. Too much of anything is not a good thing. Again, dutiful pasture management is key.

When soil lacks the protective cover of vegetation, rain (or irrigation water) pounds its surface, dislodging small particles and washing them away, compacting the uppermost topsoil, and realigning surface soil particles so that the soil hardens and crusts.

reducing soil porosity and causing a textural change to the soil surface.

It is quite remarkable how much instantaneous pressure a tiny raindrop or water drop from an irrigation source exerts on the microscopic soil particles it impacts. Raindrops pounding the soil surface resemble miniature meteorites crashing into the earth. These pounding droplets compact the uppermost zone of the soil and, on a microscopic scale, force the tiny clay, silt, sand, and humus particles to align themselves in response to the impact pressure. And as rainwater leaches nutrients and humus from the soil surface into its depths, the upper soil becomes depleted, causing it to crust (further, if calcium leaches down, the upper layer becomes magnesium enriched, which causes it to harden). This hard crust, like a shell, is somewhat impervious to water absorption and prone to surface runoff.

A thick grass and residue cover on the soil surface reduces the force of the falling water droplets, and the cattle's trampling hooves break up the soil's surface, preparing the soil for the next rainfall.

Take Responsibility for Water

Much of the agricultural world complains of droughts, erosion, diminishing water tables, depleted groundwater, flooding, water-use restrictions, and high water-use costs. But should these conditions really surprise us?

Consider how dramatically a mismanaged pasture rotation can decrease water absorption and increase runoff and evaporation rates. Add to this the vast acreages of tillage agriculture that are particularly prone to these problems because they lack a thick soil cover of grass and residue, don't enjoy the benefits of livestock grazing, and have exposed sun-baked soils, among other challenges. Many of these problems are our own doing or are at least greatly magnified by our management influence.

If we stop blaming fate, bad weather, climate change, and bad luck and recognize our role in creating the problem, then we are no longer helpless and can take action. By simply managing our land to make greater use of each raindrop that falls, we can easily double or even triple the effectiveness of every storm. Capturing a conservative extra 4½ percent of the water normally lost to evaporation and runoff by altering our pasture management and adding it to the meager 9 percent absorbed by the soil translates into a 50 percent increase in water infiltration. How much irrigation water would it take to achieve the same effect, considering that irrigation water is lost to evaporation and runoff at the same rate as, or at an even higher rate than, the average rainfall? If you compensate for water shortages with technological quick fixes such as irrigation systems, soil aerators, fertilizer dependency, equipment operation, maintenance costs, and a great deal of daily labor, what is the cost?

The Disadvantages of Irrigation

Irrigation is one of the most misused tools in the agricultural world, creating even more problems than it alleviates. Our production-oriented mind-set (crops and livestock are strictly managed to maximize yields, which translate into tons of forage or grain to the acre or pounds of beef per cow) often overshadows the hidden costs of irrigation, which exceed the high costs of equipment investment, repair, maintenance, labor, water, and pumping, as well as the opportunity cost of an investment that could otherwise be spent on infrastructure development, cattle purchases, and land rentals. In the end, though volume per acre with irrigation may look great, net income per acre suffers dramatically. And unfortunately, irrigation, much like fertilizer application, creates a vicious cycle of dependence.

Irrigation is one of the most misused tools in the agricultural world.

An irrigated field is an artificial environment. Because water is supplied artificially, the plants come to depend on it: Drought-tolerant plant species no longer prevail, but are replaced by moisture-loving species that adapt to the irrigated environment. Because water is continually supplied at the soil's surface, the grass roots are not encouraged to burrow deep for water, and without further irrigation the shallow roots are no longer able to meet the plants' needs.

Many irrigation methods are quite damaging to the soil; unless they are carefully dispersed, the large water droplets hammer the soil surface, leading to soil crusting and compaction. Delivering too much water at once due to inappropriately sized water nozzles or insufficient sprinkler moves leaches nutrients and causes small-scale flooding and runoff that further harm the soil. If the irrigation water does not rotate around the land quickly enough, the shallow-rooted plants will be starved for water, causing wilting and unnecessary stress to the plants, which in turn affects production and pasture health.

It is tricky to calculate the amount of water required and how often it should be delivered because watering needs vary from one soil type to the next and from day to day based on evaporation rates. Irrigation-dependant crops and soil are not very tolerant of overwatering.

Even the water itself can be a problem. Unlike rainwater, most irrigation water sources contain nutrients or salts or both that affect soil fertility and build up salt in the soil. Some water jurisdictions that supply municipal water for agricultural purposes chlorinate the water to avoid separating residential and agricultural supply pipelines. Chlorine is a powerful disinfectant that kills any bacteria in the water, but it also attacks the soil organisms, particularly the bacteria and other soil microbes responsible for organic matter decomposition and nutrient recycling, thereby further reducing the soil's vitality. The chlorine in the water may also build up additional salts as they bond with nutrients in the soil (sodium, potassium, calcium, magnesium).

Does this mean that irrigation is bad? Like most technology available to us, it is merely a tool and, like any tool, must be used judiciously with a great deal of awareness about its effects in order to avoid creating unintentional negative consequences. Using it is certainly not as easy as simply turning on the water, eyeballing how much you need, and deciding all is well. Clearly, maximizing your rainfall and soil moisture through good pasture and soil management practices must come first, before irrigation is even considered. If you choose to use irrigation as a tool to manage your pasture production, you must learn how to calculate your irrigation needs.

Calculating Your Irrigation Needs

Irrigation is measured as inches of water applied to and stored in the soil. How much water you can supply before damaging your soil depends on the soil type, its moisture storage capacity, and even the effective crop root depth. How often you need to supply water depends on the evapotranspiration rate, which is the combined process of plant transpiration and moisture returning to the atmosphere through evaporation. A discussion of how to calculate evapotranspiration rates follows on page 199.

The soil zone extending from the surface to the base of the effective root depth is called the *moisture storage zone*. Once this zone is depleted of its available water, crop wilting (drought stress) begins. On average, the effective root depth of most pasture grass is only 1½ feet, though alfalfa's effective root depth extends to 4 feet, giving it a larger soil moisture storage zone from which to draw. If, however, the soil is compacted, drains quickly (as is the case with a highly porous, coarse-grained soil such as sand), or has fertility problems, or if the subsoil drains easily or is excessively coarse (gravel), then the effective root depth may be even less. The chart at right shows the amount of water that can be stored in the moisture storage zone in various kinds of soil.

The amount of water available from the moisture storage zone also depends on the crop. Only 50 percent of the water in the moisture storage zone is available to most plant species, though many garden crops use even less. If you want to replenish the moisture storage zone in a silt loam in which pasture grasses are growing, for example, use the following calculation:

SOIL MOISTURE STORAGE CAPACITY BY SOIL TYPE

Soil type	Inches of available water per foot of soil (in/ft)
Sand	1.0
Loamy sand	1.2
Sandy loam	.5
Fine sandy loam	1.7
Loam	2.1
Silt loam	2.5
Clay loam	2.4
Clay	2.4
Organic soils	3.0

Ted W. Van der Gulik, ed. *BC Sprinkler Irrigation Manual.* Vernon, BC. Irrigation Industry Association of British Columbia. 1989. Pages 2–4.

CALCULATING MAXIMUM WATER APPLIED AT ANY ONE TIME

(plant root depth) × (water storage capacity of soil type) × (percent of moisture available to crop) = (maximum irrigation water applied in a single application)

For example:

Effective grass root depth = 1.5 feet

Water storage capacity in silt loam = 2.5 inches/foot

Percent available to crop = 50

1.5 feet × 2.5 inches/foot × 0.50 = 1.875 inches of water

In this example, when the plants reach their wilting point, only 1.875 inches of water can be applied in a single application to fill the storage reservoir of that specific soil type without causing excess water damage to the soil and plants, runoff, and leaching.

MAXIMUM SOIL MOISTURE INFILTRATION RATE ON A GRASS-COVERED SOIL BY SOIL TYPE

Soil type	Maximum application rate for soil covered by grass sod in inches per hour
Sand	0.75
Loamy sand	0.65
Sandy loam	0.45
Fine sandy loam	0.40
Loam	0.35
Silt loam	0.35
Clay loam	0.30
Clay	0.25
Organic soils	0.50

Note: On slopes exceeding 10 percent gradient, decrease application rate by 25 percent. On slopes exceeding 20 percent gradient, decrease application rate by 50 percent.

From Ted W. Van der Gulik, ed. *BC Sprinkler Irrigation Manual.* Vernon, BC. Irrigation Industry Association of British Columbia. 1989. Pages 2–5.

How quickly you apply this water is also crucial. If you exceed the maximum infiltration rate of water into the soil, the soil will not be able to keep up with the application rate, causing runoff that will damage the soil. The chart above shows how this infiltration rate varies by soil type on grass-covered soils. The maximum infiltration rate is also the maximum irrigation application rate for each soil type.

Let's continue our example and assume that the silt loam pasture we are irrigating has a 15 percent gradient. Yet the 7.14-hour irrigation length calculated does not suit a regular daily irrigation schedule. If you want to change your moves to suit your labor availability and fit into a regular daily irrigation schedule — for instance, to match 8-, 12-, or 24-hour irrigation moves — you have to adjust your water application rates by changing your nozzle sizes to avoid exceeding the maximum rate.

REPLENISHING THE MOISTURE STORAGE ZONE

Silt loam = 0.35 inch/hour

Correction for gradient = 0.35 inch/hour × 25% = 0.0875 inch/hour decrease in application rate

Maximum application rate = 0.3500 inch/hour − 0.0875 inch/hour = 0.2625 inch/hour

Length of time to fill up the soil moisture storage capacity = 1.875 inches ÷ 0.2625 inch/hour = 7.14 hours

Evapotranspiration Rate

How often you need to irrigate a particular piece of ground (known as the *irrigation interval*) depends on how quickly the soil moisture reservoir is depleted through soil surface evaporation and plant transpiration (moisture released into the air as vapor through the leaves of plants); when considered together they are called the *evapotranspiration* (ET) rate. The ET rate will change depending on how much water can be stored in the moisture storage zone and the climate (humidity, temperature, solar intensity).

As a general rule, the smaller the capacity of the moisture storage zone (the chart on page 197 shows the storage capacity of various soil types), the faster the water evaporates. The capacity of the moisture storage zone is determined by soil type and the maximum depth from which plants can effectively draw water because evaporation slows as the thickness of the soil layer above the water increases, providing increased protection from the atmosphere. For example, sandy soils are more susceptible to evaporation than clay soils (larger spaces allow air and moisture to move through the soil more easily, contributing to an increased evaporation rate), and water at a root depth of 4 feet (accessible only to deep-rooted plants like alfalfa) evaporates more slowly than water at a root depth of 1½ feet) (the maximum effective root depth of most grasses). Your agricultural Extension agent will be able to provide you with local peak ET rates related to your specific climate for varying soil moisture storage capacities (also called the *maximum soil water deficit*).

To continue with our silt loam example, if this soil was in the relatively humid, mild oceanside climate of Vancouver, British Columbia, the peak ET rate would be as shown in the chart at right. The moisture storage zone is capable of holding 1.875 inches of water (see calculation on page 197), which, when depleted, is equal to a *maximum soil water deficit* of 1.875 inches. Thus, according to the chart, we can estimate that our peak ET rate would be approximately 0.20 inch per day. At this rate, in dry weather during the hottest part of the summer, the irrigation interval should be nine days:

1.875 inches of soil water deficit ÷
0.20 inch/day lost to ET = 9.15 days

Many farmers simply cannot achieve short enough irrigation intervals because they are restricted by labor and irrigation equipment costs. Instead, they try to make do with less irrigation equipment and longer rotations; they put more water on the pasture during each irrigation application with the idea that volume can make up for shorter intervals. Unfortunately, the long-term result of such an approach is a leached, compacted soil, wholly dependent on continuing irrigation, whose production is on a downward spiral. This reminds me of the tongue-in-cheek expression "I'm losing $50 a head, but I'll make up for it in volume," which is just as impossible in the irrigation world as it is in economics.

PEAK EVAPOTRANSPIRATION RATE IN VANCOUVER, BC

Maximum soil water deficit in inches	Peak evapotranspiration rate per day in inches
1	0.24
2	0.20
3	0.18
4	0.17
5	0.16

IRRIGATION METHODS COMPARED

Unfortunately, there is no single irrigation method available to suit all irrigation situations. The ideal cost-effective irrigation system ultimately depends on your specific irrigation needs. Your choice of system is determined by the balance you strike among the irrigation system's efficiency as it relates to the plants and soil, the initial investment cost of the irrigation system, and the ongoing financial and labor commitments associated with its operation and maintenance. Here, in alphabetical order, is a comparison of various irrigation systems to help you weigh your options.

Hand-move Sprinkler

- It has a relatively low initial investment cost of pipe and main line to service large areas.

- It's very portable; allows flexibility of location; allows you to service many small plots, fields broken by trees, undulating ground, and so on.

- It's extremely labor-intensive!

- It's a low-pressure system.

- It's prone to sprinkler-nozzle blockages from debris entering the pipes during pipe moves and hookups.

Hand-move Stationary Gun

- It has a relatively low initial investment cost of pipe and main line to service large areas, even slightly less than hand-move sprinklers.

- It's very portable; allows flexibility of location; allows you to service many small plots, fields broken by trees, undulating ground, and so on.

- It has a higher rate of evaporation and wind drift and a harder impact on soil if the nozzle/pressure ratio does not break up the spray properly.

- It's extremely labor-intensive, though slightly less so than hand-move sprinklers.

- It's a high-pressure system.

- It's resistant to small debris.

HAND-MOVE SPRINKLER

HAND-MOVE STATIONARY GUN

Pivot Irrigation

- It requires a high initial investment in equipment and costly installation of power and water access to the pivot site.

- Irrigation is usually limited to one site per pivot; the unit is not easily mobile.

- The field must be relatively level and obstruction-free, although a pivot can operate on a slope of up to 30 percent.

- It requires very little labor.

- It's a low-pressure system.

- It's relatively maintenance-free, though some farmers will disagree. If buying a used pivot, never reuse the bolts; they will cause you endless grief.

- Water efficiency and operating pressures vary with design.

  - *Spray heads mounted on laterals lose a significant percentage of water to evaporation and wind drift.*

  - *Mounting spray heads on drop pipes reduces evaporation and wind drift. (Drop pipes, extensions mounted underneath the pivot arm, decrease the height between the soil and the water spray, thereby reducing evaporation and wind drift losses.)*

  - *Mounting a spray boom with multiple nozzles on each drop pipe allows you to reduce water pressure, spreads the water application over a wider area to increase water infiltration, and reduces impact on the soil.*

  - *Multiple perforated drag hoses pulled behind the spray boom further reduce evaporation and wind drift; water seeps out directly into the ground. This is undoubtedly the most water-efficient pivot design.*

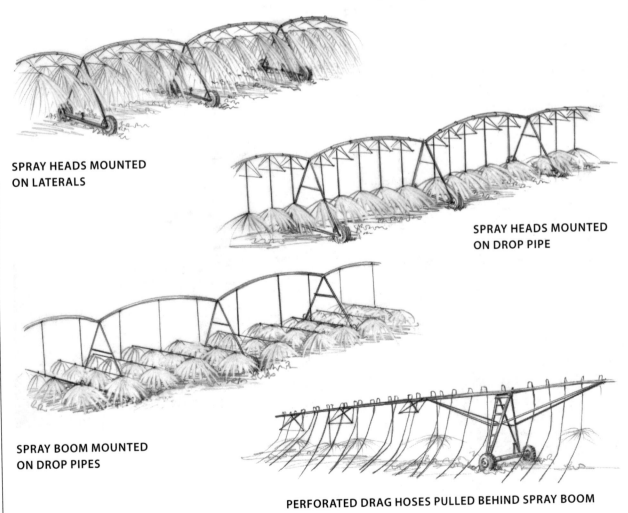

SPRAY HEADS MOUNTED ON LATERALS

SPRAY HEADS MOUNTED ON DROP PIPE

SPRAY BOOM MOUNTED ON DROP PIPES

PERFORATED DRAG HOSES PULLED BEHIND SPRAY BOOM

Hose-Reel

- It has a high initial investment cost of main line and hose reels and a tractor is required to move the hose reel and pull out the gun.

- Its gun option results in a higher rate of evaporation and wind drift and a harder impact on soil if nozzle/pressure ratio does not break up spray; it requires high pressure. It can be used on steep hillsides and across hillsides, thereby maintaining a consistent operating pressure at the nozzle as opposed to watering a hillside from top to bottom (or bottom to top), which causes gravity-related pressure increases/losses at the nozzle. It has a much wider spray arc than does a spray boom.

- It has the option of a spray boom on a cart, which utilizes lower pressure and is gentler on soil, but has a much narrower application width and is not ideal for irrigation runs on steep gradients or runs pulled across hillsides due to the risk of the cart tipping over sideways.

- It's a very portable system; it's simple to move from one field to the next and does not require extra labor investment to do so.

- It provides long reaches; its hoses are more than ¼ mile long.

- It requires very little labor, though still more than for pivot irrigation.

- The equipment is high-maintenance. It is best to use water-turbine-driven reels to turn the drum; do not use engine-driven models, which significantly increase labor, maintenance, and operating costs.

Traveling Gun or Traveling Spider

- It requires the lowest initial investment cost for larger acreages.

- Unlike a hose reel, traveling guns/spiders can be moved with an ATV instead of a tractor!

- It's very portable; it can be used to water fields of all sizes and shapes. The hose reaches up to ¼ mile.

- The cart winds up cable to pull itself forward, but the hose will flatten the grass as it's pulled behind the cart — a notable drawback.

- It requires little labor, though more than the pivot.

- It's a low-pressure system, particularly if the cart is mounted with a spinning spray boom to dispense water, instead of a typical sprinkler head such as those used on a stationary gun.

- It's relatively low-maintenance.

TRAVELING SPIDER

HOSE-REEL

Wheel-Move

- The initial investment cost (for pipe) is relatively low, though slightly more than for a hand-move sprinkler and a hand-move stationary gun.

- It's not very portable; its use is restricted to one large, open, relatively flat field at a time.

- It's very labor-intensive to disassemble and move the wheel line to a new field.

- It doesn't lend itself to use on undulating or sloped terrain.

- Its labor requirements are *far* lower than for a hand-move sprinkler and a hand-move stationary gun.

- It's a low-pressure system.

WHEEL-MOVE

Controlled Drainage/Subirrigation

Many agricultural lands are drained by ditches and tile drainage systems to deal with a high water table, particularly in spring. These can be used to give the farmer unique control over the subsurface water table by controlling the amount of water drained out of the system.

In spring you can allow these drains to function normally, but as summer heat increases, you can block them off, which will cause the groundwater to accumulate and raise the water table, allowing the plant roots to reach the water. You can use ditches in the same way by installing a series of small dams along the length of each ditch. By opening and closing a small gate or by sliding a couple of timbers down between two posts, you can raise or lower the water level. In this way, you can effectively subirrigate a field by keeping the water table higher.

In pastures and fields where controlled drainage is possible due to the presence of drainage ditches or tile drainage, it is by far the most efficient irrigation method and has the least impact on the land.

Flood Irrigation

I don't recommend flood irrigation because the water excesses required to spread water across large areas tend to cause topsoil leaching and erosion as it moves soil particles downslope. Over time this can cause serious nutrient and soil loss on the uphill side of the flooded areas as well as significant nutrient leaching throughout the flood-irrigated area.

TEMPORARY DAM IN DRAINAGE DITCH

Important Points to Know about Irrigation

Calculating your pastures' irrigation requirements is merely one aspect of maximizing irrigation benefits and minimizing irrigation-related costs. The points listed below address a number of additional important aspects of effective irrigation design and operation, such as how to protect plants and soil from unnecessary irrigation-related damage; how to apply irrigation water evenly to the plants; how to use most efficiently your irrigation water; and how to minimize the investment, labor, maintenance, and operating costs related to irrigation use.

■ On pages 197–198 we calculated the maximum application rate of water to soil, but this is not the same as the irrigation rate. Watering systems are not 100 percent efficient; a great deal of water is lost to evaporation, wind drift, and runoff. The inefficiency caused by these losses must be considered when determining nozzle sizes, the duration of water application, and irrigation intervals.

On average, sprinklers are rated at an efficiency of between 72 and 78 percent, whereas stationary and traveling guns vary between 60 and 75 percent efficiency. Pivot efficiency varies dramatically depending on where the sprinklers are mounted on the pivot system. These statistics are merely averages, however. Irrigating during windy weather, extremely hot temperatures, and the hottest part of the day will significantly reduce the efficiency of all irrigation systems and cause irrigation efficiency to fluctuate dramatically from hour to hour and from one spot in the pasture to the next.

■ Spray from sprinklers and stationary guns should overlap by at least 40 percent to achieve even water distribution. Traveling guns need to overlap by only a few feet between runs.

■ Low-pressure systems (sprinklers, traveling spiders, wheel-moves, hand-move sprinklers, and pivots) incur significantly lower pumping costs. High-pressure systems (stationary guns and traveling reels) cause high friction losses in pipes and nozzles that the pump must push against.

■ Sprinkler nozzles must be sized to break up the stream of water into small droplets (spray) before the water reaches the ground. Insufficient pressure or oversized nozzles prevents the water stream from breaking into small water particles, leading to an uneven distribution of water across the soil surface. Furthermore, the force of the violent impact caused by the water stream when it hits the soil's surface increases soil compaction and damages the soil structure in addition to dislodging individual soil particles from the ground and washing them downslope, leading to increased soil erosion.

■ Very cold irrigation water will shock the plants and cause their growth to slow for a period of time after irrigation.

Business PLANNING and Marketing

PART THREE

Chapter

14

Land and Equipment

Unless you are in the real estate business, you sell farm equipment, or you are a farm equipment contractor, owning land and farming equipment unnecessarily ties up a significant portion of your capital, which you might otherwise use to earn an income by investing in more cattle and grazing infrastructure. In many cases, the enormous land and equipment investments of conventional cattle farming practices are prohibitive to emerging cattle farmers and inhibit the profitability of many existing farms. In this chapter we will discuss viable alternatives to land ownership and explain how to identify suitable grazing pastures for your natural, grass-based cattle herd. I will also describe the minimal equipment inventory necessary for the production methods described in this book so your capital remains invested in cattle and your goal of having a natural, grass-based cattle farm can become an affordable proposition.

Acquiring Land for Your Natural, Grass-Based Cattle Enterprise

Avoid landownership at the onset of your grazing business. Landownership ties up the bulk of your capital, which does not bring you a direct income and probably comes with a mortgage. High mortgage payments put an enormous amount of unnecessary pressure on a fledgling business and severely limit how much you can invest in things that are moneymaking, particularly cattle, electric fences,

and livestock water. To make matters worse, real estate prices have become severely inflated above the land's agricultural value, especially in communities where the appeal of country living and the quality of the view drive prices ever skyward.

Locating Land Rentals

Once you shed the idea that farming requires landownership, you will find that renting agricultural land is far more economical and gives you many more options. Renting land is a tax-deductible expense and keeps your money working for you so you can grow your business much more quickly, rather than tying it up as capital that will not be recovered again until the land is sold. It also gives you much more flexibility to make changes to your business strategy because it eliminates ties to something you have to sell in order to move your enterprise to a more favorable location.

City Folks

Finding landowners from whom to rent is not difficult. More and more agricultural real estate is being bought by wealthy city people who enjoy country living but don't want the hassle of farming. These people are happy to rent their land in exchange for your herd's grazing impact, which will keep their land green and tidy-looking and prevent brush and weeds from taking over. In this way, these landowners can share the dream of farm life without doing the work.

In many cases, city landowners do not value their land for its agricultural purpose and look at land maintenance as an expense. Some people will let you graze their land for free just to keep it looking nice. Often, the prettier the view, the less it is valued for its agricultural potential, so the rent decreases accordingly. Your cattle-grazing enterprise is a free grass-mowing and horticultural service to such people.

Older Landowners

The average age of farmers in North America is approaching sixty-five. Many older farmers don't want to farm anymore, but they do want to keep their land and live on it. They are often happy to rent it, and in some cases, they are even willing to form business partnerships: You do the work while they provide some of the capital to buy cattle, build fences, and develop water sites. These older farmers are a wonderful resource for the beginning grazing entrepreneur, who typically has high energy but is low on capital.

Neglected Land

Keep your eye open for neglected land that is sitting idle. Likewise, take note if you see a brush mower clipping grass. Grazing can be an economical alternative to maintaining government lands such as floodways, rights of way, and other public areas that are currently being mechanically maintained at great expense. Cattle can even be a great asset to revegetation efforts on old mine sites and on other land reclamation projects.

Evaluating Land Potential

Try to find land within a twenty-minute radius of your home. You do not want to spend your time and money commuting. In addition, the greater the distance between your home and your grazing land, the more vulnerable you become to mishaps because it becomes increasingly difficult to oversee your herd and respond to problems (such as broken fences, frozen water troughs, and disease) quickly and efficiently, before they escalate into serious disasters. Land closer to town will give you more market options, even if the acreage is smaller. A spread of 50 acres close to town will make you wealthy much faster than 500 acres in the middle of nowhere.

Irrigated land costs more to rent and usually requires a high investment in equipment, maintenance, labor, and valuable time best spent elsewhere. (See chapter 13 for a discussion on the disadvantages of irrigation and how to improve soil moisture infiltration and moisture retention through effective grazing habits.) It is usually less expensive to achieve the same stock numbers by renting, fencing, and setting up livestock water on additional land.

When you evaluate potential rental land, in addition to your visual inspection of the grazing land to assess its grass varieties, grass density, winter grazing potential, and grazing risks such as poisonous plants, bad spring mud, and summer drought damage, ask yourself these questions.

■ Is there enough cleared pastureland to make your infrastructure investment worthwhile? A site where timber covers more than 50 percent of the land is not worth your fencing investment.

■ Does the land have a reliable year-round water source? What will it take to develop?

■ What kind of road access do you have? Can you get livestock transports in and out all year or will you be facing mud just when the cattle trucks are supposed to arrive?

■ Does the land have access to power for running a reliable energizer? Solar power works but is more cumbersome. Having power to pump water is also a big help.

■ What is the soil condition? Take some soil tests before signing a lease. In addition to your visual inspection of the pasture quality, these will give you a good idea of the land's grazing potential. Soil tests will show you what kind of soil problems you are inheriting and what it will cost to fix them before you commit yourself. Look for mineral deficiencies or excesses, low organic matter, and a problematic cationic exchange capacity (CEC). A CEC below 2.5 requires high yearly fertilizer inputs because the soil does not have a large nutrient reserve. An extremely high CEC with a serious nutrient problem is expensive to fix, but a high CEC in a balanced soil provides you with a large nutrient reserve. Beware of drought-prone sandy or gravelly soils.

Is there someone living on the land or is there a neighbor alongside it to provide a deterrent to vandalism and negligent trespassers and to notice problems during the hours of the day when you are absent?

What kind of neighbors will you have? Will you be vulnerable to dog harassment from subdivisions or pollution from pesticide users and industry?

Is there a highway nearby? Avoid land bisected by a busy highway, which can be a major hindrance to your grazing rotation. Scattered land locations make your grazing rotations very tedious and some by-laws restrict herding cattle across pavement.

Is there potential to graze adjoining land in the foreseeable future, when your business expands? This is much more desirable than having to create separate grazing cells at new locations when you want to expand your business.

What kind of infrastructure is on the land? Beware of barbed-wire-fence divisions that are costly to remove and impossible to live with because they do not suit your grazing management. It is often easier to develop neglected hay fields or cropland than it is to rent land with someone else's barbed-wire infrastructure on it.

Is the landowner pleasant to work with? There are so many wonderful people out there that it isn't worth putting up with a difficult personality, no matter how attractive the location.

Can you get a long lease? A minimum five-year lease on land is desirable to recover your fence and water infrastructure investments.

Is the lease clear? Make sure you sign a clear lease that outlines both your responsibilities to the land and the landowner's obligations and commitments. A clear *written* lease is the key to a positive, long-lived business relationship.

Each land rental adds to your résumé for the next land rental agreement. Keep the place looking tidy and organized so you can show it to other landowners to gain their confidence in renting to you. Word gets around; if you keep your rentals looking great, landowners may well begin seeking you out to enlist your land maintenance and grazing service.

Equipment

A high-profit, low-cost cattle enterprise requires only a minimal amount of equipment. Any love affair with the rest of the time- and laborsaving equipment and technology marketed to farmers is simply a vice that jeopardizes a farm's profitability.

Unfortunately, once the technology bug bites, it doesn't take long to realize our enterprise is designed around the machinery, not the other way around. You wouldn't decide what kind of house to build based on the tools you find in your toolbox, yet once you've made a financial commitment to a piece of heavy equipment and it's depreciating in your yard, the race is on for ways to justify owning it and to make it pay for itself. Equipment ownership rapidly takes over farm leadership, causing your business to flirt with disaster.

You can't just eliminate equipment; you have to replace it with effective management.

It's important that your money be tied up in things that directly make money. Unless equipment is part of a custom-work or rental business or you sell equipment for a living, machinery never directly makes money. It rusts; depreciates; and requires maintenance, fuel, insurance, repairs, and a great deal of your time just to keep around. Consider how much of your time is stolen by your equipment just to keep it working. This is time that is taken away from cattle management, planning, and marketing.

Your management program should allow you to invest money in what directly creates wealth: cattle. Money spent on anything other than cattle, land rental, and pasture-management infrastructure (fences and water) has an enormous opportunity cost because it could be invested in additional cattle to earn you a larger income. Think about how many extra cattle are represented by the costs of owning and caring for a tractor.

On many struggling farms, a large fleet of rusting equipment is what stands between making effective changes toward profitability and the hardships the farmer is currently experiencing. Sending all of that equipment down the road as quickly as possible will, in all likelihood, pay for most of the infrastructure changes that are necessary to profitably manage a twelve-month-long grazing system. It will also free up time, create a financial buffer, and force a farmer to make effective changes to management because he or she no longer has financially draining technology. You can't just eliminate equipment, however; you have to replace it with effective management.

In place of this expensive equipment and diesel engines, what are the bare essentials required by a twelve-month-long grazing-management system?

■ An ATV or pickup to move mineral feeders, bring to cattle livestock supplies such as minerals and salt, perform fence checks, and haul fence equipment.

■ An ATV trailer to carry all your livestock supplies and tools; a homemade flat-deck trailer with removable sides and a permanent storage box on the front works well. It can carry large water tubs from field to field and portable fencing gear can be stowed in the storage box on the front.

■ Electric fences, an energizer, a corral, and a water system.

■ Fencing tools: wire tightener, wire unroller, hammer, and fencing pliers. See the illustration below for an example of a homemade wire unroller. To make one, weld a short section of pipe to the apex of each vertical metal triangle so you can slide a metal bar through the pipe sections and support the wire rolls above the ground. A pin on each end of the metal bar keeps it from sliding out while the wire spools are being unrolled. The triangles should be tall enough to allow the wire spools to spin freely on the metal bar and wide enough to give the wire unroller stability so it doesn't tip over.

■ Mineral feeders for each group of cattle. Portable mineral feeders are easily constructed using big plastic drums, which are widely available in the dairy industry; milking-parlor-cleaning detergents come in them. Simply cut the drum in half and bolt it between the two cross members on the mineral feeder frame, as shown in the illustration on page 210. Bolt a painted plywood roof on top and two wooden skids on the bottom to complete the portable feeder. Metal skids slide much more easily but will rust and wear out very quickly. Metal loops bolted on the ends of each wooden skid help the feeders slide over obstacles so they won't flip when you tow them during paddock moves. A full loop prevents injury to the cattle's feet (sharp edges on the skids invite foot rot).

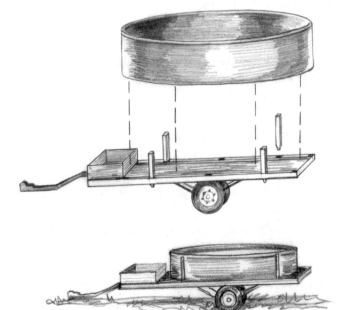

A home-built flat-deck ATV trailer fitted with pocket stakes to carry a mobile water tub during pasture moves. Solid sides and a tailgate can be mounted in place of the pocket stakes.

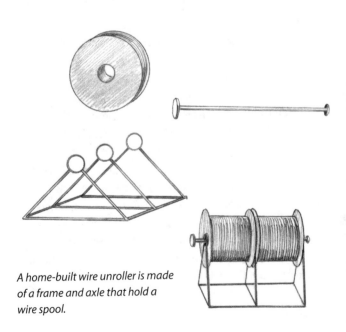

A home-built wire unroller is made of a frame and axle that hold a wire spool.

A home-built portable mineral feeder uses a halved plastic drum to contain the mineral mix.

■ Stock trailer and pickup truck. For farmers who grass-finish and direct-market their own animals, this is an addition to the list. Although I am a firm believer in custom livestock hauling, if you are slaughtering animals weekly throughout summer and fall, it will pay to own or share a stock trailer and used pickup truck to move small numbers of animals to the butcher in order to ensure a dependable, timely delivery of meat to your customers. With these vehicles you can also guarantee that the hauling is as quiet, gentle, and stress-free as possible so that the high-quality meat you have spent so long producing maintains its quality to the very end. Yet when you calculate how much a truck and stock trailer cost to own and maintain, recall that they must be used very frequently over a good portion of the year before it's worth owning them instead of renting them or contracting out your livestock hauling needs.

■ Single livestock scale. If you grass-finish beef, it may also be worthwhile to add a single-animal livestock scale to your corral to weigh animals before hauling them to a butcher or selling live, finished animals to your customers. The scale is also an optional tool to ensure good finishing weight before sending animals to slaughter.

There is little beyond what is on this list that qualifies for equipment ownership. What if you live in a heavy snowfall area? Neither grazing nor self-feeding cattle require the removal of large amounts of snow; a snow blade mounted on your personal vehicle will suffice to get your vehicle in and out of the driveway.

That's it! It certainly is a lean list compared to what's in most farmers' vast machinery yards and toolsheds. Remember: If it rusts or depreciates, it has no place in your grass-based beef production enterprise. Keep your money in cattle so you can invest your time in management.

CONTRACT WORK

If you choose to include a small drought reserve of hay, use some of it to settle new animals after shipping. Or if you feed stored feeds during a portion of the winter, contract a custom haying crew to make your feed or buy it from off the farm so you won't need to keep on hand a fleet of equipment. Store the feed directly in the fields where it will be self-fed to the cattle during the winter using portable electric fencing, as described in chapter 7. If you buy hay, have it delivered and laid out directly where you will need it later. Whichever way you choose, you still don't need to own a tractor. You can contract out fertilizing, seeding, and any other machinery-related work. It simply does not pay to own equipment unless it is something that you will use daily.

Chapter 15

Market Options

YOUR PROFIT MARGIN IS CONTROLLED BY THE three factors of volume, price, and cost:

volume × price − cost = profit margin

Before further analyzing these three factors, it is worth taking a moment to understand the commodity marketing system that has become the standard for exchanging cattle in the conventional cattle industry. Commodities, by definition, are raw or primary products categorized by a standardized measure of quality and produced in large quantities by many different producers so that items from different producers are considered equivalent. Examples include many agricultural products (i.e., grain and cattle), fuel, minerals, and foreign currencies.

Commodities are traded through a standardized marketing system in which the seller offers his product for sale, usually at a designated meeting place such as an auction or stock exchange. The price of these commodities is determined by competitive bids and offers based on supply, demand, and speculation about future value. Purchasing decisions are made almost solely on the price and category of the product, not on differences in quality or features.

This commodity marketing system has trained us to think of profitability in terms of volume because the price is established by the auction, not by the seller. Consequently, we have come to believe that the profitability of our farms is ultimately determined by how many cattle we own, how many acres we farm, and how many sales we make each year.

Yet for every extra dollar earned by increasing the volume of commodity-market sales, typically only 3 to 10 percent of it is profit because every increase in animals, bushels of grain, or bales of hay comes with a proportional increase in production costs. No wonder so many farmers are stretched so thin in the commodity-market system as they try to keep their businesses growing faster than their profit margins shrink!

The saying "A dollar saved is a dollar earned" is not far off the mark because every dollar you cut from your costs is an extra dollar in your pocket. Unlike a volume increase, which increases the profit margin only by a small percentage, every time you decrease your costs, your profit margin increases by approximately the same amount, in roughly a 1 to 1 relationship.

But what about price? If you are marketing your animals through commodity-market channels, price will vary only as much as the cattle cycle (cyclical fluctuations in cattle prices based on nationwide cattle herd inventory numbers and speculations about future cattle market prices, typically following a ten-year cyclical pattern) varies from day to day and year to year. By understanding the auction marketing system, you can influence your sales price to a small extent in an effort to sell your cattle at the upper end of the commodity price scale, such as by using to your advantage the price fluctuations of the cattle cycle, by (legally) minimizing weight losses during shipping, by influencing the sales order of your cattle relative to the other cattle being sold, and by presenting your cattle to advertise their quality (i.e., their health and future weight gain potential). Yet the influence you have on your bottom line is nevertheless relatively minimal, hence the focus on further cutting costs and increasing sales volumes.

But what if you step outside the commodity pricing world into the world of direct-marketing, niche markets, and organic or natural market avenues? How much of a difference can that make to your bottom line? Some niche markets can command a considerable premium over commodity-market prices, while other niche marketing options essentially allow you to set your own price for your product. The increase in profitability is considerable. According to Allan Nation ("Allan's Observations," *The Stockman Grassfarmer,* vol. 57, no. 8 [August 2000], 15–18), on average, for every 1 percent increase in price, you will see a 12 percent increase in your net profit without increasing your sales volume or decreasing your costs. Wow! This is why niche markets are emerging so vigorously in agriculture today.

You will always be rewarded for being out of sync with the masses.

Commodity-Market Opportunities

In the commodity market production system, the cow/calf producer sells calves, the stocker operation buys and sells stockers (or custom grazes them), and the feedlot finishes the stockers for slaughter. This is the McDonaldization of beef: Everyone focuses on a specialty and volume determines the bottom line. Yet even if you are selling animals into this rigid market environment, there are some marketing options available to you to increase your profitability.

Selling to the Late-Spring Stocker Market

Calving in summer gives cow/calf producers two options for marketing their calves to which other conventional producers do not have access. After the calves are wintered very inexpensively on cow's

milk, the weanlings can be sold in late spring to the stocker market. Because it is cheaper to grow grass than to produce hay, everyone wants to buy stockers in spring and sell them off grass in fall to avoid the high cost and effort of wintering animals. This drives the stocker prices quite high, which means you can fetch a generous premium for your weanlings during the spring market runup in addition to significantly reducing your production costs by calving in summer and wintering the calves with their mothers. You will always be rewarded for being out of sync with the masses.

Making the Most of Compensatory Gain

Alternatively, calves wintered so inexpensively on cow's milk will have a considerable amount of compensatory gain when the green grass returns because their slow metabolism, caused by their reduced nutritional intake during the winter, allows them to gain weight very efficiently for a short period of time in the spring until their metabolism speeds up again in response to the sharp increase in feed quality and quantity. Compensatory gain is discussed in detail on page 37. It is to the producer's advantage to keep calves through summer and realize those gains personally instead of passing them on to someone else. Typically, the calves will hit the ideal market weight of 700 to 800 pounds (the ideal weight for feedlot buyers to purchase young cattle for finishing) in late August to late September, so again the producer will be rewarded for being out of sync: Those animals will sell before the flood of spring-born weanlings drives down market prices in fall. This is an example of vertically stacking your enterprises; you are now in both the cow/calf and the stocker business.

Keeping Animals Longer

The other consequence of the specialization and fragmentation of the beef-production industry is that each time animals are sold to the next step (cow/calf to stocker to finishing), there are sales commissions, transport costs, health costs, and weight losses due to shipping-related stress. By keeping your animals longer, through at least two of these production steps, you save on these costs, thus adding to your profit margin.

You have to consider each production step as a separate economic enterprise. If your cow/calf enterprise cannot produce stockers more cheaply than it costs to buy them, then the enterprise is not viable, even if the advantage of incorporating both production steps makes sense in theory. And if your stocker enterprise cannot buy those stockers from the cow/calf enterprise at market value and still be profitable, then those calves should be sold elsewhere and the stocker enterprise eliminated.

The same applies to stockers if you grass-finish them. Every step of the business has to be economically viable on its own, despite how closely it is related to the next. To determine the financial viability of each step, as a planning exercise you have to theoretically sell the products of each stage to the next stage of your business at full market value, even though you don't actually split up your enterprises in this way in your bookkeeping records.

Cull and Empty Cows

Your cull cows and empty cows (unbred cows or heifers that failed to conceive during the scheduled breeding season) can give you additional market leverage if you are careful about when you sell them. Empty cattle are the least expensive animals to fatten because they are not pregnant, lactating, or growing (they have reached their adult frame size). Their cost per pound of gain is the lowest of all classes of cattle, yet because of their low auction price, their potential is often overlooked by farmers who dump them on the market as soon as they cull them from the herd.

These cull animals should be kept until after the grass greens up, and they should be allowed to fatten on grass (usually with considerable compensatory gain after the lean winter) before going to the auction. Even at cutter prices (cutters are cattle considered to be too old, with meat of a quality that is too poor to use for prime cuts; their meat is usually ground into hamburger), the additional 150 to 200 pounds — representing an increase of 5 to 7 in body condition score (see chapter 3 for an explanation of BCS) — will result in a significantly higher return on those animals. Waiting costs you little more than the additional grass the culled cows eat, which is usually in excess at that time of year.

Focusing on Stockers

Concentrating solely on stockers destined for the commodity market can also be quite lucrative, though it can leave you vulnerable to the cattle price cycle, depending on your yearly purchase-and-sales strategy. (For more information on the stocker market, see chapter 16.)

Grass-finishing animals for the commodity market is still uncommon in North America. Here, the commodity market is still predominantly biased toward feedlot-finished beef. Anyone who produces grass-finished beef has much better price opportunities by selling his animals through niche markets.

This feedlot bias is not the rule everywhere in the world. Many countries, including Argentina and New Zealand — both of which produce world-famous beef — grass-finish a significant portion of their beef (and lamb) for the commodity market, not only to produce healthier meat but also because grass-finishing is cheaper. I have recently run across some large North American enterprises that grass-finish on intensively managed, pure alfalfa grazing rotations strictly for the commodity beef market. They are doing remarkably well, and I predict that this idea will catch on, so we had best take advantage of the grass-finished niche market before it becomes a mainstream production practice.

Niche-Market Opportunities

Commodity markets, by their nature, require standardization and uniformity that all its producers adhere to in order to achieve a predictable price structure and supply network. But no one wants a product that is designed to fit everyone. We all want things that are unique and custom designed and produced to suit our individual requirements and desires.

Most advertisements try to make us believe that their products are unique and specifically designed to meet our individual needs and that using their products will distinguish us from the masses. From cars to chocolate bars, hamburgers to real estate, specialty cheeses to designer clothing, we are willing to pay more for products that are superior in quality, have a better view, look prettier, appear healthier, or appeal to our sense of uniqueness, curiosity, and imagination. The food market is no different in this respect. Yet despite aggressive advertisements claiming the uniqueness of specific brands of commodity meats, the identities of commodity producers remain anonymous to the consumer, uniquely produced products are lost in the anonymous supply network, and the standardization of commodity products to create an equitable price structure does not reward unique production efforts.

Capitalizing on Your Unfair Advantage

The niche-market producer has an unfair advantage because his or her product can be custom-tailored to a specific individual market specialty. Consequently, its uniqueness is recognized by the consumer and rewarded accordingly. In the niche-market world, we are rewarded for producing something unusual; it pays to be different.

Unlimited Potential

Niche markets are currently undersaturated. There simply is not enough supply to fill the demand for unique products. For example, the organic industry is one of the fastest-growing industries in North America, constrained only by its supply of organic products, not by customer demand.

When direct-marketing natural, chemical-free, hormone-free, or wild-grown beef, your market options are limitless. You get to create your own market for what you can produce, set your own price (within reason, of course), and target a market that guarantees you an income that meets your personal life-quality and financial expectations.

Knowing Your Target Market

If you have a good idea, do your marketing homework, and produce a quality product, you are almost guaranteed to sell everything you can produce (or even more than you are able to produce) in the niche-market environment as long as you know exactly what your target market is before you design your production system. If you don't know where you want to be, how will you know what you have to do to get there?

After all, whether you are producing for organic wholesalers (certified organic beef), restaurants (featuring, for example, "grass-finished natural beef

from the Arctic meadows of the lush, pristine Yukon River Valley"), health food stores (grass-finished organic beef from Alberta, for example), ethnic communities (for instance, kosher grass-raised beef), or season-specific meats (such as natural, pasture-raised spiced beef, sausages, tenderloin, corned beef, and rouladen — every cultural heritage has favorite holiday recipes that call for beef), or you are direct-marketing to your neighbors (by the side or the cut in fall or all year long from a farm meat outlet), your target market dramatically influences your production conditions, breeding season, finishing strategy, slaughter arrangements, and much more.

You can sell just about anything in the commodity market, but among niche markets, the production and marketing constraints are specific to the target market and anything you produce that does not meet these constraints becomes unmarketable except through conventional commodity-market channels. You simply have to know what your market goal is before you allow yourself to be lured by a seemingly endless number of opportunities. Each unique option comes with its own specific set of rules and production constraints.

Choose Three for Stability

To be financially stable, you should have at least three separate enterprises on your farm that do not rely on the same market cycle, buyers, suppliers, and weather conditions. Having multiple sources of income gives you financial buoyancy to survive market collapses, production problems, and other misfortunes. This is not to say that you need to diversify into three unrelated enterprises such as grain, beef, and a market garden; if you do, you might very well be spreading yourself too thin. The commodity-market producer might choose cow/calf, stocker, and some direct-marketed grass-finished sales for financial buoyancy. The same basic infrastructure and skills are needed for all three, but they have different buyers, their market cycles are offset by several years (e.g., cow/calf prices typically do not bottom out until two years after finished cattle prices do and stockers tend to be most profitable when cattle are cheapest to buy — that is, at the bottom of the market cycle, when other cattle enterprises are struggling to make money), and they have different weather-related vulnerabilities, giving the producer different options to market products in the face of a crisis. Note that direct-marketed beef prices are relatively unaffected by the ten-year cattle price cycle on the commodities market, but sales volumes may fluctuate unexpectedly based on contracts changing or buyers going out of business, which is discussed further below. Yet direct-market sales will be particularly profitable during the low end of the ten-year cattle price cycle if the animals used for grass-finishing and direct marketing are purchased from off the farm as stockers or weanlings. In this scenario, they will be particularly cheap to purchase while the direct-marketed sales price remains fairly constant (leading to a high profit margin).

Niche-market producers tend to be less vulnerable to market cycles because they are typically dealing with the end-product consumers, and the cost of beef changes very little at the consumer level throughout the ten-year cattle price cycle. Nevertheless, restaurants change their menus or change hands, suppliers find better deals, health food stores find other suppliers, and even neighbors can be fickle customers. You want markets that are independent of one another and products that require the same skills and infrastructure. Never limit yourself to a single niche-market outlet or a single buyer; sell to at least several independent restaurants or health food chains that have no connection to one another.

In a direct-marketing situation, diverse markets give you bargaining power and the confidence to present what you proudly produce ("This is what we have, and this is what it costs"), whereas a single market outlet makes you desperate during price negotiations and vulnerable to altering your production to meet the buyer's demands. You need the confidence resulting from diverse market exposure to make outlets want to buy your product at your price.

Determining Your Target Market

Deciding on your target market is a process of determining what your land, climate, and growing conditions can do best and then custom designing your product to the clientele you can access from your location. Do your homework so that you can answer the following questions.

- What kind of community can you access?

- Does it have health food stores, and is there an awareness of and demand for organic products?

- What kinds of local restaurants are there?

- What is the income level of the population base to which you have easy access?

- Are there any ethnic communities in your area? What are their meat preferences?

- Do you have to transport your product over long distances to a market or will you be selling locally so your clientele knows you and your farm?

The ideal market opportunity is created simply by tailoring a system to produce a product that suits the market to which you have access.

Small Communities

A small, rural, resource-based community is less likely to demand certified organic products; natural products will suit it just fine because cost is a bigger concern. Typically, many residents will be equipped with deep freezers (because of the relatively large population of hunters) and will be more inclined to buy meat by the side or quarter. A seasonal meat supply by the side or split side will not intimidate them.

In a smaller community where your customers can get to know you directly, it is easier to build a reputation for quality and production integrity. Selling your meat as *certified* organic, which is possible only if your land, production methods, and livestock are verified and subsequently certified by an independent organic association (with its extra certification costs), is not likely to be necessary. This market seems well suited to meat labeled "naturally produced" and supplied on a seasonal basis by the half or quarter. Your customers will be local restaurants, grocery outlets, families, and individuals. Although there likely will be fewer high-end restaurants interested in high-priced cuts, small communities are an excellent market for selling large volumes of ground beef with a distinctive marketing label, especially during the summer barbecue season.

Large Communities

Larger communities are less accustomed to a seasonal meat supply, so your meat will likely be sold in smaller portions over a longer season. Determine your slaughter time and meat storage accordingly. Large communities usually have plenty of commercial freezer space that you can rent, allowing you to extend your marketing season for selling meat by the cut, variety packages, or box programs (a system of pre-selling a regular purchase of farm products with a predetermined value, with selection determined by the seasonal availability; this is described in detail on page 260), or for supplying grocery and health food stores. Customers in larger communities are usually more willing to pay for marketing distinctions; people will accept a higher price for your organic or certified-organic meat.

If your farm is local, you can easily build a reputation for quality without obtaining independent certification. Your local label and your customers' ability to see your farm will inspire the same confidence in your quality and integrity, but at a lower cost to you and with far less paperwork than is necessary to certify through an organic association.

The more urban your community, the more customers will thrive on coming to your farm to purchase meat and enjoy the farm experience. The farther people are removed from the land, the more they value being able to experience the country, especially if they have children. A farm outlet with value-added products will be welcomed and eagerly sought. The additional agritourism opportunities in such a community are endless.

Ethnic Communities

Determine the demands of your local ethnic communities. They will be eager and loyal customers if they can find a reliable source for their specialty meats (such as kosher-slaughtered beef, grass-finished lamb, and organ meat).

The Need for Certification

Wholesalers, restaurant chains, and other third-party buyers have their own unique requirements. They are even more adamant about a year-round meat supply, thus your production, slaughter, freezer storage arrangements, and price will have to allow for those needs. They also will likely want your product to carry an independently verified label (certified organic beef, certified Angus beef, certified natural beef) as a guarantee of quality because both the third-party buyers and the customers they sell to have no contact with your farm and do not know you.

The farther your meat travels, the more likely it is to require a certified label as part of your marketing strategy and the more your customers will rely on the label's description. Even the most basic label will inspire interest and confidence in your product if it is produced locally. As soon as your product travels, however, it must carry a label that distinguishes it further; provides an assurance of clean, natural, healthy production conditions; and appeals to a customer's curiosity and desire to be unique, in order to compete against other products. What's more, the farther your product travels, the more your market will begin to resemble commodity sales because of all the additional transport costs, wholesaler commissions or price margins, storage fees, and so on, which normally stay in your own pocket when you sell to a local niche market.

The Restaurant Market

Restaurants love to feature natural or organic meats that are high quality and stimulate the imagination of their customers. For instance, buffalo burgers are a hot commodity in roadhouses, and five-star restaurants sell for top dollar a tiny portion of quail meat along with two baby carrots, three peas, six noodles, and a sprig of mint for decoration.

When marketing to restaurants, use your label to emphasize the uniqueness of your product and to appeal to a customer's imagination. You must be able to supply fresh or frozen meat in smaller quantities over a long season or even year-round. The better the restaurant, the more likely it is to feature a high-quality seasonal specialty on its menu, especially if you have the opportunity to market directly to the chef and to spark his or her imagination with the uniqueness, attractiveness, and seasonality of your product.

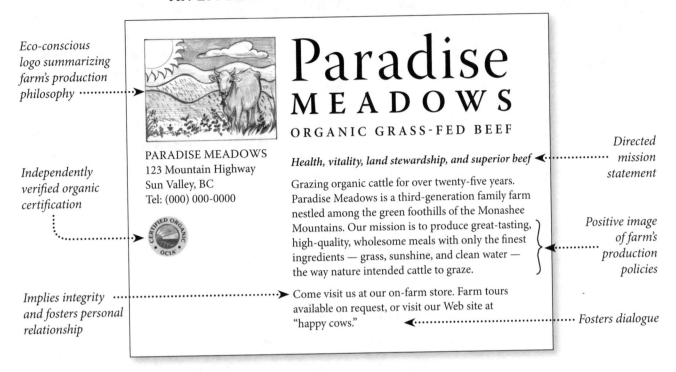

Eco-conscious logo summarizing farm's production philosophy ·········

Independently verified organic certification

Implies integrity and fosters personal relationship

Paradise
MEADOWS
ORGANIC GRASS-FED BEEF

PARADISE MEADOWS
123 Mountain Highway
Sun Valley, BC
Tel: (000) 000-0000

Health, vitality, land stewardship, and superior beef

Grazing organic cattle for over twenty-five years. Paradise Meadows is a third-generation family farm nestled among the green foothills of the Monashee Mountains. Our mission is to produce great-tasting, high-quality, wholesome meals with only the finest ingredients — grass, sunshine, and clean water — the way nature intended cattle to graze.

Come visit us at our on-farm store. Farm tours available on request, or visit our Web site at "happy cows."

Directed mission statement

Positive image of farm's production policies

Fosters dialogue

Retail Markets

Health food stores, groceries, and other food outlets vary greatly in their requirements. The larger the chain, the more likely purchases are controlled by wholesale buyers, but some stores will nevertheless be happy to feature seasonal meat with or without freezer storage.

It is best to talk directly to the owner, manager, or meat department manager about your product; the closer you get to the top, the more likely the person you talk to will have the interest and authority to create exceptions in the store's buying policies to suit your product, instead of requiring you to adjust to meet the store's needs.

Never discount the possibility of selling to large food chains. Many grocers are open to carrying natural and organic meats (today, many large chain stores even feature an organic section) if the population base in the community is wealthy enough and health-conscious. Usually the bigger chains prefer to feature the same specialty meats in all store outlets.

You can overcome this obstacle by partnering with additional natural-beef producers or by forming marketing cooperatives with other natural producers, the way TK Ranch Natural Meats in Alberta has done with tremendous success. Some grocers in individual outlets, however, will make exceptions to promote unique or exceptional local produce, even from very small suppliers, particularly because many of these grocery stores have a mandate to expand their organic selection to cater to the growing organic market. It all depends on the demographics of the community, the individual attitude of the store manager, and the innovative marketing techniques of the individual producer.

By-products

Never underestimate the value of selling your by-products as dog food. Organic dog food is always in demand, especially in larger, wealthier communities, where "certified organic" and "natural" dog food is a rapidly growing trend.

By researching your market demographics, you will be able to identify your marketing opportunities and what you must do to reach that market in terms of production, advertising, and supplying your customer base. For example, determine if your target market requires you to have organic certification, kosher slaughter practices, or access to a frozen meat storage facility. Decide how long your slaughter season will need to be. Determine if you need a state/provincial or a federally inspected slaughter facility for your intended customers to be legally allowed to buy your meat and whether you will market lots of hamburger-grade cattle or only younger animals with higher-grade cuts. Clearly, these determinations will affect your production plan. Knowledge is power; once you know what your potential customers are looking for, you will be more capable of finding a way to meet their demands.

THREE CASE STUDIES

FOLLOWING ARE THREE CASE STUDIES of different-sized communities presented to illustrate how to identify the unique marketing opportunities specific to each and to demonstrate how to tailor your production and marketing strategies to take advantage of these opportunities. In each case study, I have paired two communities on either side of the Canada/United States border with similar marketing opportunities and similar population bases. The first focuses on Helena, Montana, and Grande Prairie, Alberta, to explore the wide range of marketing opportunities found around a midsized community and to provide a sample of the unexpected marketing opportunities that can be found by digging below the surface in and around your trade area. The second study examines the marketing opportunities presented to producers living within reach of two similar large urban areas: Seattle, Washington, and Vancouver, British Columbia. The third illustrates the marketing opportunities unique to producers farming in two similar rural agricultural areas, one outside of Lewistown, Montana, and the other in Kindersley, Saskatchewan.

Helena, Montana / Grande Prairie, Alberta

Producers farming somewhere in the vicinity of Helena, Montana, or Grande Prairie, Alberta, have a wide array of market opportunities unique to the specific character of their communities, both of which are medium-sized (population approximately 26,000 and 44,000, respectively) with a wide range of incomes so that a large portion of each community has the luxury of being able to pay a little more for good-quality meat. The large number of health food stores and growing organic sections in grocery stores also attest to an increased awareness of and keen interest in healthy and organic foods. Yet producers in these areas are not limited to the population base of these individual communities. Both lie in trade areas that serve a much larger population base that includes many other communities, large and small, if producers are willing to extend their marketing efforts to within a reasonable 100-mile radius. Both areas have a diverse economy and lie along major transportation routes that host substantial numbers of tourists and a considerable amount of commercial traffic. For example, Grande Prairie lies near the beginning of the Alaska Highway, which over 150,000 tourists travel each year. This area also has a considerable amount of mining, oil, and gas exploration and serves as the supply hub for an even larger region of mineral and oil and gas exploration to the north. Likewise, Helena lies along the Lewis and Clark Trail on the eastern edge of the Rocky Mountains, which adds up to considerable tourist-related marketing opportunities. Helena is also surrounded by mining and oil and gas exploration. (See below for more on the marketing opportunities related to the mining and oil and gas industries.)

These two communities have similar direct-marketing opportunities. The strong seasons here make the public keenly aware of the seasonality of nature. In addition, with access to so much wilderness and large wildlife populations, there is a large population of hunters in both communities — deep freezers and a seasonal meat supply do not intimidate the customer base. Therefore, selling meat by the quarter or side once a year to private individuals and families would work quite well.

Because direct-marketed meat does not have to travel far to reach a sizable customer base in the 100-mile radius of either of these communities, it is reasonably easy to establish a rapport with customers to guarantee your integrity and meat quality and invite them to visit your farm to view your production practices. This direct interaction can be accomplished via farmer's markets, farm gate sales, and on-farm retail outlets, by requiring customers to come to the farm to pick up their meat orders, and by hosting open-house events and farm tours.

A natural, grass-fed, or organic label on the meat definitely goes over well in these communities without requiring organic certification by an independent organic association if you limit meat sales to this direct contact with customers. Additional natural or organic meats such as pork, turkey, and lamb raised in the same basic farm infrastructure also easily find a market among this clientele. In addition, the high

dog population in the many nearby towns suggests a ready outlet for by-products as natural or organic dog food.

Note that organic certification will enable you to expand your direct-marketing efforts to include the large number of health food outlets and many restaurants found in each area, particularly if you can also extend your slaughter throughout the growing season and into the early winter, sell meat by the cut instead of by the side alone, and organize a frozen meat storage facility to extend the meat delivery season over the months when grass finishing is not possible. These health food outlets and restaurants serve not only the local population, but also the large number of tourists passing through the area each year, to whom local, grass-finished organic beef is very attractive.

Organic certification will enable you to expand your direct-marketing efforts.

Compared to urban centers and the big tourist destinations, these medium-sized working communities host fewer high-end restaurants that want the most expensive premium cuts; nevertheless, there is an abundance of restaurants and burger outlets that would benefit from featuring local meat specialties, depending on your marketing prowess in pointing this out to them. Further, vast amounts of ground beef will easily find a home during the summer barbecue season or year-round. Profit can lie in paying special attention to marketing to restaurants serving both tourists and the high volume of commercial traffic along the main thoroughfares. The number of tourists passing through also suggests a wonderful opportunity for agritourism and an on-farm restaurant or other eatery.

The mining and oil and gas exploration in both of these regions suggests another alternative marketing opportunity to the innovative grazier. Many of these exploration camps and mining towns are remote, and a company usually meets all of its employees' food requirements. These companies need meat and plenty of it as long as it's red and it's supplied regularly. While phoning the secretary at the head office will likely get you nowhere, a viable marketing strategy is to get to know the people in charge of these operations (owners, managers, suppliers, and in some cases the chefs that cook for the camps) and present them with a thorough proposal to fulfill their meat needs.

I know several company owners in the mining industry who routinely work and live in remote locations and have expressed a keen interest in grass-fed organic beef. In one case, an owner went to great trouble to supply his inaccessible northern camp with grass-fed organic beef and other organic meats and produce that had to be transported by truck and then bush plane for hundreds of miles to reach his remote camp. (I also worked in one remote camp a number of years ago where all the food-purchasing decisions were left to the camp chef, a five-star chef who worked for the heli-ski lodges in winter. We were encouraged to voice our wildest food requests, which stopped short only of flying in caviar from Russia.)

The lesson here is that many of the people who live and work in these industries and are separated from their families and the comfort of their homes expect certain luxuries and want to eat well to continue to be productive and happy in these remote settings. Every scrap of food they consume must be shipped to them, often at considerable expense and inconvenience, or someone must go to town to buy the groceries, usually with a generous budget in hand — and possibly drive right past your operation. These camps and mining towns are right outside the doorstep of these two communities. It's up to your marketing efforts to spark their imagination about particularly worthwhile eating possibilities and to convince them that your beef would add to their lives and that your business is capable of reliably supplying them with meat to fill their needs.

Suitable land for grazing is relatively inexpensive and readily available in the vicinity of either of these two communities as compared to many agricultural areas near cities in the mountainous regions farther to the west. Consequently, marketing grass-finished, certified organic beef to wholesalers supplying beef to these large markets is another viable opportunity to producers in these areas, particularly because the climate is favorable to both cow/calf production and grass finishing, which gives you the option of producing your own calves or purchasing stockers for grass finishing. An ideal option for acquiring stockers is to partner with other certified organic livestock producers who farm either locally or farther to the east, where a particularly dry climate or long distances to direct markets may limit their ability to grass-finish large numbers of cattle but allow them to produce calves very cheaply. (These producers are similar to those mentioned in the third case study, below — they are farming in rural, agricultural areas far from any urban centers.)

The well-established conventional cattle market, the ready supply of large numbers of cattle, and the large land base and low production costs of these two communities present two unique opportunities in the commodity marketing system. Producers can either engage in low-cost stocker grazing using the sell/buy economics discussed in the next chapter or use the production practices outlined in this book to manage their cow/calf herds to sell calves (or produce their own stockers) for the conventional cattle market. (Producers pursuing this option should read the discussion on the cattle cycle on page 223 for more information about how to manage your herds in a counter-cyclical marketing strategy that allows you to benefit from the cattle cycle rather than falling victim to its cattle price fluctuations.)

Seattle, Washington / Vancouver, British Columbia

Farming in the vicinity of a large metropolitan center like Seattle, Washington, or Vancouver, British Columbia, is ripe with opportunities. Both cities have an enormous population base with many ethnic communities and cultural heritages, a wide range of income levels, and a strong interest in healthy natural and organic food sources. Here, a large diversity of meat products easily finds a market.

Like most large urban centers, however, these communities are accustomed to a constant supply of food sources year-round. Slaughter in your enterprise must be spread over a much longer season, you must have freezer storage to extend the season, and you must be aware that direct-marketing customers tend to prefer smaller portions of meat spread throughout the year. Box programs and other scheduled marketing options certainly work well here. The large urban base also suggests possibilities for agritourism and on-farm eateries. The number of hamburger joints and high-end restaurants suggests markets for both substantial volumes of ground beef and expensive cuts with descriptive, health-conscious marketing labels.

If you choose to deal with sizeable wholesale-supplied health food stores, restaurants, and other third-party organic buyers, you may need to take the additional step of certifying your products. Yet the opportunity of the enormous local market at your doorstep makes it quite possible to develop your reputation and to be highly marketable without dealing with a certifying agency.

Agricultural land in the Seattle and Vancouver areas is fairly expensive, making it a scarce commodity; yet access to so many diverse high-end markets allows you to make a good income from a small number of cattle on a relatively small amount of land. With some careful planning and by using multiple plant varieties, the warm wet climate ensures excellent grass-finishing pastures almost year-round. The proximity to other cattle markets suggests that you should utilize your expensive land strictly for grass finishing by purchasing your animals as stockers or finishing-class animals, rather than maintaining a cow/calf herd as well.

Lewistown, Montana / Kindersley, Saskatchewan

Farming a long way from town, somewhere out on the open prairie in the rural, agricultural areas near Lewistown, Montana, or Kindersley, Saskatchewan, presents its own unique opportunities. Although direct-marketing options are fewer so far from urban centers, the really big advantage here is the low cost of real estate and land rental, making it feasible to farm more acreage and grass-fatten many more animals.

Being part of a large-scale rural agricultural area also means that federally inspected slaughterhouses are within easy driving distance and transportation costs are relatively low. This puts producers in the unique position of being able to produce economically substantial volumes of grass-finished organic beef for orders nationwide and, depending on export restrictions related to mad cow disease, for export to urban organic markets throughout North America and overseas in Europe and Asia. Many of these markets currently cannot find sufficient organic beef to meet their demands: European buyers, for example, have had to look as far as Argentina and New Zealand because there isn't enough certified organic beef produced in Europe or North America to fill their orders. These large markets are inaccessible to small farms typically found in the vicinity of big urban centers because these operations simply do not have enough land to consistently produce large quantities of beef and their higher production costs make it less feasible to supply meat to wholesale organic buyers.

The sizeable orders of certified organic beef for grocery chains, restaurants, and wholesale buyers worldwide will likely need to be supplied year-round and certified organic. Year-round slaughter for hamburger is always possible, and frozen meat storage facilitates a constant supply of more expensive cuts even in the winter months, when grass finishing is not possible. Being part of a large-scale, rural farming community also makes it feasible to team up with other local producers or to create organic marketing cooperatives. Another option is for northern producers to partner with producers in southern latitudes (or vice versa), who can grass-finish during different months meaning that your combined grass-finishing season is capable of filling a notable quantity of organic meat orders year-round.

The dry prairie winter facilitates a twelve-month-long grazing season, making these areas ideal for cow/calf and stocker operations. In addition to supplying your own grass-finishing market with young stock, you can market organically produced stockers to other grass-finishing areas nationwide, particularly to producers in areas that have higher production costs (to those dealing with high land costs, more difficult wintering conditions, land in urban vicinities, and areas with restricted land availability). This opportunity will increase dramatically as organic certification becomes more rigorous, restricting calves born in nonorganic operations from being finished and marketed as certified organic meat.

Farming a long way from town presents its own unique opportunities.

Both Kindersley and Lewistown are in areas that have abundant access to commodity cattle sales and also experience dry, low-snow winters that make twelve-month grazing so much easier. Producers in these areas are thus ideally positioned to branch out into commodity stockers to increase market diversity. Organically raised commodity stockers can always be rerouted into your grass-fed organic markets as you find additional marketing opportunities. What this area lacks in high-end market diversity it makes up for in low production costs.

Although the local rural community is composed primarily of other farm operations, even here, a sizable portion of organic and natural meat will find a local direct market among neighbors and the surrounding community. Many resource-based communities and rural areas are an excellent market for healthier natural or organic meat but are reluctant to buy certified organic because of the surcharge associated with certification. Local sales will rely on your reputation anyway, so certifying for this is usually unnecessary. On-farm retail outlets and agritourism are likely less suited to these rural areas because larger urban centers are several hours away and the local tourist industry is less developed. Yet it's still feasible to realize a weekly or biweekly slaughter and meat delivery to private individuals, restaurants, health food stores, and grocery outlets located in distant cities.

These three case studies illustrate different marketing strategies, production practices, and meat distribution styles that allow producers to take advantage of the market opportunities unique to each area. They also show just how many marketing options there are to choose from in any given community. Every individual farmer will have slightly different production and climatic considerations that accentuate his or her production advantages, as well as slightly different marketing opportunities waiting to be explored from her doorstep. Nevertheless, the process of matching production advantages to market opportunities to develop a marketing strategy for your individual target market is the same regardless of your location.

Stocker Cattle

Building your business around stocker cattle (calves between weaning and the final pre-slaughter finishing stage — 350 to 850 pounds — that are grown on pasture instead of a feedlot) destined for the commodity auction markets can be quite a lucrative business option. If your business is structured incorrectly, however, basing it on stocker cattle can also break you during the price fluctuations of the cattle commodity market (the cattle cycle). This chapter focuses on a business strategy that gives you financial buoyancy throughout the cattle cycle: sell/buy economics.

Traditionally, stocker businesses buy their stockers in spring, as the pastures green up; graze them throughout summer; and sell them to feedlot buyers at the auctions in fall. In theory, this makes good sense because you don't need to put up winter feed for these cattle, your labor for the operation is required for only six months, and your profit margin and yearly income is the weight gain less expenses. The cattle cycle, however, wreaks havoc with this type of buy/sell business strategy. When the cycle is on an upswing, your stockers gain value during the time you own them. But after the market peaks and the price falls again, your expensive spring stockers may well fetch less than you paid for them by the time you sell in fall. It is no joke that many stocker graziers with this kind of business management make money only five out of every ten years in the ten-year-long cattle cycle. Ouch! The sell/buy strategy yields consistently better results.

The Cattle Cycle

Cattle prices in the commodity business tend to go through a market cycle of ten to twelve years, despite the fact that beef prices in the grocery stores do not seem to change very much at all. Over time, humans tend to accentuate positive memories and suppress negative ones. This is a natural coping mechanism that allows us to live in the present instead of being controlled by events in our past. Unfortunately, however, this tends to affect adversely our decision making around cyclical events that occur over long stretches of time (perhaps it even causes the cyclical nature of many commodity markets) because recent positive experiences tend to outweigh distant negative memories. Thus, it is hardly surprising that so many farmers get stung badly with each cattle market collapse. Often, their cattle production and marketing decisions are based on their decisions related to cattle prices from the previous year, and disregard the vague lessons they learned ten years ago about the cattle cycle and how best to respond to price fluctuations without falling victim to the cycle's lure.

In a nutshell, the cattle cycle is determined by market optimism, the price of feeder corn in Kansas and feeder grain in the prairies, natural drought cycles, the flood of cattle sold in desperation during feed shortages, and the changing volume of the cattle supply. Here is roughly how it works.

As cattle prices begin to increase, people involved in all parts of the cattle industry start buying animals

THE CATTLE CYCLE

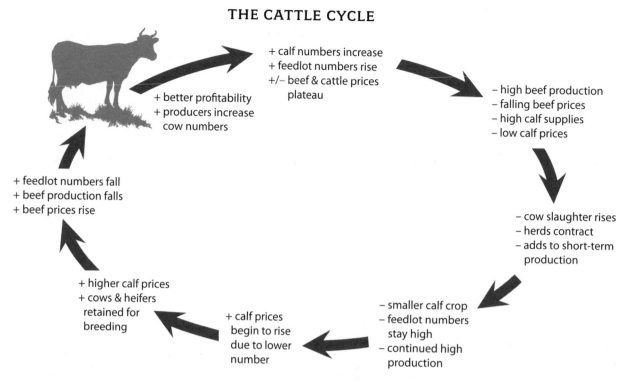

+ calf numbers increase
+ feedlot numbers rise
+/− beef & cattle prices plateau

+ better profitability
+ producers increase cow numbers

− high beef production
− falling beef prices
− high calf supplies
− low calf prices

+ feedlot numbers fall
+ beef production falls
+ beef prices rise

− cow slaughter rises
− herds contract
− adds to short-term production

+ higher calf prices
+ cows & heifers retained for breeding

+ calf prices begin to rise due to lower number

− smaller calf crop
− feedlot numbers stay high
− continued high production

The cattle cycle is based on the law of supply and demand. Each full market cycle typically takes ten to twelve years to complete.

based on market optimism. As investors compete to buy cattle to feed in the feedlots, the prices of both stockers and feeder-class cattle (cattle in the finishing stage when they are being fattened for slaughter) are driven upward. The high prices look encouraging, motivating farmers to buy breeding stock or increase their herd sizes so they too can take advantage of the good fortune. As heifers are held back from the feedlots to increase herd sizes or sell as breeding animals, the number of cattle being sold as feeders is reduced, driving up even further the prices of feeder cattle.

At some point, the production of all these new and enlarged herds comes online and begins supplying cattle to the market. The flood of cattle causes the feeder prices to decline, following the inevitable law of supply and demand. As the supply of feeder cattle exceeds the demand for cattle at the feedlots, the prices inevitably begin to drop.

Despite the falling prices, investors continue to stay in the game because they believe that feeder prices will recover. Everyone believes that cattle prices will remain the same or continue to rise. Hindsight continues to drive up breeding-class prices because investors want to get in on the good prices

they saw last year and the year before — they don't mind losing some of their earnings in the hope that their good fortune continue — but at some point their losses deplete their previous earnings and they begin losing some of their initial investment.

In short, investors follow the principle "just one more good round before I quit," which often causes Las Vegas gamblers to lose all of their gains as well as their initial wager. At this point, investors cut their losses and sell out, or get out of the game, so to speak. As the number of departing investors causes the cattle supply to increase and the demand for cattle to fall, feeder cattle prices drop in sharp market adjustments called *price breaks*. Optimism continues to drive prices upward again after each price break, but never as high as before, and each break is followed by yet another, lower price break.

Breeding-class price cycles tend to lag two to three years behind those of feeder- and stocker-class animals because of continuing market optimism among the cow/calf producers. At some point, however, calves stop bringing a profit at the auction yards, which causes their producers to begin reducing their herd sizes and rerouting replacement heifers back to the feed yards. This further increases the supply of feeders and causes feeder prices to drop even more. Because feeder-class prices have fallen, replacement animals are no longer desirable and prices plummet even further as cow/calf producers flood the market with more animals, cutting back their herds or selling out. This creates a vicious spiral of falling prices.

But every storm comes to an end. At some point, enough investors have left the game; enough cow/calf producers have gone broke, retired, or died; and herd sizes have shrunk enough that the demand for feeder cattle again exceeds the much reduced supply of available cattle, causing prices to begin rising again. The market cycle starts all over.

UNDERSTANDING THE CATTLE CYCLE

- The cattle cycle last 10 to 12 years, typically with a 6- to 7-year herd expansion phase and a 3- to 4-year herd liquidation phase.

- Cyclical price peaks occur when beef production bottoms out and cyclical price lows occur when beef production peaks, following the law of supply and demand.

- During the past few cattle cycles, feeder prices have usually bottomed out in years ending in 5 or 6, such as the market lows of 1975–76, 1985–86 and 1995–96, and feeder prices have usually peaked around years ending in 0 or 1. Yet while the herd liquidation phase starting in 1996 should have been over by 2000 in response to high cattle prices, abnormally dry weather in both Canada and the United States forced cow/calf producers to prolong this phase to a full 8 years, twice as long as usual, despite relatively high prices in the United States. (Throughout this prolonged liquidation phase, the Canadian cattle market also endured prolonged low cattle prices due to U.S. border closures over concerns about mad cow disease.)

  Consequently, based on U.S. beef herd inventory numbers, the North American beef herd expansion phase did not truly begin until 2004. With a 6- to 7-year herd accumulation phase, the next cyclical feeder price low will theoretically happen in 2009 or 2010, with the next feeder price peak occurring in 2014 or 2015, causing the rule of thumb stated above to shift by approximately 4 years. While the cattle cycle will undoubtedly continue its cyclical fluctuations, at the time of this writing it remains to be seen if the cycle will follow this predicted 4-year offset. If it does, we will see feeder prices bottoming out in years ending in 9 or 0 and peaking in years ending in 4 or 5.

- Breeding-stock prices tend to lag 2 or 3 years behind feeder prices.

- The bottom of the cattle cycle is typically marked by a terrible market price wreck when either drought or feed prices upset the already volatile market pessimism. Cattle flood the market due to a major drought or high feed prices resulting from shortages (the cost of feeder corn in Kansas being the most powerful influence), accompanied by a reduction in feedlot demand. These price wrecks usually last only as long as it takes to fatten one set of steers for slaughter: 90 to 100 days. Unfortunately, wrecks put a nail in the coffin of an already doomed situation.

- Because of the slow turnaround of the breeding cycle, it is *always* too late to try to respond to today's market prices. It takes 3 years to expand beef production at the slaughterhouses from the time cow/calf producers first decide to expand their herd's beef output to the time the increased cattle numbers finally reach slaughter-readiness. During this time, cattle slaughter numbers actually decrease because producers must first retain heifers for breeding that would otherwise be destined for slaughter. Then, when cattle prices start to drop, cattle slaughter will actually increase for several years because all the extra calves from the expanded herds and all the excess heifers culled to reduce herd sizes in response to low cattle prices must still be grown and fattened for slaughter.

  You have to predict the cattle cycle and position yourself to take advantage of market highs and lows by being ready to sell or buy opposite to the trend.

COUNTER-CYCLICAL MARKETING STRATEGIES

In an ideal world, keeping your eye on the predictable cyclical fluctuations of the 10- to 12-year cattle cycle and not letting your emotions over last year's prices get the better of you when making decisions should allow you to use the cattle cycle to your advantage; you can work contrary to what the bulk of cattle producers are doing. As I've said before, it pays to be out of sync with the masses. A number of counter-cyclical marketing strategies that allow producers to benefit from the predictable, long-term cyclical fluctuations of the cattle cycle are described below.

- When prices are at rock bottom and replacement animals are dirt cheap, increase your herd size. Prices can only go up from there.

- If market prices are strong, drop your herd size and sell as many animals as possible to take advantage of these prices. This is the time to build your financial reserves against the lean times. Unused or excess grass resulting from your herd reduction can be leased to other producers or it can be baled as hay to be stored as a drought stockpile or sold to other producers. You can rebuild your herd again when replacements are cheap.

- If you have stockers, a rising market is the time to buy lower-quality animals because they will still make you good money but will cost less.

- When the market cycle is on the decline, you want to buy the best-quality stockers because they will be cheap and will hold their resale value much better than the flood of poorer-quality animals on the market.

Does it work this way in the real world? Perhaps for those of us with nerves of steel, but I wouldn't stake my financial success on outguessing the cattle market. The market volatility from 2000 to 2004 related to drought and mad cow disease demonstrates just how unpredictable and nerve-wracking the ride up and down the cattle cycle can be. It is always easy to figure out the market in hindsight, but for most of us, a safer, more solid approach to financial success in the cattle cycle is usually advisable.

Sell/Buy Economics

To keep your business financially viable, you need cash flow. It doesn't matter so much what your net worth is doing during a market low; tax departments and bankers can lose sleep over that. What matters above all else is that your cash flow is adequate to pay your bills, debts, and taxes; make your essential purchases; and bring food to the table. Sell/buy economics ensures that you always maintain adequate cash flow, regardless of what the market is doing.

In a sell/buy market strategy, rather than waiting until the following year to buy your next set of stockers, you buy your replacement stockers immediately after you sell the ones you have so that your replacement stock is always cheaper than the animals you are selling. This difference between selling income and replacement cost generates the cash flow to pay your expenses and guarantee you a living. Your initial purchase cost when you bought your first set of stockers is your price for getting into the game, essentially a capital investment, and will be recovered years later, when you retire. The only important factor, then, is what the prices are when you initially enter the game and when you decide to sell out for good. In the interim, you are sure to have good cash flow each year regardless of whether each individual group of cattle makes or loses money (as long as its replacement group is always cheaper) or how the value of your inventory fluctuates over the course of the cattle cycle.

By replacing your stockers on the same market (or within a few weeks, at any rate), you will spend less than what you earned selling your previous animals. In a rising market cycle, you have to buy as close as possible to the sale date of your previous animals; in a falling market, you can afford to wait a little longer because replacement animals grow less expensive.

In order for this system to work, you have to keep stockers year-round, not just over the summer. Lightweight stockers bought in fall will typically be cheaper than stockers sold during the spring market runup. If your operation is structured for twelve-month-long grazing, these lightweight stockers can be grazed to meet their maintenance requirements, but with minimal weight gains, right through the winter. They recover their weight on grass in the spring as compensatory gain. With this approach, there is no need for expensive stored feeds or additional equipment during winter, so the cost of weight gain is still your cost to grow grass. Because the stockers are very light in winter, you minimize the amount of grass required during this time of year but maximize the grass consumption in summer as the animals grow and experience compensatory gain.

You can achieve further financial buoyancy by staggering your stockers so that you have multiple sale dates throughout the year. Each group must be replaced immediately in this staggered fashion. Multiple sale dates help reduce the risk of experiencing a poor sale.

Stocker Logistics

In chapter 2, I discussed how to choose a high-quality animal for your cow/calf herd based on low-maintenance and high-fertility characteristics. Yet in the commodity stocker business, the greatest opportunity lies not in buying the highest-quality animals, but in taking mediocre-quality animals and improving their quality through stress-free handling, quality grass, and high-quality management procedures.

This is not to say that emaciated, sickly animals or animals that have been recycled through the sale barn several times because they are difficult to sell are the way to go. Rather, you are looking for slimmer, rougher-looking, lower-priced animals that will blossom in your care. They still should be healthy and reasonably well nourished to respond to vaccinations, fend off transport and sale-barn stress, and acclimate to your farm. Upgrading rougher-quality stockers that have had tougher, less pampered lives is your value-adding opportunity in the price-per-pound commodity market. If your stockers are destined for a grass-finished direct-marketing scenario in which your livelihood and reputation are based on meat quality, upgrading rougher animals is less desirable.

Settling In

Settling freshly purchased animals into the farm is the most difficult challenge of any stocker operation. Often, they are freshly weaned, have been transported great distances, stood long hours in stressful sale barns, come from many different locations, and are not used to your food and water conditions. Because they come from so many different homes and are not accustomed to each other's bacterial loads, your vaccination and worming practices are particularly important. Designing your receiving strategy for these animals is the most crucial part of your whole operation. Your management strategy should be designed to minimize all stresses.

When the stressed calves arrive, process them immediately. A little bit of extra stress at this time will do less damage to their immune systems than will introducing stress later, after they have begun to settle down. Remember that stress does not simply disappear overnight, even if an animal seems calm. It takes weeks and even months for a nervous animal to become tranquil and relaxed. Once it finally settles down, its weight gains will improve dramatically; you don't want to disturb the process and rekindle the animal's nervousness. After the initial processing is complete, these calves are supposed to think they've living in paradise.

Processing

Processing should include a vaccination program, branding, and castrating. Tagging is optional: Only animals that are treated need a record, but then the record must be complete and easily accessed so that treatment can be diligent. Marker pens or spray paints, color-coded to identify a sick calf's treatment history from a distance, are a great help in facilitating your treatment regimen.

If you purchased steers, it is worthwhile to double-check each calf while it is in the chute so that you can be sure you have no bull calves in the group. This will allow you to run your steer and heifer calves together to increase your herd size and grazing efficiency. While in a feedlot situation this produces a significant amount of stressful riding due to cramped conditions and boredom, in a pasture-based operation in which the animals are moved daily to new grass and have lots of space, their minds are otherwise occupied; typically, the riding will be minimal after the initial introduction and the benefits of combining your herds will be significant.

Worming using pharmaceutical chemicals should be delayed because of the severe stress these chemicals have on the immune system. If, however, you use natural surfactants (such as Shacklee's Basic H soap, described on page 162) to worm, you can do this immediately in the water tubs in the receiving paddock/training pen. During processing, you can also dust the animals with diatomaceous earth to eliminate external parasites. (See chapter 10 for more on dealing with pests, parasites, and disease.)

Receiving Paddock/Training Pen

After processing is complete, settle the calves in your receiving paddock, which ideally should be surrounded with a permanent barbed-wire, mesh, or rail fence to contain animals that are not accustomed to electric fencing. Because the receiving paddock also functions as your training pen, where you will teach the newcomers about electric fences, build an electric drift fence (a cattle barrier that does not form an enclosure; the cattle are forced to walk around the open end of the drift fence if they want to reach the other side) to partially divide the paddock so all your animals have to come into contact with it on their way between food and water. This receiving/training pen should be directly adjacent to the processing area and hospital pen, allowing you easily to sort out sick animals, treat them, and subsequently quarantine them in the hospital pen to avoid infecting the remaining herd. This pen should also be relatively small so that sorting out sick animals during this crucial initial period is as simple and stress-free as possible.

Calves should be introduced to their mineral supplements in the receiving/training pen. These should include selenium to relieve stress and to address a deficiency in any animals coming from selenium-scarce areas. A separate free-choice access to diatomaceous earth is a great addition, particularly if your mineral feeder has buffet-style compartments

A RECEIVING/TRAINING PEN AND SICK-PEN MODEL

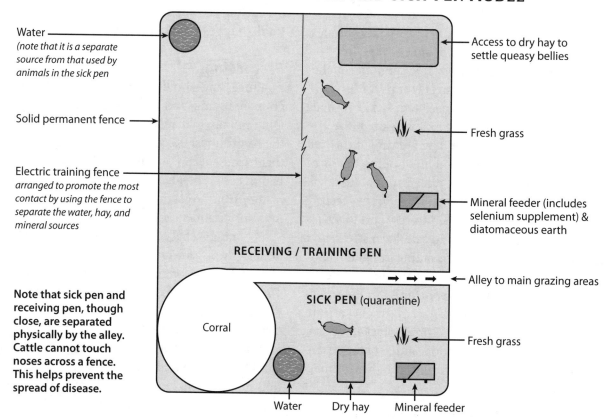

Water
(note that it is a separate source from that used by animals in the sick pen

Solid permanent fence →

Electric training fence
arranged to promote the most contact by using the fence to separate the water, hay, and mineral sources

Access to dry hay to settle queasy bellies

Fresh grass

Mineral feeder (includes selenium supplement) & diatomaceous earth

RECEIVING / TRAINING PEN

Alley to main grazing areas

Note that sick pen and receiving pen, though close, are separated physically by the alley. Cattle cannot touch noses across a fence. This helps prevent the spread of disease.

Corral

SICK PEN (quarantine)

Fresh grass

Water   Dry hay   Mineral feeder

for separating ingredients. This will help any calves with a calcium deficiency or internal parasites.

Spending a few days in the receiving/training pen when they first arrive will teach calves to respect the electric wires before they are moved out into the grass rotations. Feed them dry hay during this time to settle their nervous stomachs and solidify their runny manure. For the small amount required, it is usually most cost-effective simply to buy the hay. When it is delivered, have it loaded directly into feeders or laid out right in the training pen so you can set up portable panels or an electric wire and the cattle can self-feed. It will save you from requiring any equipment to handle the hay.

Within four to seven days, the calves should be ready to leave the training pen; your observations will tell you when the calves have ceased running into the training wire. Although you want to keep the calves close to the corral and hospital pen during the crucial first two weeks, it is equally important to start a pasture rotation on fresh grass as soon as possible to get the calves away from any oral-fecal contamination and to reduce disease pressure. Fresh growing grass will contribute far more to the calves' health than will dry hay.

Health Checks

During the first few weeks, you will have to do at least two health checks per day on your new arrivals. Make sure your water trough stays clean; wash the tubs periodically with a disinfectant.

Start the animals' grazing rotation in the immediate vicinity of the corral/hospital pen area so that sorting out sick animals continues to be as stress-free as possible. To avoid disease transfer, keep them on a separate rotation for up to a month before merging them with the rest of the animals on your farm.

Animals that are treated should be kept in the hospital pen for at least a week after their symptoms disappear so that you can monitor relapses and avoid infecting other animals. If possible, the hospital pen should be subdivided to create a small pasture rotation for the quarantined calves, with each subdivision accessible from the corral at all times to facilitate treatments. Here they can eat fresh grass instead of hay, oral-fecal contamination is minimized, and sick calves heal faster.

With cattle arriving from so many sources in such a high state of stress, I can't emphasize enough how important your hospital program is. Remove calves from the herd at the very earliest signs of trouble, treat them immediately, and keep them quarantined until well after they are healed. It is far wiser to treat early on if you suspect an illness than to wait in hope that the problem will solve itself.

Introducing Your Lead Animal

Immediately introduce your lead animal to the new arrivals. Because this experienced cow or steer knows the ropes, it will be crucial in teaching your new arrivals about gates and pasture moves. It will also help considerably in teaching your stockers to follow calmly during pasture moves. This is important because it helps to keep the animals much calmer and easier to manage than if you drive them from behind — and a calm move is much easier on gates and alleys, preserving your infrastructure. The large numbers of animals typical of many stocker operations makes calm moves vital so that herds can be merged into as few groups as possible and because sorting is common in a staggered marketing strategy.

An experienced lead animal is an excellent tool for teaching new stockers about gate locations and how to follow calmly during pasture moves.

AN EFFECTIVE ALLEY SETUP

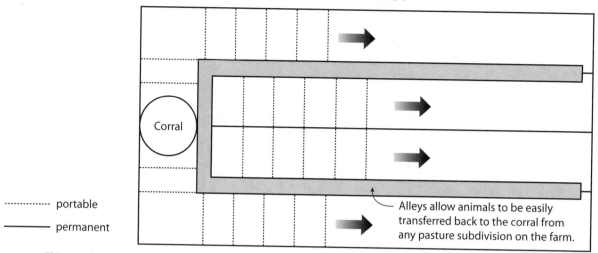

portable ············
permanent ──────

Alleys allow animals to be easily transferred back to the corral from any pasture subdivision on the farm.

This sample pasture layout illustrates how every portable electric-fence subdivision (dashed lines) between the permanent pasture divisions (solid lines) is connected to the alley network (in gray) leading back to the corral.

Building Alleys

For a stocker operation, it is particularly important that alleys leading back to your corral area are accessible from all of your major pasture divisions. This is crucial when you start sorting calves for market or for further treatment. Such alleys will also serve to transport your stocker herd easily from one end of the pasture rotation to the other to facilitate your grazing management. They should be wide enough to handle the largest herd on your farm in order to guarantee a smooth flow, especially for stockers, because they are more nervous, impatient, and excitable than older animals. Adequate space prevents mishaps and allows you to move your cattle calmly and efficiently. Please refer to page 230 for a detailed discussion of building alleys for your cattle.

Sorting and Separating

A few days to a few weeks before the animals approach their ideal market weight, the cattle should be sorted into separate weight groups that will all be sold on the same day. This avoids stressful sorting just before transport and allows the animals that are about to be sold to be grazed in a leader–follower system, one pasture slice ahead of the main herd, or set-stocked in a separate pasture to maximize gains before sale day. Set stocking is a grazing technique in which a small group of cattle is given access to a large pasture without a grazing rotation but at such a low

stock density that the animals can be very selective and graze only the best-quality forages while leaving the rest. Plan your group sizes to maximize your cattle truck loading capacities.

Before the sale date arrives, make sure your truckers know your procedure so that cattle prods stay home and there's no whooping, hollering, or banging gates on sale day. The day before loading, walk through all the steps so the gates are ready, the fences are secure, and the loading chute and pens are in good order. On sale day have your lead animal quietly follow you to the loading corral with the sale group in tow. Keep everything absolutely calm and silent. Eliminating the stress to your animals is the only legal way to avoid shrinkage; the difference this can make to your profit margin is quite remarkable. *Shrinkage* is the loss of livestock weight between the farm and the weigh scale at the auction and is caused by moisture losses due to evaporation, sweating, dehydration, urination, passing manure, and diarrhea. It is exacerbated by long hauling distances, long waits in the corral or auction yard, hot weather, and any form of stress including handling stress, nutritional stress, and disease. The last thing you want to do is lose all that hard-earned weight (and your profits) just before the auction. Besides, calm animals will gain better and stay healthier for the buyers too, who will probably thank you in the long run.

Chapter 17

Grass-Finished Beef

GRASS-FINISHING BEEF IS THE ART OF CREATING the healthiest and most tender and flavorful meat on the market. Although many of your preparations, infrastructure, and handling practices are identical to those used in the stocker business, the primary difference is that you are taking your calves right up to their finished slaughter weight. This requires even more diligent attention to nutrition and stress management.

The Art of Grass Finishing

If you are grass-finishing cattle for the commodity market, your choice of animals is less crucial. Because the emphasis is on weight, you can upgrade lower-priced, rougher-looking, slimmer animals through your quality management program. If, however, you are planning to sell directly to consumers, restaurants, or wholesalers in the natural or organic meat market, you have to carefully sort through your animals to eliminate hard-to-finish, nervous, high-maintenance individuals because their temperament will make their meat tough and ruin your overall reputation for producing flavorful, tender meat.

In addition to slaughtering finished steers and heifers, you can also butcher young cows up to five years of age with confidence about meat flavor and tenderness. These cows are actually more flavorful than the much younger steers and heifers typically butchered for their prime cuts (a marketing advan-

tage), and they are particularly quick and easy to grass-finish because they are no longer growing. In some countries, such as France, much older cows are slaughtered for their meat cuts (and not just for hamburger) because of the tremendous flavor. If these older animals are handled stress-free during their lifetime, their meat can still be extremely tender. Meat flavor is directly related to age: The older the animal, the more flavor it has. In France, specialty-labeled meats are never slaughtered before they are at least thirty months old, just to ensure good flavor.

Slaughter Weight and Frame Size

The ideal slaughter weight is determined by cattle's frame size. For instance, smaller-framed cattle such as Angus and Galloway typically reach finishing weight at roughly 1,000 to 1,050 pounds (the weight range is slightly less for heifers), starting around sixteen months of age. This means that you can usually start slaughtering them before their second winter.

On the other end of the scale are the larger-framed continental breeds such as Simmental and some of the tropical breeds such as Brahman. Because of their larger frame size, they won't reach the same degree of fatness (grade) until they are in the 1,400-pound range or, in some cases, even considerably heavier than that. They require an additional six to nine months to put on the extra pounds.

High-maintenance, leggy individuals may take even longer to reach their ideal finish grade for slaughter. If you get impatient and slaughter an animal too soon, you will end up with little more than tough, lean, hamburger-grade meat.

Meat Quality and Stress

Meat quality depends on the stress level of the animal. Stress causes the pH of meat to rise, which in turn causes the meat to begin breaking down almost immediately after butchering. Short-term stress leads to toughness and flavor loss in meat. Long-term or extreme stress causes the pH to rise so much that the meat becomes soft, mushy, dark, and sticky. The carcasses of these animals are called *dark cutters*. Although the high pH seems to prevent the meat of dark cutters from becoming tough, it begins graying almost immediately and has a very limited shelf life.

Nervous, stressed cattle never become much more than rubbery, unsavory hamburger. Calm, low-stress handling techniques are an absolute must. Calm animals gain weight much more quickly, and their meat is more tender and more flavorful, grades higher, looks more appealing, and has a longer shelf life.

The key is a calm, quiet routine. Cattle love predictability. If they have to constantly figure out a changing routine or adjust to inconsistent handling styles, they will become more alert and more attentive, which are manifestations of stress. It can take weeks or even months to fully calm cattle after handling problems are corrected, and they will remain nervous and alert long after the short settle-down time that follows a brief stressful event.

Train your cattle so that pasture moves are always quiet and happen at a walking pace (remember, cattle naturally walk much more slowly than you do). Teach them not to run through gates to new pasture. Running (theirs or yours), even a short dash through a gate to new grass, always creates nervousness and excitement in cattle. Remember that on a subconscious level, they associate running with predators and escaping from danger. Train your cattle to follow you quietly to a new pasture and whenever possible, use a lead animal (see chapter 9 for more on low-stress handling).

Gaining Weight before Slaughter Improves Meat Tenderness

Grass-finished animals must be gaining weight at the time of slaughter to ensure meat tenderness. Meat becomes tough through the nutritional stress of entering winter-maintenance mode or through a metabolism slowdown to compensate for a poorer quality or lower quantity of feed.

You should always grass-finish on your absolute best-quality pastures to maximize gains before slaughter. Seeding alfalfa, clover, and other legumes into your finishing pastures will greatly improve the grass quality and maximize weight gains to create a more tender grass-finished meat product. The rich grass/legume mix will also maintain a higher nutritional value later into fall weather, allowing you to extend your grass-finishing season beyond the finishing season of a pure grass pasture. Pure stands of clover, milk vetch, and other bloat-safe legumes make ideal grass-finishing pastures, and after they have been frost-killed to eliminate the risk of bloat, pure stands of alfalfa also work well to grass-finish.

Never move animals with empty stomachs from a poor pasture to a rich pasture.

Grazing Alfalfa and Preventing Bloat

Daily gains on straight alfalfa can easily rival or even exceed those achievable on high-grain diets in feedlots. No discussion about grazing pure stands of alfalfa is complete, however, without understanding alfalfa's bloat risk.

The plant cells in alfalfa leaves are degraded and fermented very easily by rumen microbes. If a diet is too rich in alfalfa, rumen microbes are able to multiply unusually quickly, producing a frothy slime that traps fermentation gases and prevents an animal from being able to burp up the gas, causing bloat. While mild cases are resolved on their own, particularly if the affected animal is made to walk about to help relieve the gas, if severe cases are not detected and treated early enough, the rumen can become so enlarged that it presses on the diaphragm and the animal is prevented from breathing or dies of heart failure. Death can occur in as little as an hour. If you are grazing pure alfalfa, it is worthwhile to keep bloat treatment materials on hand from your veterinarian, but in an emergency, to relieve the gas, you can puncture the animal's rumen with a knife and treat the wound later.

The more lush and rich the alfalfa, the more rapidly it is fermented, thereby increasing the risk of bloat. Dew or rain on alfalfa also increases the bloat risk, so it's best to wait for plants to dry before moving cattle onto alfalfa pasture. The risk of bloat decreases as the plant matures to full bloom, as plants dehydrate with age, hot weather, and after they die. (A word of caution about "low-bloat" alfalfa varieties: Even though they are somewhat less prone to causing bloat than other varieties, they nevertheless pose a significant risk and must be grazed with just as much caution as other varieties.)

Frost also initially causes a substantial increase in the risk of bloat because water freezing in the plant cells literally breaks open the cell walls, causing the plant material to become even more easily fermentable. Once the plant has been killed by frost and has had a chance to dehydrate, the risk of bloat is greatly reduced. This usually takes a week or so. For a period of approximately two weeks, however, from the time of the first killing frosts until the alfalfa has died down and dried, alfalfa cannot be grazed, even by cattle already accustomed to grazing it, because of bloat risk.

You must be very careful in introducing cattle to an alfalfa-rich or pure alfalfa pasture. Moving cattle from dry pastures into lush alfalfa stands is particularly dangerous. Carreful planning, however, can make the transition safe. Some commodity-marketed, grass-finishing operations have even perfected grazing 100 percent pure alfalfa stands through the summer months, despite the risk of bloat, by briefly introducing cattle to alfalfa for a short period each day, starting with no more than an hour of alfalfa grazing in any single grazing period and slowly increasing the cattle's daily exposure to alfalfa. Within a week the animals are ready to remain on alfalfa for the rest of the season.

To keep the cattle from gorging themselves when they are first introduced to alfalfa, you should turn them out on alfalfa only after they have filled up on grass from another pasture. Moving cattle to a fresh field of alfalfa after the end of the morning grazing period, when the cattle are most full and after the morning dew has lifted, will significantly decrease the risk of bloat. It is not possible to include non-alfalfa fields in the rotation without having to go through the transition period all over again.

When the fall killing frosts arrive, remove the animals from the alfalfa pastures for a couple of weeks until the alfalfa has died and dried out. Alternatively, a bloat inhibitor can been injected into the water source during this transition period to allow you to continue grazing the pasture until the frost-killed alfalfa has dried sufficiently to be bloat-safe again.

Incidentally, bloat can also occur simply by moving cattle from a dry, poor-quality grass pasture to a lush, rich grass pasture that has no alfalfa simply because of the sudden increase in fermentation produced by the lush grasses in the rumen of an animal unaccustomed to this rich diet. Never move animals with empty stomachs from a dry, poor pasture to a lush, rich pasture; wait until they have grazed their fill, or stuff them full of hay and watch them carefully, perhaps moving them back and forth between the two pastures until they grow accustomed to the rich food.

Mineral Supplements

Your mineral supplement program is also crucial to a successful grass-finishing program. A mineral- or vitamin-deficient animal will become nervous and nutritionally stressed. Kelp meal used as part of your mineral supplement seems to have the added bonus of being a natural meat tenderizer as well as adding flavor to the meat. See chapter 7 for a detailed discussion about supplementation and using kelp meal as a mineral supplement.

Grazing Options to Maximize Nutrition during Grass Finishing

Sorting out the finishing-class animals from the main herd is most suitable when you're grass-finishing large groups of animals or finishing-quality pasture space is limited. The finishing-class animals can graze the best-quality grass by adopting a leader–follower grazing style in which the finishing-class animals are grazed ahead of the main herd to give them access to the choice grass. Alternatively, if you have available high-quality grass-finishing pastures, such as legume stands, the finishing-class animals can be separated from the main herd for grass finishing and moved to a separate pasture rotation in the high-quality pasture.

The drawback to separating the finishing-class animals from the main herd is that you have an additional herd to manage, complicating your management and reducing the grazing efficiency of your herd. If grazed separately, the group of separated animals must be large enough for the cattle to retain their sense of group security. They must also be separated long before their slaughter date to overcome the anxiety of being removed from their original herd to avoid any adverse affect on their meat quality. During the last two weeks prior to slaughter, their gains can be maximized by switching from a pasture rotation to *set stocking*, which means that the animals are given a large area of choice pasture at an extremely low herd density to maximize their intake of only the highest-quality grass.

If the pasture quality is high, grazing the finishing-class animals along with the main herd will work just as well in a daily-pasture-move system that leaves a tall grass residual after grazing. The lower nutritional intake that results from having to share the grass with so many other animals is counterbalanced by the increased grazing efficiency caused by grazing in the competitive environment and security of the larger herd. You do, however, lose your ability to set-stock these animals unless you waste valuable grass on the main herd or cause stress to the slaughter animals by separating them at the last minute.

Slaughter

The trip to the slaughterhouse can make or break all your careful grass-finishing efforts. The lowest level of stress during slaughter is achieved by having the animals killed on the farm by a mobile slaughter unit. An animal simply walks off the pasture into a holding facility, where it is killed. This setup is rarely available, however, because it is far more cumbersome for the butcher for reasons of hygiene, temperature control, and waste disposal. Because of these

concerns, it's also very difficult to get meat inspected in a mobile slaughter unit. Because it is illegal to sell uninspected meat (except in a very limited number of locations and in small amounts in Canada; see the discussion about meat inspections on page 239), without this inspection your market access is limited to selling only live animals to customers who will use mobile units to custom-slaughter their own animals (check local regulations about live animals sold for slaughter in this manner).

The trip to the slaughterhouse can make or break all your careful grass-finishing efforts.

In most cases, your livestock have to travel. Try to find a slaughter facility less than three hours from your farm and absolutely no farther than eight hours. If the trip is too long, you will have to settle the animals for ten days or more before butchering to avoid stress effects on the meat quality.

Make sure loading and trucking are as quiet and stress-free as possible. Troubleshoot your loading facility the day before loading: Eliminate loud metal-on-metal noises in both the loading chute and the trailer, dampen rattling trailer gates, and provide a nonslip trailer floor. Don't crowd animals in the trailer, and when you begin the trip, drive as though you were transporting newborn babies. Cornering too fast, braking too suddenly, or hitting bumps and ruts can ruin your meat quality. (Remember, most stock trailers are not equipped with shocks, so a ride that feels smooth to the driver in the truck can be extremely jarring to the animals in the trailer.) If you hire out your transport, make sure the driver understands how crucial this is to you. This is not the time to be shy about your demands.

Don't be afraid to shop around for a slaughterhouse; the differences among them can be considerable. Besides looking for a facility that meets your meat-cutting criteria, you need one that handles the cattle gently and quietly before they are slaughtered. Slaughter employees who holler, bang gates, and use cattle prods will ruin all your grass-finishing efforts to create a tender meat product. Spend some time at the facility to oversee the sorting, handling, and killing of your animals until you develop a rapport with the butcher that ensures your animals aren't stressed before slaughter.

Slaughtering an animal while it is still experiencing the stress of trucking will result in tough meat. After the trip, it takes up to an hour to calm cattle's elevated heart rate and adrenaline rush related to short-term stress. Ideally, animals should be settled in individual pens at the slaughter facility, with access to clean water and good-quality hay prior to slaughter. They should, however, be slaughtered within six hours after transport, and *never* later than twenty-four hours after transport. After this, meat quality will begin to be affected due to the animals' stress at being separated from their herd on the home farm and being held in small confinement pens in an unfamiliar environment with unfamiliar water, feed, smells, sights, and sounds.

The Health Benefits of Grass-Fed Beef

One of the first things people tend to notice about grass-finished beef is the fat color. It is significantly more yellow than the fat in grain-fed beef due to a carotenoid (fat-soluble plant pigment) called *beta-carotene,* which the grazing animal incorporates into its fat and meat. Beta-carotene is used by the body to make vitamin A and is also a powerful antioxidant protecting cells from free-radical and oxygen damage. It also plays a role in preventing cancer-cell growth and cardiovascular disease. Grain does not contain a significant amount of beta-carotene, resulting in whiter fat. Meat with yellow fat is considered unappealing, and its price is consequently marked down by the grain-driven feedlot industry, primarily because that industry cannot produce it. Once customers are educated about the cause of this color difference and the health benefits associated with the yellow fat, however, they will quickly prefer it (much as people have come to seek out free-range chicken eggs for their rich yellow egg yolks).

Fat Content

The benefits of beef raised on pasture do not stop with beta-carotene-rich fat. Grass-fed beef is four to six times lower in fat than grain-fed beef and contains only half the saturated fat, making it as lean as poultry and wild game. This reduced-fat content also makes it significantly lower in calories than grain-fed beef.

Beneficial Acids and Vitamins

Omega-3 fatty acids, important to the healthy function of the brain and linked to lowering blood pressure, fighting depression, and reducing cancer, are produced by living green plant leaves. By contrast, a diet too rich in omega-6 fatty acids, which are found in the seed heads of plants (grains), has been linked to obesity, diabetes, immune system disorders, and cancer. To function well, our bodies require a balance between these two fatty acids (we don't want to altogether eliminate the omega-6 fatty acids), but for many of us, a diet high in grain (including grain-fed beef) has knocked that balance completely off-kilter. Grass-fed beef is three times higher in omega-3 fatty acids than grain-fed beef; in fact, the balance between omega-3 and omega-6 fatty acids in grass-fed beef is almost the ideal ratio required by our bodies.

Grass-fed beef also contains five times more conjugated linoleic acid (CLA), which can prevent and even reduce cancer-cell growth, and four times more vitamin E than grain-fed beef. (Vitamin E helps form red blood cells, is a powerful antioxidant that protects cell membranes and other fat-soluble body tissues, and aids in the prevention of heart disease and cancer.) Omega-3 fatty acids, beta-carotene, CLA, and vitamin E, all produced in living green plant tissues, are stored in the cow's meat and fat in forms that our bodies can easily access. But the meat and fat of a cow whose diet consisted of hay, silage, or overly mature grass (dead plant materials) does not contain all these beneficial ingredients, even though the absence of grain in her diet lowers the omega-6 fatty acids in her meat and fat. Supplementing a cow's diet with grain while on pasture will also eliminate all these grass-diet benefits and raise the omega-6 fatty acid content of the meat.

Only grazing green leafy grasses (and the dormant grazing reserve during winter grazing) will provide these health benefits. Grazing corn in the leaf stage will also provide these benefits because corn is a grass variety; once it forms cobs, however, the corn becomes a grain.

Impact on Herd Health

When we consider all the health benefits we derive from grass-fed beef, it should not be surprising that cattle grazed on a green pasture diet are also much healthier. They too seem to experience immune-boosting effects, a healthier heart, fewer tumors, lower blood pressure, longer life, and a host of additional benefits.

Because pasture-fed cattle are not raised in confinement, they are not subject to the manure concentrations, cramped living conditions, and high disease pressures associated with feedlots. Raising them in a pasture reduces or even eliminates the use of toxic pharmaceutical pesticides to control parasites and all but eliminates residues of high doses of antibiotics used on cattle in feedlot conditions. If these chemicals and antibiotics are not used on the cattle, they can't be passed on to us.

Given the premiums paid for healthy grass-fed beef and the lower cost of producing beef on grass, there is no excuse for producers to use growth hormones. Although some nonorganic grass-fed producers might still use hormones, most grass-fed beef should be free of them.

Environmental Benefits

And what about the environmental benefits of grass-fed beef? In a twelve-month grazing scenario, no air-polluting heavy equipment is needed. Large manure concentrations that cause air and water pollution are prevented by rotational grazing, whereby grazing cattle spread manure across the pastures and soil microorganisms recycle it back into the ground without creating odors or polluting runoff.

Raising grain, corn, and other tillage crops for animal feed exposes thousands of acres of bare soil to the harsh, oxidizing effects of atmospheric oxygen. This leads to evaporation of many of the soil elements, including carbon, which oxidizes to form carbon dioxide (CO_2), one of the greenhouse gases most responsible for global warming. Unfortunately, two-thirds or more of the grain produced in North America is grown for animal consumption, not for human food.

Grass-fed beef raised and finished in a pasture grazing rotation without the use of grains has the opposite effect on CO_2 levels. The grazing process, which causes the plant roots to continually die back and deposit their carbon in the soil, allows the plants to draw significant amounts of CO_2 from the atmosphere and deposit it in the soil as organic carbon (humus and dead plant organic matter), thereby reducing its levels in the atmospheric. The pasture sod acts like a skin preventing the soil from losing its carbon deposits to the air through oxidization. Pasture cows are a first line of defense against global warming. Perhaps grass farmers should be demanding environmental tax credits!

GRAIN-FED CATTLE AND *E. COLI* RISK

E. coli bacteria occur naturally in the digestive systems of cattle and other ruminants. On a grass diet, the stomach pH of a cow is around 6.4, which is only slightly acidic (nearly neutral). *E. coli* living in this mildly acidic environment inside the grass-fed cow's digestive system are not adapted to surviving in a highly acidic environment. If they were to accidentally contaminate our food or water, our own highly acidic stomach environment would protect us from infection because *E. coli* cannot survive in such conditions.

When cattle are fed grain, however, their stomach acidity increases to a pH of 5.8 to 5.3 (remember, the lower the pH score, the higher the acidity). Over time, this forces the *E. coli* bacteria in a cow's stomach to adapt to highly acid conditions.

Once these bacteria have adapted, they will continue to thrive if they enter our acidic digestive systems through contaminated food or water. This eliminates our natural defense against *E. coli* in the event of contamination from a grain-fed animal. *E. coli* contamination from grass-fed cattle thus poses a much lower health risk to humans than *E. coli* bacteria from grain-fed cattle.

Lessons from the Slaughterhouse

Another real market opportunity for grass-finished beef producers is the sale of organs from grass-fed animals. Consumers are hungry for healthy sources of organ meat such as beef liver, kidney, and heart.

Although grain-fed organs often end up as dog food, organs from healthy grass-fed animals are often the very first cuts of meat to sell out for human consumption in direct-market grass-fed beef enterprises. Grain-fed livers and other organs are often unfit for human consumption because they are forced to process pharmaceutical chemicals and antibiotics introduced to livestock in feedlot environments and because the process of digesting large amounts of grain produces high acidity, leading to livers that are riddled with abscesses from dealing with toxins resulting from these acid conditions. Indeed, if feedlot animals on high-grain diets are not slaughtered before their second birthday, they often die of liver failure soon after this time.

By contrast, the organs of pasture-raised beef do not encounter these problems, and consumers jump at the opportunity to consume vitamin- and nutrient-rich organ meats when they come from a healthy source.

Meat Tenderness

Two key factors affecting meat tenderness are the carcass finish (amount, character, and distribution of external, internal, and intramuscular fat on the carcass) and the amount of stress experienced by the animal during its life and especially during its last days and hours. Stress sources (including handling stress, nutritional stress, climatic stress, environmental stress, social stress, disease stress, and other management-related stresses, all of which are within the control of your management practices) cause cattle muscle fibers to tense, making the meat excessively firm.

Meat requires a certain amount of intramuscular fat to make it tender; if it has insufficient intramuscular fat, it will be dry, tough, and flavorless. Thin animals, such as those slaughtered during a drought or after a long winter, have depleted the intramuscular fat in their meat to supplement themselves through the period of feed scarcity. Likewise, animals that are slaughtered when they are too young have not deposited sufficient intramuscular fat in their meat to make it tender.

Marbling refers to the deposition of fat in small bands between the muscle fibers, appearing as small streaks of fat within the meat. Marbling has become a common way to gauge the carcass finish of grain-finished beef cattle: Grain-finished meat tends to show a high incidence of intramuscular fat when the grain-finished animal reaches its finished slaughter grade. Contrary to public opinion, however, it is not the marbling that makes meat tender and flavorful, but rather intramuscular fat deposited within the meat fibers on a microscopic level.

Meat from grass-finished cattle is leaner and consequently shows less marbling between the meat fibers than grain-finished beef — yet it contains sufficient intramuscular fat within the meat fibers so that the meat will be as tender as heavily marbled grain-finished beef. To understand how grass-finished beef can be lean yet flavorful, we have only to consider how lean wild game meat is, even from the fattest young buck harvested at the ideal time in the fall, and how tender wild game meat can be if it is harvested in a stress-free manner outside the rut season, and how fatty grain-finished farmed game can be, though it may be no more tender or flavorful than properly harvested wild game meat from an animal of the same age and species.

Interestingly, some beef breeds do not marble at all, even on a grain diet, despite the fact that they deposit intramuscular fat within the meat fibers. For example, the meat of Piedmontese cattle is reputed to be among the most tender, despite its absence of marbling.

Because meat becomes more tender when it is aged (hung) in the cooler before being cut and wrapped, ideally, it should be hung for at least fourteen days. In addition, attention should be paid to cooling the meat: Cooling too quickly causes the muscle fibers to shrink, making the meat tough. Unfortunately, the rapid-cooling standards of most slaughterhouses (and government regulations) are based on preventing spoilage in grain-fed beef. Because grain-fed beef contains more fat, which acts as an insulator, it does not cool as rapidly. By contrast, grass-fed beef cools much more quickly; thus, if it is to chill at the same rate as grain-fed beef and maintain its tenderness, ideally it should be cooled more slowly.

The longer shelf life of grass-fed beef is related to the high occurrence of vitamin E and other antioxidants in an all-grass diet. These prevent the fat from becoming rancid and the meat from losing color too quickly. Grass-fed beef maintains its color for up to thirty days; grain-fed beef loses its color after a few short days unless it is treated with an antioxidant, as some grocery stores do for their displays. This can be discerned by cutting into the meat: Rather than the outside retaining its red color while the inside loses it, the inside should retain its color longer than the outside because it is physically protected from oxidation by the external layers of meat!

Cutting and Wrapping

When cattle are slaughtered, typically only 50 to 62 percent of their live weight ends up hanging on the rail. The rest is head, hide, guts, blood, and other waste material. This *dress percentage (hanging weight* of the carcass as a percentage of live weight) varies from breed to breed and even from one animal to the next. Finished steers yield an average of 62 percent, and heifers up to 2 percent less. Cows and bulls show even greater dress percentage variability depending on the age and degree of carcass finish, with well-fleshed cows and bulls with a BCS of 7 averaging around 50 percent and 55 percent, respectively. Breeds such as Hereford and Simmental have heavier hides, feet, and viscera and therefore a lower dress percentage than breeds such as Angus and Limousin.

Grass-fed animals should yield slightly higher than grain-fed animals of the same weight and frame size because grass-fed meat is lower in fat (muscle fiber weighs more than fat). If you are selling meat directly to consumers as wholes, sides, or quarters, consider selling split sides instead of front and hind quarters. All the prime cuts are in the back half of the animal; the cuts from the front half are tougher and harder to sell. By selling split sides, all the cuts

from a side are divided up equally and you are not left with vast numbers of hard-to-sell front quarters or faced with the difficult task of pricing front and hind quarters differently.

When the carcass is divided into its various cuts, many bones are left over, fat is trimmed, and there are some cutting losses and shrinkage. The yield from hanging-carcass weight to cut-and-wrapped meat is called the *percent cutability,* or *carcass yield.* This yield depends on how much of the meat is deboned based on the types of cuts you choose. The rule of thumb for average percent cutability is 76 percent, although this varies quite dramatically with how the meat is cut (as mentioned above), as well as carcass finish (grade), breed, and age and class of animal. This means that a 1,150-pound steer yields the following amount of meat after cutting and wrapping.

1,150 lbs × 62% × 76% =
542 pounds of meat (organs extra)

live weight × dress percentage × carcass yield =
cut-and-wrap yield

Smaller Portions

The 14-ounce, 2-inch-thick steak is a thing of the past. The average modern consumer is looking not for volume, but for smaller, healthier, more flavorful portions. Smaller-sized steaks and roasts are far more popular, meaning that smaller-framed breeds are easier to market, which suits the pasture-based producer very well because these animals are lower maintenance and thus cheaper to produce.

Deboning and Muscle Groups

Another meat-cutting option to diversify marketability is to completely debone the animal and cut the meat into muscle groups, as many hunters and some Europeans prefer. Rather than the meat cutter cutting these up into individual steaks, the muscle groups are left whole and wrapped individually and the final cutting is left to the customers, who can cut their own steaks or use the meat for stir-fry, stews, or roasts, according to individual preferences. This gives customers the option of varying steak thickness based on cooking demands. Meat cut in this way stays fresh longer in the freezer because there are fewer cuts across muscle fibers that can cause the meat to slowly lose moisture and dry out.

Wrapping

Ideally, meat for the freezer is double-wrapped in brown butcher paper. Some people wrap the meat in plastic wrap prior to wrapping it in butcher paper to further seal in moisture and extend its freezer life. Cheap plastic wrap, however, degases some of the chemical components of the plastic into whatever it is covering. At best, this chemical leaching can affect both the taste and the smell of the meat.

Another attractive option for freezing meat destined for store shelves and grocery outlets (where meat presentation is half the battle) is to freeze the meat on trays before vacuum-packing it. This gives the meat a crisp appearance, whereas vacuum packing before freezing will collapse the soft, malleable meat into an unattractive blob. Steaks that have a crisp shape are appealing to customers.

Inspections

It is critical that you inform yourself about inspections. Every region has different standards; the rules vary dramatically not only from one region to the next but also according to the market to which you are selling. Meat sales are also regulated on a local level; municipalities and farmer's markets in both the United States and Canada may have further regulations about where and when meat may be sold, even if it meets the inspection requirements of the state. It is important that you research your federal, state, and local regulations before designing your marketing strategy.

In Canada some of the following are rules: Meat traveling across provincial borders or being exported out of the country must be slaughtered at a federally inspected slaughterhouse. Federal inspections require a federally licensed government veterinarian to be on staff at the slaughterhouse at all times. The permanent inspector's wages along with the more stringent building codes of a federal facility make killing, cutting, and wrapping meat in a federally inspected facility the most expensive option.

Provincially inspected meat does not require the same degree of stringent building infrastructure and the meat inspections are often carried out by a licensed local veterinarian who is not required to be on staff at all times. This considerably decreases the costs of provincially inspected facilities, making cutting and wrapping more affordable. It's important to note, however, that meat from a provincially inspected facility cannot be exported beyond the boundaries of the province in which it was slaughtered.

Grass-fed animals should yield slightly higher than grain-fed animals of the same weight.

Meat sold commercially, such as to restaurants and health food stores, requires at least a provincial inspection because these outlets cannot sell uninspected meat. Most municipalities also require that meat sold within their boundaries have at least a provincial inspection. This is not always the case, however, particularly in some very small communities.

Beyond municipal boundaries, the rules vary dramatically. Some regions insist on a provincial inspection; in other areas you can still direct-market uninspected meat by the whole, side, or split side for private sales outside of municipal boundaries, though these lax regulations are rapidly becoming more stringent and the number of allowable uninspected meat sales is both limited and subject to interpretation. This inspection exemption is meant for small hobby farms that butcher a very small number of animals, but as numbers climb to those of a sustainable business, the window of opportunity closes.

If you live within a regulated region, you have to slaughter at a provincially inspected facility in order to direct-market your meat. Even if you live outside the regulated zone, you still require a provincially inspected facility for meat that you intend to sell within inspection zones — a situation that often affects farms located just outside city limits that want to sell in town.

All meat sold in the United States must be inspected. There are four different types of slaughtering plants, which are determined by the meat inspection they receive: First, federally inspected plants must comply with USDA standards. All meat that crosses state or national lines must be federally inspected. Once inspected, it can be sold to anyone, whether by the whole carcass, side, or cut.

Second, there are federal-state cooperative inspection plants. These are slaughterhouses where the USDA is responsible for the inspection but state meat inspectors are contracted to perform the inspection according to federal standards. This is a federal inspection and receives the USDA stamp.

Third, there are non-federally inspected plants. These are state-inspected plants that comply with state inspection standards, which must be equal to USDA standards. All meat with this inspection standard may be sold and transported only within the state in which it is inspected; it cannot cross state lines. Some states do not have state-inspected plants, operating instead only under the federal inspection status. Other states have only state-inspected plants and therefore cannot export meat to other states.

Fourth, there are custom-exempt plants, which are exempt from the federal meat inspection status, though they still must meet health standards. Custom-exempt slaughterhouses do not sell meat; they may only custom-slaughter and process meat for the owners of live animals for the exclusive use of these owners and their households, non-paying guests, and employees. This meat may not be sold to another party by the owner of the live animal and must be labeled "Not for Sale." Mobile slaughter plants fall under this category, except for a single federally inspected mobile unit serving the San Juan Islands (though Montana's legislature has also recently approved a measure to authorize state inspection of mobile slaughter plants).

Selling "On the Hoof"

If an inspected plant is either unavailable or too far away to be economically viable to your marketing strategy due to high shipping costs, in some cases you can use a custom-exempt slaughterhouse by selling live animals to your customers, which the custom-exempt slaughterhouse will then process as your customers' live animals. Customers will either deal with the meat cutter directly or make arrangements through you. This option works only for direct-marketing beef to private individuals who want a whole, side, or split side of beef and who buy live animals or partial shares in live animals. It is not an option for selling meat by the cut.

Unless you charge a price per pound of live weight, which will be to your disadvantage, if you sell live animals to your customers for custom slaughtering, you must somehow legally recover the difference in carcass weight and carcass yield through innovative pricing schemes and charges related to the handling and delivery of the live animals. This can be accomplished by selling a live animal for a base price and then charging a delivery fee that varies according to the carcass hanging weight. It is vitally important that you check your local regulations before engaging in any kind of live beef sales destined for immediate custom slaughter to ensure that your operation, how you market your animals to your customers, how you adjust your price to accommodate the differences in carcass weight, and the number of animals marketed in this way comply with federal, state, and municipal laws.

Certified Organic Meat

Certified organic meat must be processed at a facility certified for organic slaughter. In most cases, this means that the slaughter facility must provide separate meat storage and hanging space to separate certified and uncertified meat, and certified organic animals must be killed and butchered at a different time to prevent mix-ups and contamination. Usually this is accomplished by slaughtering organic animals on a designated day of the week after the facility is thoroughly cleaned according to organic certification standards.

Pricing Your Meat

Pricing your grass-fed beef can be a challenge because there is no commodity standard to which you can refer. You must take into account your cost of production per pound of beef produced and your price must allow you to meet the financial goals you have set for yourself (see chapter 21). Yet your price is restricted by what your target market will bear. The fewer people there are between you and the customer who eats the meat, the higher your price margin can be.

Direct-marketed sales by the whole carcass, side, or split side are usually calculated on a price-per-pound basis, based on the dress percentage or hanging weight. Because customers will have their meat cut according to their preferences (variable deboning depending on cuts selected), this method does not affect your income structure, whereas selling based on carcass yield makes your income unpredictable.

Your price per pound will depend on your cost of production per pound of beef plus the desired profit margin percentage that you set for yourself during your financial planning. It will be based on how many animals you slaughter and the expected dress percentage (50 to 62 percent) of each class of animal. Researching what other, similar operations in your area are charging for their direct-market sales will usually give you a rough indication of the price range the market will support.

Pricing for Specialty Markets

To price meat by the cut for wholesalers, restaurants, farm retail outlets, or box programs (discussed on page 260), you will have to work with your butcher to determine the exact carcass yield and the breakdown by weight for each of the cuts. See page 294 for a chart that shows a typical breakdown of meat by the cut for an average 1,150-pound steer. From this you can determine if the price structure you choose for all your cuts adds up to a profitable average price per pound of meat produced. Don't forget to add the cost of freezer storage to your cost of production!

A trip to the meat counter at the grocery store to look at the prices will usually tell you what your price structure should resemble. Raise or lower the grocery price structure by an equal percentage to reach a price that your target market will bear and that fulfills your financial goals. The risk is ending up with wasted undesirable cuts. The grocery store price structure will help here because it is designed to sell out all cuts equally. The most expensive cuts of direct-marketed meat tend to sell quickest.

Pickup and Delivery

I recommend that your customers pick up their meat directly from the meat-cutting facility on a designated pickup day, allowing you to meet them to receive payment before the meat is released from the safety of the freezer. You can't run a business on promises of future payment and you certainly can't afford the time, fuel, and frustration of tracking down payments after the meat has been picked up.

It is crucial to provide an order form clearly defining a payment structure to ensure payment upon delivery. A down payment on orders will guarantee customers their meat and ensure that they'll actually pick up meat they've ordered. Orders without a financial commitment are unpredictable and make it difficult to plan sales.

If you don't put a value on your time, neither will your customers.

Although I don't recommend delivering to private sales, if you choose to deliver, charge an additional delivery fee. Don't underestimate the cost of your vehicle, fuel, and time. What often seems like a negligible service can quickly cut into your profit margin. If you don't put a value on your time, neither will your customers. You want to avoid the scenario of wasting valuable time by waiting for a late customer to arrive home or spending an hour loading an overcrowded deep freezer while you have a load of meat thawing in the back of your vehicle.

Meat sold by the cut will usually be sold to stores, wholesalers, or restaurants, or to retail customers through an on-farm retail outlet equipped with freezers. Large orders, scheduled orders, or box programs should all be guaranteed in advance by a down payment or sales contract so that volumes don't change or demands for certain cuts aren't canceled at the last minute.

Transporting Meat

Transporting meat for scheduled orders, meat delivery, or even a retail business can be accomplished in a number of ways without incurring the cost of a freezer truck. Small orders can easily be transported in coolers for considerable distances if the cooler is chilled and packed inside the freezer prior to traveling. I have routinely traveled two days or more with a sealed fifty-five-quart cooler packed to the brim with frozen meat without losing any meat to thawing, despite an outside temperature of 68°F/20°C. No ice is necessary; the frozen meat will do a better job of keeping the cooler cold. Just be sure that the entire cooler is filled and do not, under any circumstances, open it before reaching your destination. This method will allow you to accomplish short trips for local deliveries.

For longer travel distances or in warmer temperatures, you can carry a deep freezer in the back of a pickup truck. Again, the large mass of meat in the deep freezer will keep the meat frozen. For smaller volumes of meat, fill up the additional space with ice. If you must access the freezer multiple times during the delivery run, use a converter from 12 volts to 110 volts to plug the deep freezer into your vehicle's power outlet.

Long-distance and large orders can easily be delivered by renting space in one of the thousands of freezer trucks that are constantly traversing the country. Many have extra room and welcome the opportunity for increased business. A number of smaller food outlets even have small freezer trucks that run only part time; these either have underutilized space on their regular routes or can be hired — with or without a driver — on days they don't make deliveries.

Cooking Grass-Fed Beef

Because grass-fed beef is leaner than grain-fed beef, to preserve its tenderness and flavor it should be handled slightly differently during cooking, and to prepare it you should use different recipes from those used for grain-fed beef. To help your customers enjoy the experience of cooking and eating grass-fed beef and ensure their continued satisfaction with your meat, this information should be passed on to them either directly on the meat labels or in pamphlets accompanying a meat delivery. To educate your customers, it is also worthwhile to include in your customer newsletter your favorite beef recipes along with the cooking differences between grain-fed and grass-fed beef.

When beef is cooked too fast, the high heat destroys the structure of the proteins, which makes the meat tough and chewy. The slower the beef is cooked, the more it will retain its tenderness and flavor. Fat has an insulating effect during the cooking process: It slows heating as the meat adjusts to the cooking temperature. Because of its lower fat content, grass-fed beef requires lower cooking temperatures and shorter cooking times to cook through. Along with a shorter cooking time, grass-fed beef should be cooked using 15 to 25 percent less heat. Roasts usually cooked at 325°F/160°C should be turned down to 275°F/135°C and checked a few minutes sooner. Even barbecuing should be done at low heat. Meat from breeds that do not marble (for example, Piedmontese) will cook even faster at the same temperature because there is even less fat in the meat fibers.

Less fat also means that there is less fat to liquefy during the cooking process (the primary cause of meat shrinkage during cooking). Consequently, grass-fed beef, especially ground beef, will shrink significantly less during cooking. You can expect your burgers and ground beef to remain close to the same size throughout the cooking process.

 # Paradise
MEADOWS

Beef Order Form

_____ Whole (400–650 lbs)*
($2.75 per lb / $300 deposit)

_____ Side (200–300 lbs)*
($2.85 per lb / $170 deposit)

_____ Split Side (100–150 lbs)*
($2.75 per lb / $90 deposit)

PREFERRED SIZE*
❑ Small ❑ Medium ❑ Large

❑ Small ❑ Medium ❑ Large

❑ Small ❑ Medium ❑ Large

Please fill out the butcher's packaging instructions form on the reverse of this sheet

Preferred pickup date**: ❑ August ❑ September ❑ October ❑ November

Pickup occurs on the first Saturday of each month at:
"The Organic Butcher"
1234 Mountain Hwy, Sun Valley, BC
Tel: (250) 000-0000

There is no UPS delivery available on whole, side, or split sides of beef.

*All prices and weights for whole, side, and split sides are based on hanging weight, and include cutting, wrapping, and freezing. Beef take-home weights are approximately 25–35% less than carcass hanging weight due to trimming and deboning.

** We will do our best to accommodate your size and pickup preferences, but all orders are reserved on a first-come, first-served basis.

- -

SPECIALTY PACKAGES

_____ **40 lb Grilling Package** ($260)
(A premier selection of steaks, one roast, two stews, and 15 lbs of hamburger)

_____ **40 lb Family Package** ($220)
(A selection of roasts, stews, a few steaks, and 15 lbs of hamburger)

_____ **20 lb Grilling Package** ($120)
(A premier selection of steaks and 10 lbs of hamburger — this package fits comfortably into your fridge's freezer)

_____ **20 lb Family Package** ($110)
(A selection of roasts, stews, 10 lbs of hamburger — this package fits comfortably into your fridge's freezer)

_____ **10 lb Sample Package** ($260)
(Steaks and 5 lbs of hamburger to give you the organic grass-fed beef experience)

Preferred pickup date**: ❑ August ❑ September ❑ October ❑ November

_____ Hamburger ($260) (available year-round)

I want _____ lbs of hambuger in _____ one, _____ one and a half, or _____ two lb packages.
If not specified, hamburger will be wrapped in one lb packages.

PACKAGING INSTRUCTIONS FOR WHOLE, SIDE, AND SPLIT SIDES ONLY

Please check your preferences below (*asterisk indicates the most popular choice):

Roasts: _____ 2 lb packages _____ 3 lb packages* _____ 4 lb packages

Hamburger: _____ 1 lb packages _____ trimmed*

_____ 1.5 lb packages* _____ trimmed close

_____ 2 lb packages

T-bone steak thickness: _____ ¾" thick _____ 1" thick* _____ 1¼" thick

Sirloin steak thickness: _____ ¾" thick _____ 1" thick* _____ 1¼" thick

Round steak/cube steak: Which do you prefer?

_____ round _____ cube* _____ include with hamburger

If you choose Round: _____ ½" thick _____ half size*

_____ ¾" thick* _____ full size

_____ 1" thick

_____ 1¼" thick

If you choose cube: _____ 1/pkg _____ 2/pkg* _____ 3/pkg _____ 4/pkg

Rib steak: _____ Rib steak (bone)

_____ Rib eye (no bone, skirt)*

_____ Rib roast

Soup bone: _____ Package separately*

_____ Grind into hamburger (fatty — keep out of lean hamburger)

_____ Discard

Quantity per package: _____ none _____ one _____ two* _____ three _____ four

Stew meat: _____ Package separately*

_____ Grind into hamburger (makes hamburger leaner)

Check all that you would like with your meat order:

❑ Heart* ❑ Liver* ❑ Kidneys* ❑ Oxtail* ❑ Tongue ❑ Sweetbreads ❑ Briskets

❑ Marrow bones ❑ Other (please specify) _____ *

Sausages available by special request (processing costs extra).
All meat is hung 14 days for tenderness, except by special request.

SPECIAL INSTRUCTIONS:

_____

_____

_____

_____

_____

LIMITED YEAR-ROUND AVAILABILITY (all prices are in Canadian dollars):

_____ Dinner sausages (10/pkg)   _____ lbs × $8 per lb = $ _____

_____ Breakfast sausages (10/pkg) _____ lbs × $8 per lb = $ _____

_____ Garlic coils (single coils) _____ lbs × $8 per lb = $ _____

_____ Salami (single coils)   _____ lbs × $10 per lb = $ _____

_____ Smokies (4/pkg)   _____ lbs × $8 per lb = $ _____

_____ Pepperoni (6/pkg)   _____ lbs × $8 per lb = $ _____

_____ Beef jerky   _____ lbs × $25 per lb = $ _____

All weights are approximate; packages will be weighed during pickup/packaging.

Inquire at our on-farm store — we have limited freezer space available for year-round meat sales by the cut, hamburger, sausages, some specialty packages, and bulk pet-food scraps.

Name: _____

Mailing address: _____

_____

Delivery address (if ordered shipped via UPS): _____

_____

_____

Telephone: _____ (home) _____ (cell)

Fax: _____ E-mail: _____

Enclosed is my cheque for $ _____ (20% of order). The balance will be due on receipt.
All sales are in Canadian dollars.

❏ I will pick up my order at   ❏ I want my order shipped via UPS
The Organic Butcher   (Note that UPS shipments must be
1234 Mountain Hwy   prepaid. Call ahead for shipping
Sun Valley, BC   charges.)
Tel: (250) 000-0000

Make cheques payable to: Fictitious Farm, Fictitious Address, Tel: (000) 000-0000; E-mail, Web site.

WE WANT TO HEAR FROM YOU (meat reviews, comments, questions, suggestions):

_____

_____

_____

_____

_____

_____

PARADISE MEADOWS

"Health, Vitality, Land Stewardship, and Superior Beef"

Organic Certification

THE ULTIMATE GOAL FOR MANY ORGANIC AND natural producers is organic certification. Without a description of the production practices, organic certification guarantees only that unhealthy chemicals and antibiotics are not in the product; it does not guarantee what ingredients are in the product. Nevertheless, it is the ticket to consumer trust in situations where the consumer has no contact with or no knowledge of the producer. Meat with this certification appeals to markets such as wholesale buyers, health food stores, and restaurants, because it is a well-known standard and therefore commands a higher price on their shelves or menus. But beware: Organic certification does not automatically bring the highest income.

Seeking Certification After the Transition

The appeal of higher prices for certified organic products in health food stores lures many producers into mistakenly thinking that organic certification is the magic bullet to quickly bail them out of financial difficulties. It is also tempting to start the organic certification process as soon as possible with the idea that the long, three-year-minimum transition process required by most organic associations will allow you to figure out all the details.

But when you eliminate synthetic fertilizers, pesticides, antibiotics, and parasiticides all at once, before your management practices (grazing rota-

tion, calving season, and winter grazing) have had time to offset the need for these products, you risk expensive difficulties such as significantly reduced production levels, mineral deficiencies in your soils, mineral and nutritional deficiencies in your animals, fertility catastrophes, and crippling animal health problems.

Proceed with organic certification only *after* your production practices run like a well-oiled machine. Once the organic certification process has begun, you no longer have the option of reaching for any of the conventional tools to bail you out while you figure out how to solve your production challenges through natural means; the only option you have if you encounter a problem is to use expensive organic inputs (fertilizers, feed supplements, parasite-control materials, and other off-farm resources), costly natural remedies, and tedious veterinary alternatives.

It is remarkably easy to lose yourself in advertisements for alternative organic products to replace the chemical and pharmaceutical products farmers are accustomed to using in conventional agriculture. In fact, instead of replacing conventional products with more expensive, labor-intensive alternatives, you have to *eliminate the need* for these products through your management strategy. In the long run, it is far more economical to slowly wean yourself off conventional agroproducts prohibited by the organic associations while your production changes eliminate the need for them. The less desirable alternative is to cut off prohibited products overnight by beginning the certification process too soon, while you tighten your belt and hope you can survive the whole ordeal.

Headaches of the Certification Process

Organic certification can be a significant additional headache in your production process, even if you exceed all the requirements of the certifying committee. It requires that you keep a tremendous paper trail of all your business dealings and the sources of all your off-farm inputs (fertilizers, feed supplements, and parasite-control materials, for example). Required for quality assurance, these records mean nothing to your everyday management strategy yet soak up hours of your time.

Some of the product restrictions implemented by the organic certification associations may be questionable when applied to your specific operation, but the associations must uphold them because their rules must cover all producers in all production scenarios in order to be fair. Your efforts tend to become focused on how to avoid or replace certain products rather than how to improve the quality and cost-effectiveness of your natural production strategy, which can have a crippling effect on how you approach your production challenges. And of course the financial cost to carrying an organic association's "certified" label either eats into your profit margin or has to be passed on to your customers.

Choosing the Best

If you need organic certification for the market you are targeting, find out which certifying association is most recognized by your buyers before making a certification choice. They are not all equally recognized, they do not all have the same rules, and each charges a different fee for the certification process (some take a percentage of your gross income, while others charge a single fixed fee or vary the fee based on the acreage farmed).

Being certified by one association does not automatically guarantee certification by another; because standards are not universal, to be certified by another association that has different rules, you may still

have to undergo a transition process. Local organic associations are usually most suited for local markets where their label is recognized. The farther you sell from home, the more internationally recognized your certifying association must be. These will vary from country to country or from one wholesaler to the next.

Also consider what kinds of products a certifying association is most accustomed to dealing with. For example, if a local association most often certifies smaller-scale vegetable, fruit, and herb farms, trying to certify a large-scale livestock operation will be far more difficult for everyone involved than if you were to work with a larger association that is familiar with both your product and your scale of production. The rules will be more clearly defined for your particular product and production scale and you will face fewer political and bureaucratic obstacles because you won't be the odd one out.

Solving the Most Common Challenges

The challenges discussed in the following pages are the most common barriers to producers achieving organic certification for their livestock, even if their pastures already meet organic production standards. Overcoming these challenges is particularly difficult because organic alternatives to the conventional practices and materials most often used to address them are not readily available. These difficulties must be overcome not through addressing symptoms but through preventive management techniques and a comprehensive production plan that eliminates the root causes of these production obstacles.

Pharmaceutical Parasiticides

For most livestock operations trying to achieve organic certification, parasites are the biggest stumbling block. Until you find a way around lice, worms, warble flies, and other internal and external parasites, all your certifying efforts will be for naught. Parasites, then, are the measure by which you can gauge whether your livestock operation is ready for organic certification. As soon as you can eliminate cattle oilers (self-treatment scratching posts that dispense an insecticide-oil mix onto the cattle's skin and hair when they rub against the device), fly tags (insecticide-impregnated ear tags that slowly release insecticide, which the cattle inadvertantly spread over their faces and bodies as they rub together or groom themselves), and systemic parasiticides (absorbed or injected into the body of the cow), your operation will be ready to enter the organic certification process.

Good management is the only cost-effective solution to reducing your farm's medicine chest.

This is easier said than done. Parasites are the indicators of flaws in the production system and target any animal whose health is stressed. The livestock equivalent of weeds to poor soil fertility and poorly managed vegetation, they rise and fall to match the opportunities presented by your production system. Certainly, there are some natural alternatives (such as diatomaceous earth, neem oil, garlic, homeopathic remedies, and other products approved by your chosen organic association), but your success in meeting the organic regulations for parasite control depends primarily on prevention rather than on cure.

Reduce parasite pressure on your cattle with an effective grazing rotation (ideally with daily pasture moves) to disrupt the parasite life cycle and reduce parasite contact with your cattle. Manage livestock health to give your cattle the greatest resistance against parasite predation; sickly and weak animals are always primary targets.

Good soil fertility and mineral supplementation address livestock's nutrient deficiencies, an effective grazing rotation maximizes the quality of their nutritional intake, and a rigorous culling program eliminates the weaker, more parasite-susceptible animals. A small parasite load is not a problem; only when the parasite load begins to affect cattle health and meat quality does a problem develop. Instead of trying to eliminate parasites altogether, systematically eliminate their opportunities and advantages through effective management (see chapter 10 for more information about managing pests, parasites, and diseases).

Antibiotics

Although there are plenty of homeopathic, herbal, biodynamic, and other natural remedies to cure cattle of disease, prevention is much simpler than the tedious and often expensive alternatives to antibiotics. As for parasite control, rigorous attention to your grazing rotation, animal health, nutrition, and herd stress, careful herd selection and culling, and striving to match your production cycle with nature's seasonality are key to preventing disease pressure.

For example, a summer calving season timed to take advantage of benefits afforded by nature's seasonality all but eliminates the scours and coccidiosis that plague so many conventional calving seasons. An effective grazing rotation that reduces dust, flies, and overly mature grasses whose seeds can irritate cattle's eyes also reduces summer disease pressures such as pinkeye. Winter-grazing rotations all but eliminate oral-fecal contaminations that are the source of so many diseases in winter feed areas and encourage animals to sleep in fresh snow every night rather than returning to manure-laden bedding sites. Finally, a carefully planned shipping, handling, and receiving program manages shipping stress.

Good management is the only cost-effective and practical solution to reducing your farm's medicine chest. Again, the focus should be on eliminating the opportunities and advantages of the disease and on maximizing the animals' natural immunity, rather than on curing disease occurrences with quick fixes.

Synthetic Fertilizers, Especially Nitrogen

The drop in grass production associated with eliminating synthetic fertilizers can be a real kick in the teeth. Test your soils and address your soil fertility issues before beginning your organic certification process. It will be much less costly and less demanding to find sources for many of your soil's missing minerals if you balance them *before* switching to organic fertilizers.

Pay special attention to the calcium–magnesium ratio. Some trace elements are extremely difficult or even impossible to find in organic form. Once your soils are balanced, you can begin the organic certification process, but you will need an effective grazing rotation to maintain the balance of soil fertility.

Managing manure and urine distribution in your pastures is your primary defense against nitrogen deficiencies, which are much more difficult and expensive to manage with organic nitrogen fertilizers. Consider some of the biological fertilizers that build nitrogen in the soil, but watch out for microbial fertilizers that burn up the precious humus layer. (See chapter 11 for more about these fertilizers.)

Compost may sound appealing, but it can be very tedious and expensive to spread it over large acreages and it is hard to find or produce in volumes sizeable enough to maintain pasture fertility. A twelve-month-long grazing program that grazes right through the snow season will distribute manure evenly throughout the fields, allow soil microbes to compost in each little manure pile, and prevent the inevitable leaching and evaporation of nutrients and urine in your feed area and manure collection site.

Hay

Feeding forage in an organic production system can be a nightmare if you haven't weaned yourself off stored feeds and implemented an effective winter-grazing program. Producing your own hay means that some fields will not see the benefits of herd migration and there will be significant nutrient transfer from the hay fields to wherever the hay is fed. Even if you feed it in the fields during winter, manure and urine will be concentrated at the feed sites, thereby recycling the nutrients unevenly.

Organic hay is expensive to purchase. Organic hay producers cannot use synthetic fertilizers to increase their volume and replace nutrients exported from their farms with each hay sale, so they have to charge more to make ends meet. Your best option is to implement an effective twelve-month-long grazing season.

Sourcing Organic Cattle

As organic regulations addressing where cattle are born and raised grow increasingly tighter, a significant new market will emerge to supply organic grass-finishing operations with organic calves. Now, however, there are few sources for organic calves; in many cases it is more realistic to produce your own if you are planning to market large numbers of them.

It's important to be aware of the production methods used by your certified organic cattle suppliers. If the animals have been raised on a heavy diet of organic grain, you may be inheriting more health problems down the road than you would have with many nonorganic grass-raised cattle.

Fences

Treated fence posts are becoming tightly regulated by organic certification associations. Unfortunately, in a livestock operation it is hard to tell cattle to leave an ungrazed strip around the fence posts to avoid consuming grass that is potentially contaminated by the chemicals used in treating the posts. Building fences with untreated posts that fall over in five years seems a waste of time. Metal posts are an expensive option.

You may be able to find or make posts from local hardwoods such as cypress, ironwood, ironbark, hawthorn, acacia, or eucalyptus. Many of these woods are extremely dense and rot-resistant and therefore do not require chemical anti-rot treatments. Some electric-fencing companies actually sell eucalyptus or acacia electric-fence posts (for example, Insultimber posts are sold by Gallagher Power Fence Systems). Because many of these hardwoods are nonconductive, they allow you to nail or wire-tie electric wire directly to the post without insulators.

Buffer Zones along Farm Boundaries

To prevent chemical residue drift from conventional farms, roadsides, and industrial lands, buffer zones along farm boundaries are usually required by certification organizations. Although they can be planted with tall bushes and trees (such as blackberry bushes and sea buckthorn) or allowed to return to the wild brush that naturally takes over ungrazed land (creating wonderful natural wildlife and bird retreats), these strips can quickly become a significant weed reservoir.

Some boundaries are clear enough to grow nonorganic crops around the farm perimeter, but it is not easy to dedicate steep, rocky, or bushy land to another use beyond cattle grazing, and such areas are impossible to navigate with a mower. If the bushes are not thick enough to choke out undesirable weeds in the buffer zone, you could graze the strips in a number of ways and still meet organic certification requirements.

- If you are allowed to maintain a split herd of organic and nonorganic cattle, you can graze your noncertified cattle around the perimeter.

- Because your bull herd is used solely for breeding and not directly as meat, it may not need to be certified. These bulls can graze around the farm perimeter outside of the breeding season.

- Rent the perimeter strip to someone with horses or other cattle; the strip will be easy to graze in a rotation as long as water is available alongside it.

- Graze the strip with noncertified sheep if your fences will contain them. Alternatively, if you integrate noncertified sheep into your cattle herd, they will be able to walk underneath the single-wire pasture fence to graze the strip while the cattle are contained within the certified organic pastures. If you use this method, you should not exceed a hundred sheep in a single group of cattle; a larger number of sheep might wander indiscriminately all over your farm rather than remain with the cattle for protection.

Dynamic Marketing

MARKETING YOUR PRODUCTS IS AN ART. You win the interest and loyalty of your customers by painting a picture of the mouthwatering quality of your beef — an image that should inspire your customers' imagination, fill them with positive emotions, and give them confidence in the quality of your meat.

The image created by your marketing expresses all the joy, effort, and pride you have invested in your beef. An uninspiring marketing image makes the very same product appear dull, negative, untrustworthy, bland, and unappetizing. Imagine viewing a black-and-white photograph of gray vegetables, steely tomatoes, washed-out shrimp, and a filet mignon that look like a lump of lead. Would such a colorless picture inspire the same gourmet dining experience as the full-color version — and would it inspire a customer's loyalty?

Developing a Profile of Your Target Customers

To develop an effective marketing strategy tailored to your target customers, we can look to the marketing strategies of the radio industry. The owner of a country radio station once explained to me that stations develop profiles of their prospective listeners for every time slot. For example, daytime country music radio programming may be designed to target the profile of the forty-year-old female listener. While it does not exclude other listeners, the adver-

tising, song selection, and dialogue are designed to attract this audience and keep it in its comfort zone. Every time slot is tailored to a different profile of age, sex, profession, and income level depending on who is most likely to be listening during that particular time slot and what his or her particular desires, passions, concerns, weaknesses, and needs are. This is the fundamental method of building a loyal clientele for each time slot. Marketing meat may not be the same as marketing music, but the strategy is the same for building a loyal clientele by tailoring marketing to client profiles.

Many direct-marketing beef ads are written by men emphasizing the things they are passionate about when they produce their beef, yet men are not typically the ones who make decisions about where to buy meat. In many family households, women make the grocery shopping decisions. They are the ones whose concerns, interests, priorities, and needs must be met if they are to become your loyal customers. Glowing descriptions of cowboys chasing cattle on pure, lush rangeland simply will not inspire them in the same way as a description of your meat's health benefits, remarkable flavor, cooking ease, and positive impact on their lifestyle and culinary interests.

By contrast, advertising your meat at a steakhouse demands an entirely different focus. Here the main red-meat-eating customers are men dining socially; they are the ones who are more likely to order steak and burgers from the menu. You are trying to sell the steakhouse an image that will make your pasture-raised beef stand out above the chicken wings, fish and chips, and lamb chops when their customers look at the menu.

A fine dining restaurant attracts a different clientele. Here, people are coming for a memorable dining experience. You want to sell an image that will entice customers' taste buds with pretty images, rich flavors, succulent tenderness, unique and rare ingredients, and unsurpassed quality. Know who the chef is before you talk to him or her, and know who the clientele is. What kind of people are you targeting? Have you seen the menus or eaten at the restaurants to which you are marketing?

If you are marketing to a grocery or health food store, spend some time walking the aisles to see what kind of clientele it is catering to before developing your marketing strategy. Homemakers, direct-marketing customers, steakhouses, fine dining restaurants, family restaurants, burger joints, health food stores, wholesalers, and grocery chains all speak to different clienteles; so should your marketing strategy.

Assume nothing; if it isn't on the label, it probably isn't in the package.

Demystifying Product Labels

The new awareness of organic products and health foods has swamped the grocery store shelf with a multitude of new words. This new age of food is a marketplace where every product strives to distinguish itself by advertising the one special quality that makes it more delicious, nutritious, or somehow a cut above the rest. All of these words can be terribly confusing to the consumer: How was the beef raised, who guarantees what, who is the authority listed, what is the difference between seemingly identical terms, and what on earth are all these strange-sounding names in the ingredients list?

This confusing flood of labels, advertisements, and claims is no less daunting for the producers, who must determine the difference between beef from their steer standing knee-deep in the pasture and a certified organic steak bought at the health food store, the differences among all the certifying agencies, how to access all these new and obscure markets, and how to make their food products seem special in all these new niche markets.

Here I will try to clarify the labels. Unfortunately, as will become obvious, their interpretation is subjective due to a lack of standardization. Despite their good intent, blanket terms are so broad that they leave enough loopholes for a cow to fall through. Without a more thorough description, the good reputation of the producer to rely on, or more knowledge about the beef described, I would hesitate to draw any health- and nutrient-related conclusions from such labels. Often what is left out is far more significant than what is included. No one leaves out facts that would distinguish his or her product as healthier, more nutritious, or better tasting. Assume nothing; if it isn't on the label, it probably isn't in the package.

On the other hand, once the name is incorporated into the package label, there isn't always room for the kind of detailed description we'd like to see. The labels do grab your attention, however, persuading you to pick up the packages and see what's written on the back. If the product is distinct in some way, the back of the package should tell the rest of the story, including, perhaps, an indication about the production methods used, the producer's philosophy, and any other clues about how this specific product is healthier, more environmentally sustainable, or otherwise unique from its competition.

If the label says that the beef is organic, don't assume that the animal was raised in lush, knee-deep green grass and never saw a tractor or a kernel of grain. If you don't find that information somewhere on the package, it's not because the marketer forgot to mention it; it's because he or she can't lay claim to it.

Alberta Beef/Montana Beef

These two examples of location-specific marketing labels state only that the beef on your plate comes from Alberta or Montana, with the implication that because Alberta and Montana are "cattle country," producers there should know what they are doing when it comes to growing good beef. They are successful marketing images found frequently not only in grocery stores, but also on many restaurant menus. The mental imagery conjured up by these labels associates beef with the tall prairie grass stretching to the horizon, where lie the pristine, snowcapped Rocky Mountains. It does not draw the consumers' attention to the enormous grain-finishing feedlots that play such a significant role in the Alberta and Montana agriculture sectors. Both provide great examples of how a well-presented, location-specific description can be a strong marketing tool while in fact they tell the consumer only where the animal was raised and offer nothing about the methods used to produce the beef.

Antibiotic-Free Beef

Beef labeled in this fashion is, as the phrase implies, free of antibiotics. We have to trust in the good faith of the producer that he or she means the animal was free of antibiotics during its entire lifespan and not just for a withdrawal period.

It is worth mentioning, however, that raising cattle without the aid of antibiotics is no small feat. The number of cattle treated with some form of antibiotic during their lives is staggering due to the stressful conditions of crowded feed pens, mud, fecal contact, weaning, shipping, injury, flies, and a multitude of other common scenarios leading to disease. Beef that is labeled "antibiotic-free" does imply that the producer is conscientious enough in his or her production system to avoid using these medications. Unless it's specifically stated, however, the extent of this conscientiousness is unclear and it certainly does not lay claim to the health benefits of beef raised and finished exclusively on grass.

Argentina-Style Steakhouse/ Argentina-Style Beef

This marketing language, used by restaurants, is not to be dismissed lightly. Argentina has a world-renowned reputation for its amazingly flavorful and tender beef, due in part to its unique grilling technique, but perhaps more because its beef production system is primarily grass-based. Most of the cattle are finished not on grain, but on a well-run grass- and legume-finishing regimen. The high standard of Argentina-style beef is a remarkable success

story by which we can measure our own beef quality and production.

Breed-Specific Beef

Including information on beef breed is part of marketing a picture to distinguish a product from the rest on the store shelf. Of course, because we all like to know we are eating something special, the more we can visualize the beef we consume, the better it will taste to us. Knowing the breed helps paint in the mind of the consumer a lovely picture of cattle and where they come from. Ask any chef: Visual appeal is an integral part of fine dining. An appealing mental picture of where dinner came from is just as crucial as the ingredients themselves in making a good dinner become a memorable experience. We should remember, however, that there is as much variety within a breed as there is among breeds. Rather than making a judgment about the beef based on this breed information alone, we should read the label to find out more about the production methods that produced it.

Certified Organic Beef

This means that the producer has followed the rules set out by a certifying organic association such as the Organic Crop Improvement Association, Quality Assurance International, the California Organic Farmers Association, or the Certified Organic Associations of British Columbia (for a complete list of certifying bodies, please refer to the National Organic Program Web site; see resources, page 360). Each association outlines these rules along with various regulatory and enforcement procedures to ensure that they are followed. In the United States, each of these certifying agencies must follow a mandatory national standard outlined by the National Organic Program (NOP). Canada has its own Canadian National Standard of Organic Agriculture, but at the time of this writing, following it is entirely voluntary, not a codified federal regulation.

The "certified organic" label is not a guarantee of healthy, nutritious food. It is primarily a guarantee of what is not in the product or what was not used by the producer to grow or raise the product: certain chemicals, hormones, pesticides, herbicides, and synthetic fertilizers. Certifying agencies also have certain rules about the procedures and conditions of various production systems, but the "certified organic" label does not guarantee how the animal was raised.

For example, I have seen certified organic beef herds that spend nine months of the year in "spacious" feed pens consuming finishing rations that are extremely high in organic grain — the equivalent of an organic feedlot. The beef was certified organic *grain-fed* beef with none of the health benefits that grass finishing brings to the table.

Remember that the term "certified organic" is as subjective as many other labeling terms; if you want more than a guarantee of what is not in a certified organic product, you have to rely on a more comprehensive label or more details about the methods used to produce it.

Chemical-Free Beef (Our Cattle Don't Do Drugs)

This is a common label, and while it does not specify what things are included or excluded in raising the animal, it does imply that the producer has tried to eliminate as many chemicals as possible.

Unfortunately, unless the beef is certified organic, there is no guarantee of just what chemicals have been eliminated and whether this claim also extends to indirect sources of chemical use in the beef production system. Does it mean that antibiotics, hormones, vaccines, externally applied pesticides, synthetically fertilized feed sources, synthetic mineral supplements, and herbicide- or pesticide-sprayed feed sources were not used? Without a clear standard for the term "chemical-free," its true meaning ultimately is determined by the producer's interpretation.

Free Range

This label is most common in the white meat (that is, chicken, turkey, and pork) and egg industry, where the bulk of birds and pigs are raised in confinement on feed rations. The free-range claim is that the birds and pigs were raised with access to space for exercise and to their natural evolutionary food supply of grass, bugs, and grubs. The resulting rich yellow yolks in eggs from free-range chickens are but one example of how this change in environment sets apart these birds from those producing the pale yellow yolks of eggs of the more mainstream production methods. The label's popularity is also growing in the beef market.

Free Run

This term, reserved primarily for labeling eggs as well as the meat from chickens, turkeys, and other poultry, is used by producers to distinguish their birds from those raised in small, crowded battery cages. Free-run animals are supposedly given the space to move about for exercise and to interact with one another. The label makes no further claims

CONVENTIONALLY PRODUCED BEEF

This is not a designation that we ever see on our store shelves, yet it is worth taking a look at "conventional" production to understand the origins of the beef industry and what the status quo is in mainstream agriculture.

Conventional production, combining the elements of grain, auction barns, and feedlots, is responsible for the typical steak on the grocery store shelf. It is characterized by an assembly-line production style in which all players try to cut the greatest costs from their production specialty in order to make a living from a commodity market, and in which volume of production is the main variable that each farmer manipulates in order to remain financially viable.

The production chain is broken into many parts, with each successive cattle owner specializing in one aspect. It is large-scale, mainstream agriculture, commodity priced and driven by global market supply and demand. The prairie grain fields and the lush corn belts are the pulse of the cattle feed industry and big equipment and diesel exhaust are the engines that make the cogs of the production chain go round and round. It is from this production scenario that niche opportunities for the organic, natural, and grass-fed meat industries emerge.

about feed sources, what the animals' "run" looks like, whether agrochemicals are used, and so on. The assumption is therefore that the birds did not have access to grass, bugs, and grubs, for if they had, it would be advantageous to label them "free range."

Grain-Fed Beef

Grain feeding represents the status quo for most beef production in North America, though it did not become the mainstream of the beef industry until the early 1900s, when the tallow industry became important. At that time, the fat from beef animals was rendered into oil and became a valuable secondary by-product of the beef industry. Grain finishing produces a much greater amount of visible internal, external, and intramuscular fat (including marbling) on the finished animal compared to grass-finished animals. By feeding grain, producers could create and sell more valuable tallow. With the emergence and growth of the petroleum industry, the tallow market has all but disappeared, but cattle continue to be grain-fed in the belief that the high-fat content of the meat is the decisive factor in tenderness and flavor and thus in profitability. This is not entirely true. While intramuscular fat is important to meat tenderness and flavor, the excess fat produced by grain finishing does not result in a more tender, flavorful beef product than grass-finished beef of the same finishing grade produced in a carefully planned finishing program. (See chapter 17 for more about how grass-fed beef should be handled and cooked to preserve its tenderness and flavor.)

Flavor is a subjective quality that depends entirely on the consumer's palate. I find grass-finished beef more flavorful than grain-finished beef. I also find that the almost colorless fat found on grain-fed beef has very little flavor when compared to the flavorful yellowish fat produced by grass finishing.

Like grain finishing, successful grass finishing is a deliberate and carefully planned process. Plucking a random steer from dry range conditions that do not meet the strict conditions necessary for successful grass finishing will not produce comparable results, yet beef raised in this way is often labeled "grass-fed" only because it was not finished on grain, which explains why grass finishing has developed its undeserved bad reputation in the grain-finishing industry.

Grass-Fed Beef

Grass-fed beef refers to the production system used in raising the animal. It does not guarantee any freedom from chemical residues, which must be addressed separately. The "grass-fed" label claims that the animal ate primarily grass or lived a good deal of its life on pasture. *Grass-finished beef* or *pasture-raised beef* also refers to an animal's diet and should therefore be synonymous with the term *grass-fed beef.*

In my mind, these labels imply that the animal did not eat any grain in its lifetime. Because these terms are not standardized, however, this is not necessarily the case. Like *grass-finished* and *pasture-raised,* the term *grass-fed* does not actually guarantee that grain was not fed to the animal at some point, such as during the finishing stage of the animal's life. There is no standard that determines what percentage of the animal's life is spent on grass or prevents producers from feeding grain to an animal at some point in its life.

It is worth remembering that most cattle spend the bulk of their lives on pasture and are not fed any significant grain until the finishing stage, when they go to the feedlot. In addition, some producers supplement their cattle with grain during the finishing stage while the cattle continue to live on pasture, and thereby use the "grass-fed" label because the animal lived in a grass environment all its life. In this case, the grain is still essentially the finishing ration and the grass contributes only roughage. (The human and animal health benefits of finishing cattle on an all-grass diet are discussed more specifically in chapters 4 and 17.)

Grass-Finished Beef

To me, *grass-finished* implies that the animal was fed on grass during the finishing stage of its life; grass is the food that fattened the animal in preparation for slaughter. Although essentially the same as *grass-fed,* it refers specifically to the final finishing stage of the animal's life and indicates that this stage occurred on a grass diet.

Once again, however, there is no standardized usage of this term in the industry. Some farmers grass-finish with a little bit of grain added to the diet at the very end of the process, believing that the small amount and short duration of grain in the diet will not have a negative effect on the healthiness of the beef, but rather will improve flavor and tenderness. Not only is this grain unnecessary if you design the correct grass-finishing program to suit your animals, farm, climate, and market, but even a small amount of grain can negatively effect the nutrient makeup of the beef when it reaches the consumer's plate. (See chapter 17 for information about the ill effects of excess omega-6 fatty acids.)

Hormone-Free Beef

This claim suggests that no growth hormones were used on the animal during its life span and that the consumer is therefore protected from hormone residue. It makes no other claims about how the animal lived or what it ate or about the use of any other materials restricted by certified organic production. Unless the label specifically mentions that the beef is free of antibiotics, pesticides, herbicides, and other agricultural chemicals or pharmaceuticals, the meat may still have been exposed to these during its production.

Natural

This is the most difficult label to define. It claims that the meat was raised in the most natural conditions. Now, to me this means that the animal lived as it would have in its original wild, natural environment. But different farmers or meat sellers interpret the word *natural* very differently. To some, it may mean the cattle were raised without chemicals, hormones, synthetic vitamins, synthetic fertilizers, synthetic feed additives, or antibiotics. To others, it does not preclude that an animal was finished on a grain ration, perhaps even in a feedlot ("natural, grain-fed beef" is not an uncommon advertising label). To still others, it may refer to any combination of these conditions or that the animal was raised outside on grass or on range as opposed to in a feedlot, or simply that it was raised humanely.

Ultimately, the label means very little without the producer's definition. There is no standard that must be followed in labeling a product "natural," and use of the word is at the discretion of the producer.

Organic Beef

In October 2002, the National Organic Program (NOP) was enacted as federal legislation in the United States to provide the organic industry with codified standards. This legislation restricts the term *organic* to certified organic producers (see Certified Organic Beef on page 253). Only manufacturing facilities and producers with less than $5,000 in sales per year may sell products labeled as "organic" to consumers or retail establishments without requiring organic certification. In addition, they must not imply that the operation has been certified; it is unlawful to sell uncertified products labeled "organic" for further processing by a certified organic party. These exempt producers must maintain an audit trail and comply with the organic production standards set forth by the NOP.

Unfortunately, not all countries have federally legislated organic standards. For example, at the time of this writing, the Canadian National Standard for Organic Agriculture is only a voluntary set of guidelines that have not yet been passed into a codified federal regulation. In countries without a national standard, any beef labeled "organic" without being certified falls into the same category as beef termed *natural*. Without a codified national standard, products carrying the organic label are still subject to the interpretation of the individual producer.

Range-Fed Beef

The "range-fed" label implies that the animal spent most of its life on range or free ranging as opposed to being either in the feedlot or in a pasture. I have most often seen the term applied to cattle that have spent some or all of their lives grazing in mountains, on extensively managed land, on community pasture, or on extensive grazing leases. It is difficult to say what the nutrient value or finishing technique of these animals was without knowing more about the cattle; the label does not specify anything about production practices. It does, however, catch the consumer's attention by painting a picture of the life of the animal. From there it's up to the rest of the label to provide further information.

Your Advertising Strategy and Label

In the niche-market world, opportunity lies in creating a unique individual product (only you and perhaps a few others produce it) that addresses the specific needs of the clients (who feel as though the product was designed specifically for them). Your label is the instrument to communicate to your customers the distinct nature of your beef. Unless all your meat sales are through a farmer's market, where you have the opportunity to speak to customers directly, your label must speak for you. It is what will catch customers' attention and provide the opportunity for them to want to know more about your product and to purchase it.

Study Cheese

A great place to learn about how to distinguish your product is in the specialty-cheese aisle at the grocery store. A typical label on a specialty European cheese, for instance, describes the flavor, alludes to the uniqueness of the cheese-making process, emphasizes the freshness of the ingredients and herbs, and often includes a picturesque description of the hand-milked, breed-specific dairy cows that graze on the lush pastures on mountain slopes of some pristine, attractively named valley in the Alps of a certain country.

It's all there: distinctive flavor to suit the consumer's distinguished, individual taste; the care and uniqueness of the production process; a brief hint of the natural, healthy conditions and ingredients used to make the product special; and a description of the lush location that makes it all possible. The implication is that this type of cheese simply cannot be made anywhere else in the world. If you like the taste of a specific cheese, you will remain loyal to that brand because there is no other brand exactly like it. Unlike commercial products A, B, and C (you have no idea what sets them apart other than their name), this cheese leaves no doubt in your mind that it is unique and irreplaceable.

Be Positive and Don't Invite Comparison

A successful label is built on describing what your product is, not what it is not. Describing what your product is not is a negative approach that invites comparing and choosing good over evil. It puts you in a position of having to prove yourself to your customers; you are consequently on the defensive instead of able to speak glowingly of the flavor, health, and quality of your product.

Your brief opportunity to connect with a potential customer should be spent marketing just how desirable your product is, not defending why someone should pick your product over someone else's.

FRONT

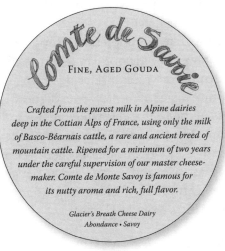

BACK

The front and back of an effective cheese label (fictional) that describes the product's distinctive flavor and the special quality of the location where it was produced. It suggests that it was made with care; a natural, healthy process; and quality ingredients.

Arctic Pastures Organic Beef

Our Scottish Highland cattle are raised on grass, sunshine, and a fresh mountain breeze to bring you the tender flavors of the lush meadows of the Yukon River Valley. Our lean, grass-finished beef is high in omega-3 oils and rich in vitamins and minerals.

Arctic Pastures Organic Beef

Our cattle are raised without hormones, antibiotics, or harmful pesticides. We use environmentally friendly production techniques to grass-finish our cattle without the use of fattening grains and crowded feedlot pens. Our lean beef is higher in omega-3 oils and richer in vitamins and minerals, bringing your family a healthier, more environmentally friendly beef alternative.

By comparing your product to others, you are also subconsciously criticizing customers who have been choosing grain-fed, conventional beef, which puts them on the defensive.

Instead of inviting comparison, your label should draw in the customer and open the door for him or her to ask you why grass-finishing, natural, pasture-raised, high omega-3s, low-stress handling, grazing migrations, rolling green meadowlands, and all the other descriptors you use on your label are so important. Once a potential customer comes to you with questions about your label, the battle to make a sale is already half won; you'll have time to elaborate and get into the inspiring details.

Consider the different impressions produced by the two similar beef labels above: The first is designed to be eye-catching and mouthwatering, giving customers a positive image of the flavor, tenderness, and healthfulness of the beef and leaving them wanting to know more. The second label plays on the fears of consumers and is judgmental of other beef producers. Using criticism and scare tactics to persuade customers to buy from you only leaves your product vulnerable to criticisms and judgment.

Keep It Simple

The key to successful marketing is a positive, brief, eye-catching label. Too much information is a turn-off. Remember that you will have the opportu-

nity to answer questions once you have established a rapport with your customers. You are selling an image of how much better their quality of life will be because of their choice to buy your meat. This simple concept is what makes some beer commercials successful: lots of friends, good times spent in nature, happy people, and no worries — an enviable life associated with the use of that specific product.

Be an Ambassador

You are the ambassador for the image of enjoyable flavors, healthy benefits, and good living that your niche-market beef is trying to promote. Be confident about your quality and price. Look successful and happy; a positive, upbeat attitude will inspire your customers to want to share in your good fortune of being able to eat such tasty, tender, healthy meat. This means you have to be sold on your own beef. Make sure that you eat the very best of your own produce; if you truly love it, your enthusiasm will be contagious and your advertising will naturally focus on your enjoyment. Talking about the beef you enjoyed for dinner last night will win you far more customers than preaching about feedlots, grain, hormones, and pesticides.

Marketing to Businesses

Different markets require different marketing strategies. Marketing to businesses such as health food stores, wholesalers, and restaurants is very different from dealing directly with your end consumer. They will require a different marketing strategy from that necessary to woo the public at the farmer's market.

The public want to know that your product is good for them, that it will be flavorful, and that it is unique. While these points are also important to businesses because they are looking for products that will attract their customers, ultimately the deciding factor between signing a contract with you or someone else is your reliability and professionalism as a business partner. The quality of your meat products is only one small factor in what makes a business want to contract with you.

To succeed in contracting with businesses, you can't just show up with your hands in your pockets and say, "You want some beef?" A well-planned business proposal, prepared in advance, not only advertises the attractiveness of your beef, but also imparts to businesses confidence in your ability to reliably and regularly produce and deliver to their door your wonderful product. Simply put, businesses need to be able to make money. What good is a fabulous beef product if it isn't reliably delivered? A business's profitability depends on your promise to supply a product in the amount, time frame, and seasonal duration promised by your business contract.

Thoroughly inform yourself about a business and its needs before you approach the owner or representative. Arm yourself with confidence, professionalism, and a concrete offer to begin your negotiations. The way you present yourself, your manner of talking, your dress, and the impressions you create in the first few seconds when you walk through a business's door or talk on the phone with its representatives all have a greater impact on the success or failure of obtaining a meat contract than the actual meat you are selling. The question any business will ask is "Can this person deliver on his or her promises?"

Market Exposure

In the next few pages I present a list of a number of innovative avenues through which to direct-market your meat. Following this is a list of advertising strategies designed to reach out to the community to raise interest in your beef and attract new customers. With all the personal interactions you develop in your customer base through your direct-marketing efforts, there are a number of value-added marketing opportunities that naturally present themselves. A short list of these opportunities appears at the end of this chapter. These three lists are just a beginning to get your creativity flowing. As you come to know your community and gain confidence, you will surely find many unique marketing opportunities for your beef products.

Direct-Marketing Options

Here are some direct-marketing ideas to increase your customer base, add value to your meat, and increase market exposure for your grass-finished beef.

- Take a freezer full of meat cuts to the farmer's market.

- Run a burger stand at the country fair.

- Host a community event on your farm: Set up an informational booth; offer farm tours, meat samples, and catered food; and sell meat.

- Distribute taste samples at the health food store or grocery store that carries your beef.

- Approach a butcher shop or meat market about supplying it with beef.

- When approaching restaurants, first familiarize yourself with the menu and clientele (perhaps even eat a meal there) before you talk to the chef or owner.

- Always go to the decision makers at the top: chefs, owners, managers, the person in charge of meat and produce at the grocery store. They are the ones who have the authority to make changes to their buying programs. People who don't have that authority won't realize there is any flexibility and will discourage you, if they don't turn you away outright.

- Consider approaching the big grocery chains. Because these will want a steady supply of beef year-round to distribute to all their stores, consider working with a number of other producers in a marketing cooperative to guarantee volume. Some of these stores will also feature specialty meats once a year when they become available (just like some seasonal and local fruits) — especially if they are desirable enough to feature in this way. Your product image must be so desirable that buyers are willing to try to conform their sales to your supply rather than making you dance through hoops to get their business.

- Small corner grocery stores and convenience stores are often open to new ideas to bring customers through their doors. I've seen such groceries sell a full range of organic produce, including meat, as well as fresh baked goods like sausage rolls, meat pies, and other value-added natural and organic meat products. These stores can often be found in rural settings, as part of gas stations, for instance, where they serve as stopping-off places for everyone passing through, whether for gas or fishing bait or something to eat or a few extra groceries. These stores also exist in urban settings as convenience markets located on corners or in important neighborhood locations to cater to a certain demographic. I have even seen a video store that sold fresh farm eggs — an unlikely combination, but surprisingly effective at generating business for the two separate enterprises: Many people renting videos left with eggs and, finding them of good quality, returned for more, only to leave with a video rental tucked under one arm.

- Consider traveling door-to-door in wealthy neighborhoods (with the appropriate business licenses, of course) to deliver preordered meat by the cut or variety packages from a freezer unit in the back of your truck. These clients are the most likely to be interested in organic meats and many will appreciate the convenience you provide. Make sure your meat-delivery vehicle is prominently marked with your full contact information to advertise to other potential customers while on your delivery run. Remember, a prominently marked vehicle delivering an interesting product to the neighbor's house generates plenty of curiosity. And always leave a pamphlet with your meat deliveries so customers can share it with neighbors, friends, and family members who ask about your meat. Such deliveries of preordered meat every two to six months are already popular in many parts of the United States and Canada. Some deliverers travel great distances on a predetermined schedule and in some cases they don't even produce their own meat.

- Open an on-farm retail outlet or set up one along the roadside on a specific day of the week. Some producers even have unattended self-serve produce stands along the roadside, running them on a trust system — payments are made directly into a sealed lockbox with surprisingly little if any theft.

- Box programs are very attractive because they give you a predictable volume of meat sales and are prepaid so they do not require attendance at pickup locations. Customers sign up and pre-pay for a weekly or monthly meat pickup worth a predetermined value (for example, a choice of $20, $40, or $60 value) and designed to suit the needs of their family. The producer then fills the order (usually by packing it into a box; hence the name) with whatever combination of produce or meat is in season, providing variety for the customer and a predictable sales outlet for the producer. The customer picks up his or her box each week at a designated location and drops off last week's empty box. Selling meat this way works particularly well if combined with vegetable and fruit sales.

- Open an on-farm restaurant. There are tremendous value-added opportunities in your picturesque setting amid grazing cattle and green flowing grass. The farther this restaurant is from town, the better the quality of the food must be for people to make the effort to drive a distance to eat there. People are willing to travel a long way for a memorable dining experience; for four- and five-star food, they will happily make an evening out of it.

- Consider adding meats such as pastured lamb, turkey, chicken, and pork to your beef enterprise. Much of your infrastructure will serve multiple species and your existing beef customers are an ideal base for these new products.

Getting the Word Out

Getting the word out into the community is another very important aspect of your marketing strategy. By distributing information and raising public awareness about your unique beef product, you will be sowing the seeds for new customers to contact you or meet you at one of the retail locations where you sell your beef. The list below provides a sample of places where you can distribute information bulletins and order forms, followed by two additional advertising strategies to help get the word out to potential customers.

- Health food store bulletin boards

- Organic bakeries

- Naturopathic health clinics, herbalist clinics, and any other place where alternative medicine is practiced

- Local nutritionists' offices

- Fitness centers and gyms

- Hardware stores. Other than the grocery store, the hardware store is visited by the most complete cross section of people in the community and is a very sociable place of business. People spontaneously stop to talk to one another about all sorts of practical concerns, get advice, and discuss local goings-on. It is rare to go to a hardware store in a small community without running into someone you know — and hardware stores attract many do-it-yourselfers who like to barbecue.

- Apartment complexes, adult living communities, and retirement homes. Because people here typically have limited freezer space, they may be open to a box program or meat by the cut delivered monthly or every other month to a central pickup location close to their building. Such a program gives to people living in large cities an attractive link to the country. The number of people you have access to in a single location can make this market very worthwhile.

- Organic associations. They are sometimes approached by wholesale buyers and locals in the community looking for healthy-food suppliers.

- Ads on the community radio station and in the local newspaper. You can also find a way to inspire the local news channel or newspaper to run a feature about your innovative, environmentally friendly business practices, which would be a wonderful way to indirectly advertise your products.

- Nutritional bulletins that inform people where to find alternative healthy foods in their communities and around the country

More Value-Adding Opportunities

Having access to agricultural land and a herd of peaceful cattle nose-deep in grass creates an opportunity to diversify into value-added, agritourism enterprises. This picturesque atmosphere of your farm is the stereotypical country experience that many city dwellers and tourists crave, particularly if it is accompanied by some family-friendly activities.

- Create a cornfield labyrinth to draw big crowds.

- Run farm tours on the retired hay wagon or behind the ATV. They are a big hit with kids, as is the opportunity to actually pet a cow. Tame a few Scottish Highlands, Galloways, or cattle from other unique-looking specialty breeds that quickly find a way into people's hearts.

- Schedule Halloween events.

- Consider renting out small garden plots for customers to grow their own vegetables. They will get to experience the farm, and you will make contacts to sell beef. They essentially pay you for your market exposure.

- Open a bed-and-breakfast — a wonderful getaway where city dwellers can walk in lush pastures and see the cattle grazing.

Chapter

Helpful Business-Management Tools

In this chapter we explore several valuable management tools to facilitate the organizing, prioritizing, planning, and decision making needed to successfully manage a business. Just as we need tools to make our physical work efficient, these intellectual tools are simply principles, ideas, and strategies that facilitate our management and change our perspective.

Importance and Urgency

Most farms have an overwhelming to-do list of tasks, issues, and problems that must be addressed. The list always seems to grow at the bottom faster than we can work things off the top. Here is a neat little mental trick I have adapted from Stan Parsons's book *If You Want to Be a Cowboy, Get a Job* (Harare, Zimbabwe: Franfel Publishing, 1999), which helps prioritize things so you know where to begin.

Rate each item's importance and urgency as low, medium, or high, as shown in the chart at right. Note that as the season progresses, the importance and urgency of these items will change. For example, the urgency of planning your winter grazing may be low in July, but if you haven't done it by September, its urgency will become medium or high, depending on how soon the growing season ends.

Clearly, items that are marked as high regarding both urgency and importance should be done first, but many times you will find items that are extremely urgent but of little or no importance; these can sim-

ply be ignored for the time begin. The extremely important but low-urgency items can be scheduled for the earliest possible convenience without creating a panic at the expense of more urgent items.

Quite often, as the most important and urgent items are addressed, many other items on your list seem to solve themselves because they are a consequence of something else or dissolve through better management and planning. This process allows you to prioritize items that will have serious positive or negative effects on your business's success, or to schedule items for a later date if they are not urgent without worrying yourself to death over not addressing them immediately.

Without this process, each item on the list looks as important and urgent as the next. This prioritizing method safeguards against worrying yourself sick and losing yourself in a bottomless pit of work, and it prevents your list from crippling your business management and paralyzing much-needed changes or from overwhelming your personal life, recreation, and family time. Putting off family time is a rampant disease among most farmers: I know several sixty-five-year-old farmers who long ago postponed their honeymoons to deal with urgent items on their lists, but forty-five years later, their lists are even longer and they still haven't had time for a honeymoon or any other vacation. "Time for self" is an important item that should be included on your list; grouchiness and exhaustion are indications of when it becomes both very important and urgent.

SAMPLE TASKS, PROBLEMS, OR ISSUES
RATED ACCORDING TO IMPORTANCE AND URGENCY

Task, Problem, or Issue Date: July 1	Importance	Urgency
Plan and install new frost-proof water site for winter grazing in 4 months' time.	high	low
Troubleshoot livestock corral ahead of next week's branding.	medium	medium
Find missing branding iron for next week's branding.	medium	high
Develop fall beef price list.	medium	low
Fix broken loading chute to send grass-finished hamburger-grade cows to butcher on Wednesday for direct-marketed beef sales (cannot presently load from chute).	medium	high
Investigate and prevent possible scours outbreak among the newborn calves based on first troublesome disease symptoms among some of the calves.	high	high
Go on first family holiday in 4 years.	high	medium
Repair farm sign ahead of next weekend's open house.	low	high
Calculate cow grazing days for remaining summer grazing.	medium	medium
Fence along creek to keep cows out of the water.	medium	low
Plan winter grazing rotation.	high	low
Sell obsolete tillage equipment.	medium	low

low = would be nice to get to this, but it will hardly affect the business at all.

medium = the positive or negative outcome of this item is important.

high = extremely urgent or extremely important; the success or failure of the business depends on the outcome of this item.

The $5-, $50-, and $500-per-Hour Jobs

Often our list of tasks and our problems become so overwhelming and engrossing that we forget to step back from the daily grind and look at the business as a whole. In effect, we spend so much time working in the business that we forget to work on the business.

Our daily routine often prevents us from setting aside time for planning, management decisions, and business monitoring — as discussed on page 106, the CEO's contribution to the success of the business. Ironically, these tasks have the most influence on business profitability, yet often they are relegated to some time late at night after the kids are in bed, or squeezed into the half hour before dinner is served or when they are most apt to be interrupted by incidental phone calls, crisis management, and petty chores. Schedule time to be the CEO of your business!

Rate your jobs at $5/hour, $50/hour, or $500/hour, based on their impact on the business. This is not meant to be a reflection of what you should pay others to do work for you; it is simply a tool to help you prioritize your own tasks.

Simple chores — feeding, building fences, and servicing the ATV — are the $5/hour jobs, the business maintenance jobs. The $50/hour jobs are the daily management tasks such as going to a bull sale, selecting replacement heifers, organizing your employees at the morning meeting, and planning your day. These are the lower-level business management jobs that affect the day-to-day operation of the business.

The $500/hour jobs are the CEO's contribution to planning, long-term business management, and marketing. They are the jobs in which you work on the business and shape its future. Whenever possible, your focus and energy should be directed at these: for instance, planning your yearly calving and breeding cycle, preparing a drought strategy, planning the grazing rotation, planning your herd management, creating a marketing strategy, and designing your grass-finishing program. These jobs also include the time you spend walking through your pastures to monitor grass growth, observing how your animals are faring in your pasture management system (as opposed to the $5/hour herd health check), and monitoring the success of your management strategy. Take time weekly to tour your farm on foot. Even sitting on a hill overlooking your farm, where no one can find you and with your cell phone turned off, allowing you time to think, plan, and observe without distractions, is a $500/hour job.

How are you spending your time? If you aren't getting enough $500/hour time, you need to rethink your management strategy. If you spend your time exclusively on $5/hour and $50/hour jobs, don't expect your grass-cattle enterprise to bring you a $500/hour income. If necessary, hire someone to fill in for your $5/hour chores to give you time to do some $500/hour thinking.

Getting the Biggest Bang for Your Buck

Whenever you spend time or money, you want to get the most from your investment. Concentrate on items that will have the greatest influence on your business success. As an example, you could spend your operating budget and time spreading fertilizer, running your irrigation equipment, or improving your haying equipment and winter feeding systems, all of which will have small positive effects on your financial bottom line. Or you could start by improving your electric-fence and watering-system grid, which would help utilize your cattle's manure and urine more fully as fertilizer, improve the soil, increase soil moisture, enhance your forage man-

agement, and take you toward a twelve-month-long grazing program that no longer requires hay, silage, or winter feeding sites and equipment. That is getting the biggest bang for your buck! You also want to get as many long-term benefits as possible from a single action, instead of spending your time performing tasks such as fertilization, irrigation, and equipment maintenance that must be repeated sometime down the road.

Weighing Decisions, Solving Problems, and Handling Crises

Time and again, we are faced with complicated problems that force us to choose from many possible, equally attractive (or unattractive) options or we are faced with crises and high-pressure situations in which we have to come up with a decision under immense stress. It is very easy to let emotions rule and to make rash decisions unless we have a strategy to perform rationally.

Before exploring your options, start by reminding yourself of your goals. An emergency provides the opportunity to review your written plan for dealing with a specific crisis such as a severe drought. Write down everything; a high-pressure situation is not the time to act on a "wing and a prayer." Step-by-step writing compels you to be thorough and logical in the face of stress.

My wife drafted the following worksheet for me to help me weigh options during crises or life-altering decision-making processes. We use it frequently, particularly because I become quite impulsive and hot-blooded under stress. This worksheet is invaluable during crises and high-pressure decisions because it forces you to logically follow each option through to its consequences rather than impulsively jumping on half-baked ideas without weighing their long-term effects. I have used this worksheet to organize and guide my decision making during droughts, marketing problems, equipment purchases, career choices, and many other situations involving critical decisions. It has allowed me to isolate each option; give each a logical, step-by-step consideration; and eliminate options one by one as I work through them. (See page 343 for a blank option-weighing worksheet.)

OPTION-WEIGHING WORKSHEET EXPLAINED

Each option gets its own worksheet.

Direct and indirect expenses or costs incurred by this option. Assign a dollar value, if possible.

Direct and indirect income or savings resulting from this option. Assign a dollar value, if possible.

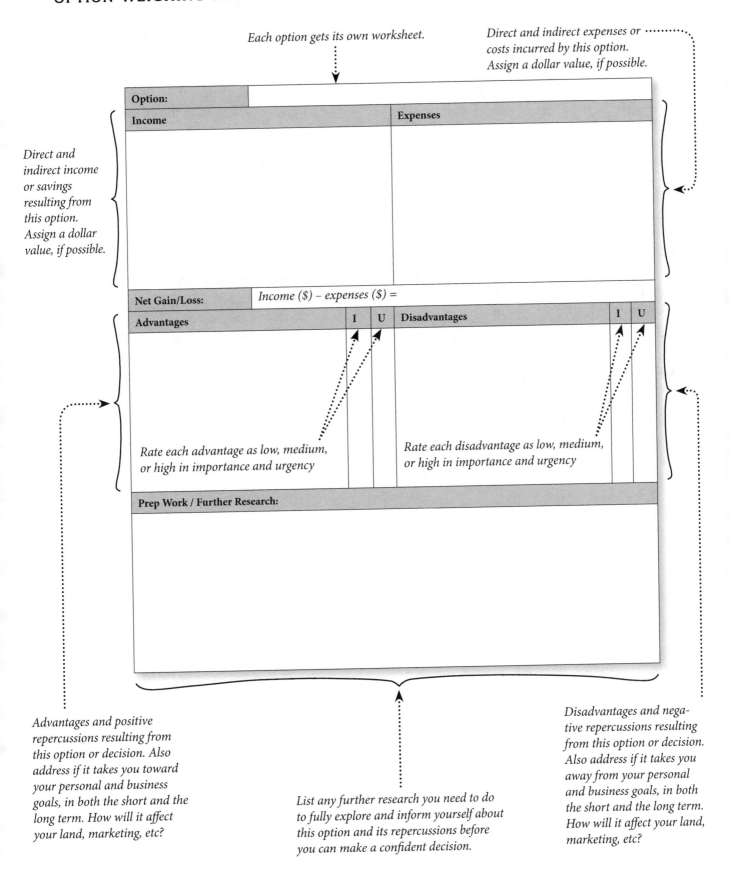

Option:			
Income		**Expenses**	

Net Gain/Loss: *Income ($) – expenses ($) =*

Advantages	I	U	Disadvantages	I	U

Rate each advantage as low, medium, or high in importance and urgency

Rate each disadvantage as low, medium, or high in importance and urgency

Prep Work / Further Research:

Advantages and positive repercussions resulting from this option or decision. Also address if it takes you toward your personal and business goals, in both the short and the long term. How will it affect your land, marketing, etc?

List any further research you need to do to fully explore and inform yourself about this option and its repercussions before you can make a confident decision.

Disadvantages and negative repercussions resulting from this option or decision. Also address if it takes you away from your personal and business goals, in both the short and the long term. How will it affect your land, marketing, etc?

10 Percent Chance, 90 Percent Work

There is no such thing as good or bad luck. Chance plays only a 10 percent role in all your experiences. Consciously or subconsciously, your efforts provide the other 90 percent. Whether you suffer drought damage or livestock death or enjoy success, the responsibility is yours. For example, while bad weather may prompt a scours outbreak in your herd, it is your feeding habits, grazing management, calving season timing, or other management decision that made your animals vulnerable to the disease or gave the advantage to the scours bacteria. Likewise, while individually you cannot stop a region-wide drought, the scale of the impact of the drought on your business is entirely in your hands: By carefully formulating and then strictly following your disaster/drought contingency plan, you can keep the drought from becoming a disaster on your farm.

Your Unfair Advantage

Every farm has an unfair advantage — some condition that distinguishes the farm, the area, or the market from everyone else's. This, however, is not an unfair advantage until you identify it and take advantage of it. If you try to over-winter calves in a high-snowfall, high-moisture climate, you may be struggling; but if you change the operation to focus on grass-finishing animals in the summer on lush high-moisture pastures, you have an unfair advantage over most producers.

A very cold, dry, prairie winter is the unfair advantage of the producer who wants to graze cow/calf pairs and stockers year-round — but only once the farm's calving season is in sync with nature's seasons. Expensive land at the margins of a large urban center seems like a curse until you consider the exceptionally high-end direct-market opportunities that such a location offers. Suddenly the land becomes an unfair advantage. Often your unfair advantage is the greatest thorn in your side until you identify it and alter your management to use it to your advantage. Think positively: Once your perspective changes, allowing you to see your unfair advantage in a positive light, you'll know your business is invincible.

Your BUSINESS PLAN: Putting Principles into Practice

PART FOUR

21

Your Goals and Market Opportunities

Custom designing your business plan begins by identifying your goals and understanding your market environment. While planning, management, and marketing are absolutely vital to the success and profitability of your natural grass-based cattle business, your goals and the expectations you have for your business give direction to these activities and provide your overall business plan with a solid foundation. Only your goals will allow you to distinguish your path from the countless other options that lead to other places. By researching your market environment before you plan your entire business, you will have a greater understanding of the market opportunities that exist in your community when you choose your enterprises and plan your production, management, and marketing strategies. You will be able to develop a business plan that is best suited to your particular market environment, takes you toward your goals, and meets your financial expectations.

Your Business Plan

Your business plan is both a blueprint for your business and a powerful planning document with which to manage your ongoing business operation and guide its future development once it is established. It is your road map that outlines your proposed products; defines your objectives; guides your financial management strategies; describes your production methods; establishes your management principles; presents your marketing plan; and provides you with a solid basis upon which to grow your business, evaluate changes, and endure adversity.

Your business plan is as valuable to an established natural, grass-based cattle business as it is to an emerging one. Thus, you will want to keep your plan updated, revisiting your goals and the financial planning sections of the plan on a yearly basis to keep your established business profitable, growing, and adapting to change. And it all starts with identifying your goals.

Starting with Goals

Specific goals are the single most powerful tools at your disposal as you begin to structure your farm's production, management, and marketing plans. They help you to develop impeccable judgment, inform your business decisions, guide you through crisis management, increase your efficiency, and are key to your farm's profitability.

No other part of your business venture is as important or has as great an impact on your success or failure as your goals. Yet goals are often the weakest part of many farm businesses, if they have been addressed at all. Even a monkey can make money during a gold rush, but farming can't be left to chance. Only well-formulated, clearly recorded goals will help you to focus, take you through the lean years, carry you through crises, help you build a fulfilling career, and make possible the stability needed to build a home, a family, and a retirement plan.

———————

When I was a youngster, my grandfather interrogated me: "What do you want to be when you're forty? You have to know; it's the only way you'll get anywhere in life." Frankly, I didn't even know which socks I would wear to kindergarten on any given day, so I humored him with a wise-sounding "Hmm . . ." and quickly changed the subject. In retrospect, I realize that he was trying to instill in me the same idea stressed by financial-planning books, motivational speakers, and bank loan officers: Only when you have a vision of where you are going will you find the path that will take you where you want to be.

———————

I used to sit at the kitchen table with my family until late in the night wondering how to find a way out of one or another financial pickle, tossing around ideas and possible solutions until everyone's stomach was knotted so tightly that even coffee couldn't soak through. It is this same kind of aimless grasping at straws that keeps farmers worrying all night and will land them in a cedar box long before their time. It is not a lack of viable options that causes grief, but not knowing how to choose the best solution.

Goals provide a clear picture of what we want our lives to look like in one, five, ten, twenty, or even forty years. Armed with these images, we are able to answer tough questions about enterprise choices, sort through options, make personal and business decisions, keep doing the things that we value, and have a reason to get out of bed in the morning.

If we don't work toward clearly defined goals, it doesn't take long for us to be up to our nostrils in muck and manure, desperately hoping for our next day off and forgetting why we got out of bed. If we don't know where we are going, how will we know if what we are doing will take us where we want to be?

Where to Begin?

We generally have a vague idea of what we want to do with our lives and farms, what we want to accomplish and what we have to change, but we don't know where to begin. Our impulse may be to jump on the first great idea that hits us so we can finally get our hands dirty: "Let's stop talking and start *doing* something." But if we jump in this way, it doesn't take long for us to realize that we have been working very hard without addressing the real issues or that a problem returns in a slightly different form despite our enthusiastic efforts to overcome it. Sound familiar?

The most difficult thing to do when we're itching for a change and desperate to start doing something is to sit down and think. The place to begin making changes is at the kitchen table with pencil, paper, and a good eraser. We must start with goals, describing exactly where we want to go before we begin brainstorming ideas for how to get there. Your business plan will address the how-to-get-there questions. Your goals are about you, not your farm; your farm comes second.

What kind of life do you want to lead? How do you want it to look? What do you value? What is important to you? Your job is to write a brief statement about the quality of life you want to lead and what kind of person you want to be.

Your Goals

In the section below, you will work through the process of identifying your goals, beginning by determining what you want from your life. From this you will develop a common set of goals that act as your guiding set of principles, values, and objectives and that express the spirit or purpose of your business as a whole.

A list of your goals should not address how you will achieve them. The specific enterprises that will be the focus of your business — how to structure your business, what markets to target, and what production methods to choose — should be addressed *after* you identify your goals and research your market environment.

I can't emphasize enough the importance of writing down your goals so you can refer to them often, preferably every day. Whether you consider your goals a daily reminder of your purpose or you prefer to wait until decision-making time to dig them out from under the mattress, they must be on paper or safely stored in your computer for three reasons.

1. Writing down your goals makes you fill in all the fuzzy parts about what you want from your life. You may know these but can't quite formulate them in your head. They are the bits you won't remember later, though they are probably the real meat and potatoes of what you want to express. Writing them down makes you accountable to them; it forces you to be thorough and keeps you from forgetting your goals.

2. When you write down your goals, you have the power to know exactly where you want to be. Doing so makes it more difficult for you to lose sight of your long-term destination during the course of the daily grind: You know that somewhere there is a piece of paper (or a computer file) reminding you that each choice you make brings you closer to or farther from what it is that you really want.

Whether you are designing your farm production plans and financial budget or feeling impatient with a cow that is refusing to enter a squeeze chute, you know the decisions you make on the spot will ultimately affect your long-term goals. You are no longer getting a job done just to make it through the day. This year's low synthetic fertil-

izer bill might not look so attractive when you recall that a long-term goal is to build up rich organic soils. And the cow that seems to deserve the cattle prod in the heat of the moment points to a flaw in your handling protocol when you remember your goal of producing stress-free, tender beef.

A goal that exists only in your mind is easy to change to suit the occasion, but when you write down your goals, you commit to them. If you start taking shortcuts, change your goals to suit the occasion, or willfully turn a blind eye to the long-term consequences of your actions, you know you are doing yourself a disservice.

3. Having written goals is a lifesaver when the real crunch comes in the form of drought, flood, market collapse, or family discord. Because they remind you where you want to go, you can emerge from the crisis still heading in the right direction. In the stress of an emergency, long-term thinking will be furthest from your mind unless you have written goals to which you can refer. Without them, you will fall victim to impulsive suggestions, rash decisions, and quick fixes that you may well regret for years to come.

Brainstorming Your Personal Aspirations

Brainstorm all the things you want from life and, more specifically, what you want your business to help you achieve personally. Write down everything that comes to mind. Be sure not to censor, evaluate, criticize, or dismiss your ideas during a brainstorming session. Doing so interferes with your mind's natural creativity. Evaluate your ideas only *after* the brainstorming session.

Describe what you want from the relationships in your life (marriage, family, friends, business interactions). Describe how you want your environment

(pastures, soils, wildlife, rivers) to look and be sure the description makes you smile. (Your dream environment and the level of connection you want to nature and the land will inform your daily decisions as you plan for your ecologically sustainable business.) Describe your financial expectations. Include your passions, favorite hobbies, vacations, foods — whatever means the most to you.

Aspirations are dreams, things you wish for. When I did this exercise, some of the aspirations I brainstormed were:

- Happiness
- Great marriage
- Nutritious, vitamin- and mineral-rich home-cooked meals every day
- Lots of wildlife that I can enjoy with my children
- Lakes teeming with big fish
- A reputation for honesty and thoroughness
- Financial stability
- No debt

Clearly, these are very personal desires, and they paint a clear picture of the kind of world I want to live in and what will make me happy.

Transforming Your Aspirations into Goals

Next transform your aspirations into goals. Rework this list until you can boil it down to seven concise sentences that express the essence of the most important aspirations on your list. Rewrite them as sentences that begin with the words "I will" or "We will." Use language that is definite, committed, and firm. Your statements should evoke mental images that fill you with emotion. For example, describe the kind of land, environment, and world you want to live in, but don't yet consider how you will reach these goals. These goals should be open-ended, providing you with a clear picture of what you want from life without limiting your options.

Consider the following examples and how they express my deepest values without limiting my options:

- I will have a healthy balance between family, career, and friends, as well as time for personal recreation.
- I will be financially stable, wealthy, and debt-free.
- I will have honest, professional, forthright business relationships.
- I will create on my land a sustainable oasis for livestock, grass, wildlife, and fish that my great-grandchildren will be able to inherit.

Keep your list and your sentences brief — and remember to write only seven sentences at the most. The list must be short enough to memorize. At least once per year you should reassess and revise your goals as your personal life changes, your business evolves, and your priorities shift. Goals are dynamic; they become more refined over time as your principles, values, and objectives become clearer through experience.

Internalizing Your Goals

The more compact and personal your goals are, the easier it is to relate to and remember them. You are far more likely to achieve your goals if you are able to memorize them because in this way they will become a part of every aspect of your life and every decision you make. Reviewing your goals frequently and memorizing them will help your subconscious to gravitate naturally toward their fulfillment. Posting them somewhere, such as on the bathroom mirror where you will be forced to look at them daily, will also help you internalize them. That is the power of suggestion! Your goals should always be at the forefront of your mind, directing the value judgments and business decisions you make.

Assessing Your Market Environment

Before you commit to specific enterprises designed to serve a specific target market, you must develop a profile of your prevailing market environment to identify potential market opportunities. Like your goals, your market research is the precursory homework that must be done before designing the remainder of your business plan. I cannot overemphasize how important it is to begin with this market research. This extra time invested up front is absolutely vital to the development of a successful business plan designed around products and production methods to suit your specific market environment. Everything you do from this point forward depends on making choices that will enable you to find a market for your chosen products. Take the time now to do your homework; it will pay off later.

Developing a Profile of Your Market Environment

Begin your market research by taking stock of the area where you live. The following list is a guide to questions you must answer to develop a comprehensive profile of your market environment. Answering them may require that you do some research.

■ What is the population of your immediate vicinity (within a 40-minute drive of your farm)? What is the population of your greater trade area (within a 100-mile radius)? Describe that population: How many live in towns? How many live in rural areas? Is it an aging population or are there many young people? Is it a transient or established population? The more specific you can be about the demographics in your area, the better informed your choice of enterprises will be and the better will be your subsequent marketing strategy.

_____

_____

_____

_____

_____

_____

■ What drives the economy of your community? Is it a particular business or industry?

_____

_____

■ What is the income-level profile of the population to which you have easy access?

_____

_____

■ Is it a tourist area? ❑ yes ❑ no

■ Are tourists or another transient population traveling through your area a potential market for you? (If yes, this would diversify your market access.) ❑ yes ❑ no

■ How much awareness of organic food exists in your community? ❑ high ❑ normal ❑ low

■ Are there any health food stores in the area? ❑ yes ❑ no

■ Is there an organic food section in the regular grocery store? ❑ yes ❑ no

■ Is there an existing demand for certified organic direct-marketed meat, or will uncertified meat (labeled as natural) better fit your community's organic awareness and income levels? ❑ certified ❑ uncertified

■ What types of local restaurants are there that might be prospective markets? Fine dining, pubs, family restaurants, burger joints?

_____

_____

_____

■ Is there an ethnic community in your area? If yes, what are its meat preferences? ❑ yes ❑ no

_____

_____

_____

■ Is commercial freezer space available in your community? (If yes, it can lengthen your marketing season to give you greater market flexibility and allow you to sell meat by the cut.) ❑ yes ❑ no

■ Will you have to transport your produce to market over long distances, or will you be selling locally to customers who personally know you and your farm or your reputation?

_____

_____

■ Will you need organic certification in order to sell to your local business customers such as health food stores, restaurants, and distributors, or will you be able to sell uncertified meat products to these market outlets?   ❑ certified   ❑ uncertified

■ Will the local population be open to a seasonal meat supply that is sold by the side or by the split side? Those in rural areas more customarily use deep freezers, whereas urban populations — especially those in gated and retirement communities and apartment complexes and transient populations made up of seasonal workers, university students, military personnel, mining or oil field workers — are more receptive to small, by-the-cut purchases that they can buy as needed during the year.   ❑ yes   ❑ no

■ What kind of value-added opportunities would suit the local population? Agritourism? An on-farm retail outlet? An on-farm restaurant? Sausage making?

_____

_____

_____

■ Are there any larger population bases within a 4-hour driving distance that you could market to through wholesale buyers, restaurants, or scheduled private meat deliveries?   ❑ yes   ❑ no

■ Are there other like-minded producers in your area with whom you can form marketing cooperatives if you decide to sell to organic or natural wholesalers that want large volumes of meat?
   ❑ yes   ❑ no

■ Is there a market for natural or organic dog food in your area?   ❑ yes   ❑ no

■ What are your unfair advantages? Often these take the form of your biggest hindrances to production until you learn to use them to your advantage (see page 266 for more about unfair advantage). For example, do you have grazing opportunities that are unique and give you an unfair advantage over other producers (e.g., lush grass-finishing condi-

tions or dry winters for winter grazing cow/calves or stockers)? Is land inexpensive where you are located? Do you have an easily accessible market at your doorstep?

_____

_____

_____

■ Will you need to produce your own calves or can you buy them?

_____

■ What are your unique marketing opportunities and how can you capitalize on them?

_____

_____

_____

The profile you develop of your community should be similar in form to the case studies in chapter 15 (see page 218), only far more detailed. Carefully analyze the market profile, and make a list of potential market opportunities. This market profile will guide you in the next chapter as you determine the focus of your business and later will inform your marketing strategy.

With your goals and marketing opportunities clearly identified, you now have at your disposal valuable tools with which to evaluate your enterprise options, production methods, management decisions, and marketing strategies. You will know whether every decision you make from this point forward in designing your grass-based cattle business takes you toward your goals and suits your particular market environment. With your goals and market research in hand, you are now ready to move on to the next phase of your business plan development — your finances — discussed in the next chapter.

Your Financial Plans

In this chapter you will choose which natural, grass-based cattle enterprise options you will focus on in your business, systematically evaluating and comparing their income potential and assessing their risk to determine which options are best suited to helping you achieve your goals and financial expectations in your specific market environment. This is a process that you will want to revisit on a yearly basis to reevaluate your chosen business enterprises against new market opportunities and alternative enterprise choices that you could add to your business as it grows or use to replace existing enterprise choices that are becoming less profitable.

The best enterprises give you the greatest potential income for your efforts.

Once you have compared your enterprise options and chosen which will be the focus of your business, you can plan the financial structure of your business. You will decide on your targeted profit margin (used to calculate your beef price) and plan your expense budget. You will also learn how to record your business's cash flow for financial planning purposes, which requires a very different set of records from those you keep for your taxes.

Finally, this chapter will show you how to calculate prices for your direct-marketed beef by the side and by the cut so that you can achieve your income expectations and targeted profit margins.

These financial management tools will not only serve to develop your business plan, but also continue to benefit the ongoing financial management of your grass-based cattle business after it is established. As you work through this section, you may want to jump ahead to subsequent chapters (chapters 23 through 30) to help you with your planning.

Determining the Focus of Your Business

Having identified your goals and completed the profile of your market environment in chapter 21, the next step in developing your business plan is to choose which enterprises will be the focus of your business. Will it focus on cow/calf pairs, stockers, grass-finishing animals, or a combination of these? Will you market through conventional commodity market channels or will you direct-market your beef to niche-market opportunities? Can you think of new markets for your beef or find innovative ways to expand your beef herd? Will you open an on-farm retail outlet, rent your neighbors' land for a stocker grazing cell, or venture into agritourism?

List the enterprises in which you are most interested. The point here is to challenge your mind to come up with as many viable options, alternative marketing avenues, and new market opportunities as possible. This is the time to let your imagination run wild; later on you will systematically evaluate ideas and calculate their income potential.

Evaluating Your Enterprise Options

The next step is to sort through your list of potential enterprises to judge how they fit with your goals and whether they suit the potential market opportunities you discovered in your market profile. Refer constantly to your goals when making such evaluations; they are an excellent screening device and will help you make sound decisions to lead to personal fulfillment and the creation of a sustainable enterprise that you can proudly pass on to your children and grandchildren.

Analyzing the Income Potential of Your Top Enterprise Choices

Next, you will evaluate your top enterprise choices for their income potential. For each enterprise, calculate the income potential in the following manner:

Step 1: List the total number of sales that you anticipate being able to make from each enterprise based on your land and your market environment.

_____

_____

_____

_____

_____

Step 2: Multiply the total number of sales by the sale price to determine the financial value of your total sales.

_____

_____

_____

_____

_____

_____

Step 3: To calculate the gross profit, deduct the cost of replacement animals (number of animals × purchase price) to maintain your herd (e.g., replacement heifers and bulls in a cow/calf enterprise, young stockers in a stocker grazing enterprise).

_____

_____

_____

_____

_____

This gross profit is your income potential from that enterprise. In the business world, this is called a gross profit margin analysis. _Do not_ factor in your expenses; they will be influenced by your production practices and by how innovative you are at finding cost-effective ways to produce your product, all of which are within your control and thus do not provide a valid means of comparing your business enterprises at this stage of your business plan. You will address your expenses later, when you plan your budget, on page 281. The best enterprises give you the most potential income for your efforts.

You can also use the Income Potential worksheet provided below to help evaluate each of your livestock enterprises. Use sales numbers, conception rates, death losses, prices, and numbers of replacement animals to arrive at your best estimate of an average year and your _average_ expectations. You will also complete this worksheet for a best-case scenario and a worst-case scenario later, in the risk assessment section on page 279.

SAMPLE INCOME POTENTIAL WORKSHEET

Based on a cow/calf to grass-finished direct-marketing operation with 258 female breeding-class animals (cows, first-calf heifers, and replacement heifers) *

Class of animal	No. of cattle	Death loss		Cows / heifers bred		No. of sales	Avg. sale weight × avg. sale price	Income from sales	No. of cattle bought	Avg. purchase weight × avg. purchase price	Purchase expenses
		No.	%	No.	#						
Cows	170	2	1	151	90	17 empty 5 cull	1,200 lbs × $0.55/lb	$14,520			
Bulls	8	½**	5			2	1,800 lbs × $0.55/lb	$1,980	2.5	$3,000 ea.	$7,500
Male calves	111	5	4			106 male calves transfer to stocker class					
Female calves	111	5	4			50 female calves transfer to replacement heifers 56 female calves transfer to female stockers					
Replacement heifers	50	1	2	44	90	38 replacement heifers transfer to first-calf heifers 5 empty replacements transfer to grass-finishing class 6 culls sell at auction 1,050 lbs × $0.60/lb = $3,780		$3,780			
First-calf heifers	38	0	0	27	70	24 first-calf heifers transfer to cow class 11 empty first-calf heifers transfer to grass-finishing 3 culls sell at auction 1,085 lbs × $0.60/lb = $1,953		$1,953			
Male stockers	106	1	1			105 male stockers transfer to grass-finishing class Death loss would be higher if stockers purchased off-farm (approx. 5%)					
Female stockers	56	1	1			55 female stockers transfer to grass-finishing class					
Grass-finishing males	105	1	1			104	1,150 lbs × 62% (dress percentage) × $2.50/lb	$185,380			
Grass-finishing females	71	1	1			70	1,050 lbs × 60% (dress percentage) × $2.50/lb	$110,250			
							Total sales income	$317,863		**Total purchase expenses**	$7,500

* 258 animals bred yearly, 222 calves born
Grass reserve adequate for 258 cow/calf pairs, 176 stocker to grass-finished animals, and 8 bulls.
All grass-finished animals direct-marketed at $2.50/lb based on hanging weight. Assume average dress percentage of 62% for steers and 60% for heifers.

Income potential = total sales income – total purchase expenses = $310,363

** fraction used here (½) shows that 1 death loss can be expected every other year.

INCOME POTENTIAL OF A GRASS-BASED CATTLE ENTERPRISE

Assuming (check one): ☐ Average Scenario ☐ Best-Case Scenario ☐ Worst-Case Scenario

Class of animal	No. of cattle	Death loss No. / %	Cows / heifers bred No. / %	No. of sales	Avg. sale weight × avg. sale price	Income from sales	No. of cattle bought	Avg. purchase weight × avg. purchase price	Purchase expenses for replacement animals
Cows									
Bulls									
Male calves									
Female calves									
Replacement heifers									
First-calf heifers									
Male stockers									
Female stockers									
Grass-finishing males									
Grass-finishing females									
					Total sales income			**Total purchase expenses**	

Income potential = total sales income – total purchase expenses =

Comparing Different Enterprises Using the Same Grass Reserve

In order to provide a fair comparison between different enterprises, it is important to utilize the same-size grass reserve as the basis of your calculations. Note, however, that different animal classes eat different amounts of grass; for example, you cannot graze on your land the same number of cow/calf pairs and stockers. Thus, when you compare the income potential of one livestock enterprise with the income potential of another livestock enterprise that would utilize the same grass resource, you have to adjust your livestock numbers accordingly, requiring certain assumptions about feed intake for animals of varying classes grazing side-by-side in a pasture:

1 stocker (400–775 lbs) = 0.5 brood cow

1 brood cow = 2 stockers (400–775 lbs)

1 grass-finishing animal (775–1,150 lbs) = 0.75 brood cow

1 brood cow = 1.33 grass-finishing animals (775–1,150 lbs)

1.5 stockers (400–775 lbs) = 1 grass-finishing animal (775–1,150 lbs)

1 stocker (400–775 lbs) = 0.66 grass-finishing animal (775–1,150 lbs)

1 bull = 1 brood cow

1 cow/calf pair = 1 brood cow

1 brood cow = 1 replacement heifer (after breeding at 16 months of age)

1 brood cow = 1 first-calf heifer

For example, let's look at the pasture required by the cow/calf to direct-marketed grass-finishing operation (with 258 cow/calf pairs) from the example shown in the worksheet above. This same pasture could instead be used to graze 752 stockers (400–775 lbs) or 500 grass-finishing class animals (775–1,150 lbs) year-round, as is shown in the calculations below:

Pasture for:
> 258 cow/calf pairs (170 cows + 38 first-calf heifers + 50 replacement heifers), 176 stockers (half year), 176 grass-finishing animals (other half of the year) and 8 bulls.
> (Note that stockers become grass-finishing animals in the same production year, so they are counted as being present for only half a year each.)

Pasture converted to stocker grazing only (400–775 lbs):
> 258 cow/calf pairs = 258 brood cows = 258 × 2 = 516 stockers
> 176 stockers × half-year = 176 ÷ 2 = 88 stockers
> 176 grass-finishing animals × half-year = 176 × 1.5 ÷ 2 = 132 stockers
> 8 bulls = 8 brood cows = 8 × 2 = 16 stockers
> Total = 516 + 88 + 132 + 16 = 752 stockers

Pasture converted to grass-finishing animals only (775–1,150 lbs):
> 258 cow/calf pairs = 258 brood cows = 258 × 1.33 = 343 grass-finishing animals
> 176 stockers × half-year = 176 × 0.66 ÷ 2 = 58 grass-finishing animals
> 176 grass-finishing animals × half-year = 176 ÷ 2 = 88 grass-finishing animals
> 8 bulls = 8 brood cows = 8 × 1.33 = 11 stockers
> Total = 343 + 58 + 88 + 11 = 500 grass-finishing animals

Complicating this conversion is the fact that stockers and grass-finishing animals are growing. Consequently, in a year-round operation, you will replace stockers and grass-finishing animals once they reach their targeted sales/slaughter weight. How quickly you can replace your animals will depend on your cattle's weight gains averaged over the course of the year, which in turn depend on the productivity of your pastures and cattle, your climate, and your management practices.

For example, at an average weight gain of 2 lbs per day, it will take 6.25 months for a stocker to grow from 400 to 775 lbs:

775 lbs – 400 lbs = 375 lbs of weight gain

375 lbs ÷ 2 lbs/day = 187.5 days

187.5 days ÷ 30 days/month = 6.25 months

Likewise, at an average weight gain of 2 lbs per day, it will take 6.25 months for a grass-finishing class animal to grow from 775 to 1,150 lbs (also a 375 lb weight gain). At an average weight gain of 2 lbs per day, over the course of an entire year, then, the grazing reserve should be able to raise approximately two batches of stockers (1,444 stockers in exactly 12 months) from purchase to sale weight, or approximately two batches of grass-finishing animals (960 grass-finishing animals in exactly 12 months) from purchase to slaughter weight.

Accounting for Risk

To create a solid business plan, you need to account for risk. Every enterprise has vulnerabilities, which is why many banks and lending agencies require you to complete a gross profit margin analysis for average, best-case, and worst-case scenarios to evaluate your income potential and its inherent risks before granting you a loan. The same risk assessment is valuable when you assess the income potential of your grass-based cattle business options. Financial stability and security come only when a business is able to withstand even a worst-case year.

Using the same Income Potential worksheet (see page 277), prepare a best-case and worst-case analysis for each of the enterprises by varying the factors that represent the greatest degree of risk for your business. These factors will differ for each enterprise and may be a combination of market prices, death losses, conception rates, the percentage of animals you are able to direct-market, and the price you can demand for your direct-marketed beef — whichever areas represent the greatest uncertainty and risk to your income. (In chapter 29 you will design your disaster/drought contingency plan to further reduce the risk associated with your chosen enterprises.)

Comparing the Enterprises

When your average, best-case, and worst-case scenario worksheets are completed for each of the enterprises you are considering, scrutinize them and decide which enterprises and market options you want to pursue. To aid in your decision, you should also compare the enterprises based on their return per acre, per capital investment, and per hours of labor (your time invested). These comparisons allow you to assess vastly different enterprises based on the common denominator that is most important to you. Making this comparison will give you a good idea about which enterprises are worth pursuing, based on their income potential. I do not recommend including a return-per-animal comparison because this focuses on maximizing productivity (at any expense) rather than maximizing profitability, as per the discussion about productivity and profitability on page 4 in this book's introduction.

For example, the return from the cow/calf through grass-finished direct-marketing operation we've been discussing might look something like this:

■ Assuming a land base of 2,000 acres: $310,363 gross profit ÷ 2,000 acres = $155/acre

■ Assuming a capital investment of $450,000: $310,363 gross profit ÷ $450,000 capital investment = 0.69 (or 69 percent) return on the dollar

■ Assuming you work full time on your enterprise, 12 months of the year (52 weeks minus 3 weeks holiday) and average 5 days/week, 6 hours/day: $310,363 gross profit ÷ 1,470 hrs = $211/hour

Afr completing the Income Potential Worksheets for each enterprise and accounting for risk as well as the enterprise comparisons based on return discussed above, you will further reduce your list of options. It is wise to maintain at least three separate enterprises or independent markets, all of which are either cattle-based or diversified into other complementary enterprises, to give your business financial buoyancy when confronted with weather challenges, market problems, disease outbreaks, and fickle buyers. You do not need three unrelated enterprises; instead, establish three separate market outlets for the same product or three enterprises that take advantage of the same resource base, basic infrastructure, and skills but provide you with access to unrelated markets to guarantee financial stability and security (this concept is discussed on page 214).

Conscious Profit Planning

The financial records that most farmers keep — little more than a financial history of the businesses — are strictly for the tax department. Unfortunately, this type of record keeping is a poor tool for financial planning. Expenses tend to rise to meet income and the various farm enterprises merge, making it difficult to compare individual enterprises, calculate costs of production for each enterprise, and identify financial drains on profitability.

You don't need to leave your profits up to chance. You can approach your financial planning in a deliberate, conscious, calculated fashion. Before planning your budget, you should set a target profit margin that automatically caps all your expenses. Doing so forces you to find innovative ways to cut expenses and compels you to rethink your production strategy so that after removing your profit share, the remaining income is sufficient to cover your expenses. This is the key to maintaining a viable business strategy.

What that profit margin should be depends entirely on your goals and financial expectations. It should provide you with a viable income, however, which is why it is so important to set your profit margin target now, based on your anticipated gross income, before you begin planning your expense budget.

Separate Your Enterprises

In your financial planning records (your yearly expense budget on page 282, and your cash flow records of budgeted monthly expenses versus actual monthly expenses on pages 284–285 and anticipated monthly income versus actual monthly income on page 286), you should split your farm into its various enterprises and keep a separate set of financial records for each. Consider each enterprise to be a stand-alone business so you can identify and address costly expenses, focus on the most profitable enterprises, and eliminate areas that are not profitable. If an enterprise does not generate a profit, it must be eliminated. The cost of equipment, labor, taxes, debt, and other expenses shared by more than one enterprise should be divided so that each enterprise carries the cost percentage that it uses.

Also, enterprises that include more than one level of production should be analyzed as though each production level is a separate business, to ensure that each portion is profitable. This should be done when you develop your expense budget (see pages 281–282) and compare your anticipated cash flow with your actual expenses and actual income (see pages 283–286). For example, if you have a cow/calf through grass-finished direct-marketing cattle enterprise, this enterprise is composed of a cow/calf enterprise, a replacement heifer (breeding stock) enterprise, a stocker program, and a grass-finishing enterprise, each of which must be viable and self-sufficient.

In these financial planning records, sell your calves to your stocker enterprise at market value, make your cow herd enterprise buy your replacement heifers at market value, and pay your stocker enterprise market value when your stockers transfer to the grass-finishing enterprise. Even if you make hay for your own animals alone, you still have a hay production enterprise and your cattle business must buy that hay at full market value. Your hay production should not be a subsidy program for your cattle enterprise (see chapter 7 for a discussion about the importance and techniques of year-round grazing).

By doing this, you will soon see the true cost of each of your enterprises. Perhaps you will find it is cheaper to buy stockers instead of producing your own in your cow/calf operation; or you may discover it is cheaper to purchase your hay supply to settle your stockers instead of producing it yourself; or you may find that hay must be eliminated altogether (to be replaced by a twelve-month grazing program) because your cattle business simply can't afford to buy it.

Budgeting Your Expenses

Next, you should list all your anticipated expenses for the year (including the dollar value) incurred by each enterprise. Each enterprise should be evaluated separately. Setting your target profit margin before you begin designing your expense budget forces you to become innovative and question every expense to see if you can eliminate or reduce it.

Keep trimming your expense budget until your business is able to meet your financial expectations. Review each item one by one — never accept an expense just because you did so in the past. Put your money into making the greatest improvements and reducing the most risk; leave less demanding expenses for another year, when there is more room in the budget. You want to get the biggest bang for every buck you spend. Ask yourself which expenses will generate the most new income and what you can do to reduce the greatest drain on your current income. Find the most costly aspects of your business and change them.

Sifting through your expenses in this manner helps you create the most positive, long-reaching changes on your farm, rather than letting your recurring operating expenses prevent you from breaking out of a production rut. For example, you may decide to spend your entire fertilizer and irrigation budget on building electric-fence divisions and develop-

ing water sites because the positive repercussions will echo through your entire enterprise and reduce expenses in the long term without having to repeat the same high fertilizer and irrigation expenses year after year.

The following worksheet is designed to help you identify your *target profit margin* and set a cap on your *maximum allowable budget for expenses* before you begin planning your yearly budget. At the bottom of the worksheet there is space to include an itemized list of yearly expenses. The sum of these expenses must be equal to or less than the maximum allowable budget for expenses calculated at the top of the worksheet in order for you to meet your financial goals. (Note that two of the spaces in the worksheet are marked with capital letters: A in the "target net income" line and B in the "total sum of budgeted expenses" line. These will be used later to calculate your meat price.)

Find the most costly aspects of your business and change them.

This budget also allows you to categorize your expenses for ease of planning by organizing them as operating expenses and farm/business improvements such as new infrastructure, more cattle, and additional land. Make sure you also include a fund to replace your equipment, fences, and vehicles, which you pay into on a yearly basis based on the life expectancy of your equipment. This will eliminate the extra expense of carrying a debt load to replace them and will spread their replacement cost over multiple years so no unpleasant surprises await you as your equipment ages. Paying into your own interest-generating fund is a lot less expensive than servicing a debt to a lender. This also provides you with some added financial stability because you can skip paying into your equipment fund in a dire emergency, something a lender would never let you do without considerable surcharges. Also include a small yearly allocation that you will put into a financial reserve as a drought fund. (See page 340 to learn how to calculate your drought fund.)

YEARLY EXPENSE BUDGET WORKSHEET

Year:_____

Calculating Maximum Allowable Expense Budget

Category	Calculation	Value
Anticipated Gross Profits ($):	from the Income Potential worksheet (use average scenario)	$
Target Profit Margin (%):	adjust as necessary to meet your financial expectations in the next row	%
Target Net Income ($):	Anticipated Gross Profits × Target Profit Margin (e.g., for a target profit margin of 25%, multiply by 0.25)	**A. $**
Maximum Allowable Budget for Expenses ($):	Anticipated Gross Profits − Target Net Income	$

Budgeting Your Expenses

Itemized Expenses:	Dollar Value ($)

Operating Expenses:

Farm/Business Improvements:

Replacement Funds & Drought Funds:

Total Sum of Budgeted Expenses:	**B. $**

Adjust expenses until Total Sum of Budgeted Expenses ($) is equal to or less than Maximum Allowable Budget for Expenses ($) at top of worksheet.

Common Financial Drains on Profitability

If you are having trouble shaving your expenses, here is a list of common issues that are often the biggest financial drains on profitability.

- Out-of-season calving

- Winter feeding

- Too many herds

- Inefficient grazing management or a poorly designed pasture rotation

- Poor marketing strategy

- Poor management priorities

Issues such as poor conception rates, high mortality, and the high cost of equipment, labor, fertilizer, weed control, irrigation, and veterinary services are usually symptoms of a bigger problem. Tightening your belt and trying to limit your budget in these areas without addressing the underlying causes will only take you a step backward in the long run. These symptoms indicate a desperate need for change in your management strategies. Such a change will eliminate them altogether without having to address them specifically. You effectively trade expenses for different management ideas. The resulting savings combined with the revenue from the sale of equipment that becomes obsolete due to changes in your management strategy will usually pay for more income-generating cattle, grazing infrastructure, and land acquisitions.

Budgeting Cash Flow: Keeping a Finger on Your Financial Pulse

Your financial planning records should allow you to track the progress of your business from month to month through the course of the year. The most effective way to do this is to use monthly record sheets that compare your budgeted monthly expenses and your actual monthly expenses. You should also use this same system to keep track of how your actual monthly income compares with your anticipated monthly income. This record-keeping system allows you to assess at a moment's glance your budgeted cash flow and your actual expenses and income so

you can keep your finger on the financial pulse of your business at all times. It is the most effective way to avoid having your expenses rise up to meet your income and swallow up your profits.

The worksheet below includes a blank monthly expense record sheet. Use the left side of the record sheet to plan your itemized monthly expense budget, filling in a separate record sheet for each month. On the right side of the record sheet fill in your actual itemized expenses as they occur throughout each month. At the end of each month you can easily see whether your budget is on target. If at the end of the month the budgeted expenses are not equal to your actual expenses, find out why! Did you forget to make an important payment? Did you overspend on something or spend in an area that was not in the budget? The net difference between your budgeted and actual expenses will tell you if you are on track, overspending, or missing a planned expenditure.

The next worksheet shows a blank itemized monthly income record sheet, which is also divided to plan your anticipated income on the left side and record your actual income on the right side so you can compare your income on a month-by-month basis. This worksheet gives you immediate feedback so you can see whether your scheduled income is going to meet your planned expectations. If not, consider this feedback an early warning sign of drought, financial crises, and slowing market cycles. Then reduce or eliminate some expenses in order to meet your financial goals: Perhaps reduce payments into your equipment replacement fund or into your drought fund, for example.

The worksheet also shows you whether your advertising strategy is successful for each of your market outlets, whether your cattle are gaining weight and selling on schedule, and whether you have any unexpected health, production, marketing, packaging, and meat-cutting issues to resolve.

If your actual income exceeds your budgeted expectations, you get immediate feedback about the success of your enterprise. Is the high income a warning sign of an oversight such as selling too many animals? If it is a sign of a premature sale date, you would need to hold over that income to the next months to meet your financial obligations then. It might also be a sign of an unexpectedly good

EXPENSE RECORD SHEET FOR THE MONTH OF: _____

Budgeted Expenses		Actual Expenses	
Name of expense	$	Name of expense	$

OPERATING EXPENSES:

Budgeted Expenses		Actual Expenses	
Name of expense	$	Name of expense	$

FARM IMPROVEMENTS:

REPLACEMENT FUND AND DROUGHT FUND:

Total budgeted expenses:		Total actual expenses:	
Net difference: +/- $			

financial year, in which case you can plan ahead and use that extra income to invest in herd increases, infrastructure improvements, land, and new income-generating expenses, or to address any bottlenecks that are currently limiting your income potential.

An unexpected increase in income should be used to increase your financial stability. For example, use some of it to build up the nutrient balance in the soil by addressing mineral deficiencies. Doing so allows you to significantly reduce your fertilizer amendment costs during drought and other less profitable years.

Extra income should also be used to add to your equipment-replacement fund, especially if you were unable to pay into the fund during some years, or it can be allotted to the financial reserve you are building for drought years. Either way, by knowing ahead of time where your income levels stand, you have an opportunity to protect your financial stability, smooth your income between good and bad years, and invest in increasing your income potential.

By planning your income and expense schedules, you can also try to manipulate the dates of your earnings and expense payments. This allows you to organize your cash flow throughout the year to avoid financial loans to tide you through the months when monthly expenses would normally have been higher than your monthly income.

INCOME RECORD SHEET FOR THE MONTH OF:_____

Income Budget		Actual Income	
Source of income	$	Source of income	$
Total anticipated income:		Total actual income:	
Net difference: +/- $			

Offsetting Your Income for Tax Purposes

Almost everyone has heard the advice about buying a new truck just before year-end to offset taxes during a good fiscal year. Yet another way offsets taxes during a good fiscal year but also enables you to store that extra income for a lean year, providing your grass-based cattle business with an additional degree of financial stability.

Most conventional synthetic fertilizers are water-soluble, causing any fertilizer excesses to be washed from the soil through the movement of ground-water. Many organic fertilizer sources (such as soft rock phosphate) are not water-soluble, meaning the unused portions remain in the soil and build a nutrient reserve. Calcium fertilizer amendments such as limestone and gypsum also last many years after they are applied. During years of exceptional income, you can build up the fertilizer reserves of phosphorus, potassium, calcium, magnesium, sulfur, and trace elements in the soil through the use of insoluble fertilizers, based on the recommendations of your soil analyses. These can be drawn upon during financially lean years, allowing you to reduce or eliminate your fertilizer expenses to help stabilize your income without affecting productivity.

The extra income can also offset taxes if you use it to develop additional water sites, add fencing, or address many of the other farm improvements that will increase your financial stability and raise your profit margin over time.

Keep a list of what to do with excess income from good years so that you can maintain financial stability through the ups and downs of the market cycle, survive the inevitable drought, fend off disease outbreaks, repair flooded crops, and survive prairie fires. It's just good risk management. After all, the average year comes around only once every decade or so. The lean years require all the financial innovation you can muster to maintain a stable income (these are the worst-case scenarios calculated to assess risk on your Income Potential worksheet) and the prosperous years (your calculated best-case scenarios) become wasted financial opportunities if you don't hoard those hard-earned dollars for a leaner day.

Determining the Price of Your Direct-Marketed Beef

The rest of this chapter provides a framework for calculating the price of your direct-marketed beef. There are two worksheets provided below to assist you with this process. The first, on pages 288–290, is designed to help you calculate your price per pound of beef sold by the whole, side, or split side based on hanging weight (also known as carcass weight). Following this, on pages 291–293, is a worksheet to help you calculate your meat prices by the cut. You will determine these prices by adjusting for your financial goals the retail price structure you find at the grocery store.

The calculations found on these two worksheets account for the cost of replacement animal purchases and any live animal sales (e.g., animals sold via cattle auction). Be sure to include all slaughter, transport, and marketing expenses in your expense budget (your Yearly Expense Budget worksheet on page 282); these expenses must be included for your meat price to be accurate. The slaughter price (also known as the kill fee) is usually a fixed cost per animal regardless of weight. Ask your butcher for a quote.

The cutting and wrapping costs will vary depending on each customer's individual cutting and wrapping preferences. Consequently, they should be paid directly to the butcher by the customer or you can add them to each individual customer's bill to account for cutting and wrapping requests. When selling meat by the cut, however, you will need to factor into your price the cutting-and-wrapping surcharge, as shown in the second worksheet. (Also see chapter 17 for more information on how to calculate your price per pound of beef.)

Calculating Your Direct-Market Meat Price Based on Hanging Weight

To calculate your direct-market price for meat sold by the whole, side, or split side, use the worksheet on pages 288–290, Calculating Direct-Market Beef Price Based on Hanging Weight.

CALCULATING DIRECT-MARKET BEEF PRICE BASED ON HANGING WEIGHT

- *Use your Income Potential worksheet for an average year and your Yearly Expense Budget worksheet to help you fill in blank spaces on the worksheet as you calculate your meat price by the whole, side, or split side, and by the cut.*

Name of item	Calculation	Value ($)
Sum of Budgeted Expenses (excluding cost of replacement animals):	Transfer line A from your Yearly Expense Budget worksheet	**A. $**
Your Target Net Income:	Transfer line B from your Yearly Expense Budget worksheet	**B. $**
Gross Profits (excluding cost of replacement animals):	Line A + line B =	**C. $**

- *Calculate the total purchase cost of replacement animals:*

Class of animal purchased	No. of animals purchased each year		Average live weight at time of purchase (lbs)		Average purchase price ($/lb)		Value of total purchases in each class ($)
Cows		×		×		=	
Bulls		×		×		=	
Male calves		×		×		=	
Female calves		×		×		=	
Replacement heifers		×		×		=	
First-calf heifers		×		×		=	
Male stockers		×		×		=	
Female stockers		×		×		=	
Grass-finished steers		×		×		=	
Grass-finished heifers		×		×		=	
Total purchase cost of replacement animals:	Sum of column listing total value of purchases in each class:						**D. $**

- *Add the total purchase cost of your replacement animals to your gross profits to calculate the total income required by your farm to cover all costs, including costs to purchase replacement animals.*

Total farm income:	Line C + line D	E. $

- *Calculate the total value of live-animal sales (e.g., auction sales):*

Class of animal sold live (e.g., via auction)	No. of animals sold live		Average live weight (lbs)		Average price for each livestock class sold live ($/lb)		Value of total live animal sales in each class ($)
Cows		×		×		=	
Bulls		×		×		=	
Male calves		×		×		=	
Female calves		×		×		=	
Replacement heifers		×		×		=	
First-calf heifers		×		×		=	
Male stockers		×		×		=	
Female stockers		×		×		=	
Grass-finished steers		×		×		=	
Grass-finished heifers		×		×		=	
Total value of live animal sales:	Sum of column listing total value of live-animal sales in each class:						F. $

- *Calculate the total value of all your direct-market sales (including meat sold by the cut) by subtracting live-animal sales from the total farm income:*

Total value of all direct-marketed beef sales:	Line E – line F =	G. $

continued on next page

- Calculate the total carcass weight of all direct-marketed beef, including all beef sold by the cut:

Class of direct-marketed animal	No. of animals sold via direct-marketing		Average live weight (lbs)		Average dress percentage of each class expressed as a decimal (e.g. 62% = 0.62)		Total lbs of direct-marketed beef in each class (lbs)
Cows (for direct-market hamburger)		×		×		=	
Bulls (for direct-market hamburger)		×		×		=	
Grass-finished steers		×		×		=	
Grass-finished heifers		×		×		=	
Grass-finished young cows		×		×		=	
Grass-finished young bulls		×		×		=	
Total lbs of beef sold via direct-marketing sales:			Sum of column listing total lbs of direct-marketed beef in each class:				H. lbs

- Calculate the average price/lb of your direct marketed beef; this is also your direct-market price for beef sold by the whole, side, or split side.

Average price/lb of your direct-marketed beef:	Line G ÷ line H =	I. $/lb

- The average price/lb of your direct-marketed beef is your direct-market price for beef carcasses sold by the whole, side, or split side. This price is based on the hanging weight (also known as the carcass weight). Cutting and wrapping are extra — your butcher will quote you or your customers a $/lb surcharge for this service. Adjust your direct-market price according to your marketing needs (e.g., to make whole beef carcasses slightly cheaper than sides or split sides):

Direct-market beef price based on hanging weight ($/lb):	from line I	I. $/lb

- If applicable, add a cutting-and-wrapping surcharge to this price:

Cut-and-wrap surcharge:	(as quoted by butcher)	J. $/lb
Direct-market beef price based on hanging weight with cutting and wrapping added to price ($/lb):	Line I + line J =	K. $/lb

Calculating Your Meat Price by the Cut

To calculate your direct-market price for meat sold by the cut, use the following worksheet, entitled Calculating Meat Price by the Cut. If you are selling many different animal classes by the cut, you should fill in a worksheet for each animal class you will slaughter and sell via by-the-cut sales (that is, steers, heifers, young cows, young bulls, and older cows primarily ground into hamburger for direct-marketing). As part of this calculation, you will need to go to the grocery store and record its retail prices for the same cuts; its price structure is designed to sell out of all cuts equally. Your butcher may also be able to supply you with a retail price structure for meat sold by the cut. You will then adjust the grocery store price structure to allow you to meet your income expectations.

CALCULATING MEAT PRICE BY THE CUT

- Use this table to calculate your meat-by-the-cut prices. It is a multistep process using your most common class of direct-marketed beef sale. You will be able to use the same price structure for all direct-marketed by-the-cut beef sales. Begin by calculating the average value of a single by-the-cut direct-marketed animal using the retail (grocery store) price structure. Then adjust the grocery price structure to come up with a price structure that allows you to meet your financial expectations. Your butcher will be able to provide you with the average dress percentage, percent cutability, and an average breakdown of meat yield for your slaughtered animals to further adjust your by-the-cut-meat prices.

Average live weight of a single animal (lbs)		Average dress percentage provided by butcher (% expressed as decimal)		Average cutability provided by butcher (% expressed as decimal)		Total meat yield per animal when sold by the cut
	×		×		=	L. (lbs)

- Calculate the value of individual meat cuts of your by-the-cut beef animal based on the grocery store price structure. Your butcher will be able to supply you with your animals' average meat yield percentage. Identify meat cuts based on your marketing strategy (e.g., roasts vs. steaks). You will adjust this price structure later to meet your financial goals.

Name of meat cut	Total meat yield from line L (lbs)		% yield / cut		Yield / animal (lbs)		Grocery store meat price ($/lb)		Value of each individual meat cut ($)
		×		=		×		=	
		×		=		×		=	
		×		=		×		=	
		×		=		×		=	
		×		=		×		=	
		×		=		×		=	
		×		=		×		=	
		×		=		×		=	
Total value of meat-by-the-cut animal based on grocery price structure:			Sum of column listing value of individual meat cuts:						M. $

- Calculate the total value per direct-marketed animal of the same animal class when sold based on hanging weight. This will be used to develop a conversion factor to adjust the prices shown in the previous section. Use the direct-market price that includes cutting and wrapping (line K from Calculating Direct-Market Beef Price Based on Hanging Weight worksheet).

Average live weight of a single animal (lbs)		Average dress percentage provided by butcher (% expressed as decimal)		Direct market price including cutting and wrapping (from line K)		Total value per direct-marketed animal (sold based on hanging weight)
	×		×		=	N. $

- Compare the value of your average by-the-cut animal (line M) with the value of an animal of the same class when it is sold by the whole, side, or split side (line N). The difference between these values will determine your conversion coefficient, allowing you to adjust your meat-by-the-cut prices so you will earn the same net income per animal as through whole, side, or split-side carcass sales.

Total value per direct-marketed animal, sold based on hanging weight (from line N)		Total value of meat-by-the-cut per animal based on grocery price structure (from line M)		Conversion coefficient
	÷		=	O.

- Now adjust the meat-by-the-cut prices used above to give you a price structure for your grass-finished beef that allows you to meet your financial expectations:

Name of meat cut	Grocery store meat price ($/lb)		Conversion coefficient (from line O)		Your price for each meat cut ($/lb)
		×		=	
		×		=	
		×		=	
		×		=	
		×		=	
		×		=	
		×		=	
		×		=	
		×		=	
		×		=	
		×		=	
		×		=	
		×		=	

- *Variety meats yield should be calculated and priced separately because it is variable and sales depend on your specific market.*

Average live weight of a single animal (lbs)		Average dress percentage provided by butcher (% expressed as decimal)		Average variety meat yield (approx. 4% = 0.04)		Total variety meat yield (all variety meat cuts) (lbs)
	×		×	0.04	=	

- *Adjust the variety meat price using the conversion factor (O). Variety meats are the liver, heart, tongue, tripe, sweet-breads, kidneys, oxtails, and brains.*

Name of meat cut	Grocery store meat price ($/lb)		Conversion coefficient (from line O)		Your price for each cut ($/lb)
		×		=	
		×		=	
		×		=	
		×		=	
		×		=	
		×		=	
		×		=	

The table below shows the meat yield, by the cut, of an average steer and includes the approximate yield percentages used to calculate the individual prices for meat by the cut for each of your animal classes.

MEAT YIELD BY THE CUT FOR AN AVERAGE STEER*

- *Note that all the ground beef and stew meat yields from each part of the animal are added together in the bottom right-hand corner.*

Average 1,150 lb steer: 62% dress percentage, 76% cutability ± 4% variety meats					
Chuck (171.2 lbs meat = 31.5%)	Blades, roasts, steaks	32.6 lbs = 6%	**Loin** (85.6 lbs meat = 15.8%)	Porterhouse steaks	18.8 lbs = 3.5%
	Ground beef and stew meat	80.1 lbs = 14.8%		T-bone steaks	9.4 lbs = 1.7%
	Arm pot roasts and steaks	34.1 lbs = 6.3%		Strip steaks	14.4 lbs = 2.7%
	Cross rib pot roasts	24.4 lbs = 4.5%		Sirloin steaks	14.7 lbs = 2.7%
Round (119.1 lbs meat = 22%)	Top round	33.3 lbs = 6.1%		Tenderloin steaks	6.5 lbs = 1.2%
	Bottom round	30 lbs = 5.5%		Ground beef, stew meat	21.8 lbs = 4%
	Tip	16.2 lbs = 3%	**Rib –** 56 lbs meat = 10.3%	Rib roasts	23 lbs = 4.2%
	Rump	7.5 lbs = 1.4%		Rib steaks	8.8 lbs = 1.6%
	Ground beef	32.1 lbs = 5.9%		Short ribs	8.3 lbs = 1.5%
Thin Cuts (110.1 lbs meat = 20.3%)	Flank steaks	3.5 lbs = 0.6%		Ground beef, stew meat	15.9 lbs = 2.9%
	Pastrami squares	2.8 lbs = 0.5%			
	Outside skirt	2.1 lbs = 0.4%	Total meat cuts (not including variety meat)		542 lbs
	Inside skirt	2.4 lbs = 0.4%	Total portion of meat used for ground beef and stew meat		233.8 lbs = 43.1%
	Boneless brisket	15.4 lbs = 2.8%	Variety meat (liver, heart, tongue, tripe, sweetbreads, kidneys, oxtails, brains)		27 lbs ≈ 4% of carcass weight
	Ground beef, stew meat	83.9 lbs = 15.5%	Fat, bone, and loss (dog food and soup bones)		144 lbs

* % yield indicates percentage of total meat cuts (542 lbs) in each cut of meat.

Chapter

23

Your Cattle Year on Grass

AN INTEGRAL PART OF CUSTOM-DESIGNING a business plan for your grass-based cattle enterprises is to plan your cattle production year to take advantage of nature's seasonality. Mimicking nature's production cycle with your cattle herd's life cycle and matching your cattle year with the yearly grass production cycle will allow you to schedule all the important events in your cattle production year and plan the timing of your business. This begins with planning your calving season because it determines all other important dates during the year, from breeding and cattle processing to weaning and grass-finishing, as well as determining slaughter and sale dates and marketing strategy.

The calving season must be timed to coincide with the peak late-spring to early-summer grass flush, which is roughly the same time grazing species give birth in the wild. This allows the cow to fatten up prior to calving and ensures her fertility for the next breeding season. In addition, calves are born during the most hospitable, disease-free time of year, cows are able to winter their calves on their milk, and cows and calves can be grazed throughout winter without feed supplements. (See chapter 3 for more on the cattle year on grass.)

In this chapter you will design your own cattle production year, customized to suit your particular climatic constraints, environmental conditions, and market opportunities. You will time your cattle production cycle to take advantage of the changing grass growth rates and grass quality over the course of the seasons and minimize feed costs, climatic challenges, parasite pressure, and predation. This comprehensive approach to planning your cattle production year is a fundamental prerequisite to creating a profitable, sustainable, and healthy natural, grass-based cattle enterprise.

Planning the Ideal Cattle Year on Grass

The best way to design your ideal cattle year on grass is to plot it on a time line or graph that shows all environmental constraints: the grass-growth curve (the changing grass growth rates over the course of the growing season; see chapter 3 and chapter 4), grass lignification (the process of grass becoming hard and unpalatable as it approaches maturity or begins to dry out or when cool-season grasses are exposed to hot weather; see chapter 4), parasite life cycles, cold, heat, dust, wet weather, and any other constraints that are predictable environmental trends in your area. These constraints make it easy to plan your ideal calving season. Once you have recorded this information, the other important livestock-handling dates will fall into place: breeding, calving, weaning, vaccination schedules, processing, slaughter schedules, and auction dates.

A Sample Cattle Year

A typical grass-based cattle production year in the Okanagan Valley in southern British Columbia might look very much like the illustration below. Notice the calving season has been timed so that cows have at least a month of predictable lush green grass to fatten on before they calve, regardless of adverse weather conditions. Fluctuations in the length of winter are normal and should be accounted for; allow enough leeway in the calving date to ensure that nature doesn't leave you vulnerable if the spring grass is late. The calving season in the illustration (June 10 to July 25) overlaps with the calving period of deer and moose in the Okanagan Valley, but theirs is shorter. Interestingly, the overwhelming majority of cows conceive during their first breeding cycle and calve out by the end of June, just as the deer do.

The grass in the Okanagan lignifies during summer due to the intense heat. Calving is timed so the calves' rumens start to become active just when summer temperatures begin to drop and the cool season grasses begin their second growth phase. Calves get the benefit of the tender second grass growth, which stimulates their delicate mouths and rumens. Their

MATCHING THE CATTLE YEAR TO THE GRASS GROWTH CYCLE

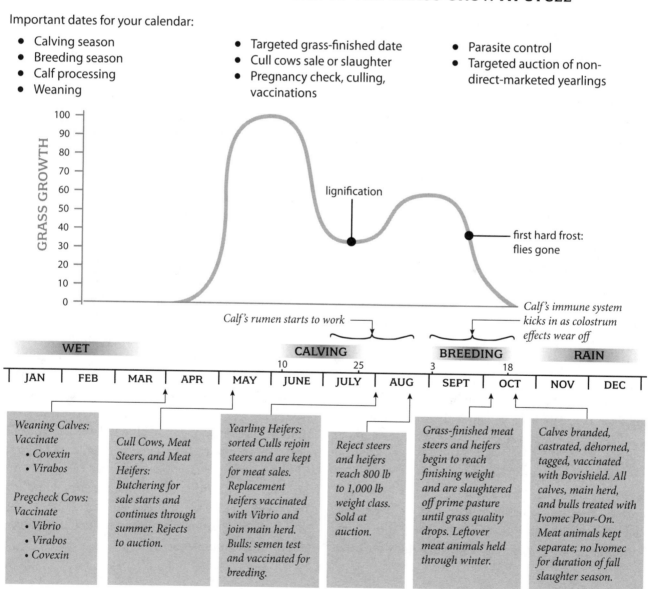

Important dates for your calendar:

- Calving season
- Breeding season
- Calf processing
- Weaning

- Targeted grass-finished date
- Cull cows sale or slaughter
- Pregnancy check, culling, vaccinations

- Parasite control
- Targeted auction of non-direct-marketed yearlings

Calf's rumen starts to work

Calf's immune system kicks in as colostrum effects wear off

Weaning Calves: Vaccinate
- Covexin
- Virabos

Pregcheck Cows: Vaccinate
- Vibrio
- Virabos
- Covexin

Cull Cows, Meat Steers, and Meat Heifers: Butchering for sale starts and continues through summer. Rejects to auction.

Yearling Heifers: sorted Culls rejoin steers and are kept for meat sales. Replacement heifers vaccinated with Vibrio and join main herd. Bulls: semen test and vaccinated for breeding.

Reject steers and heifers reach 800 lb to 1,000 lb weight class. Sold at auction.

Grass-finished meat steers and heifers begin to reach finishing weight and are slaughtered off prime pasture until grass quality drops. Leftover meat animals held through winter.

Calves branded, castrated, dehorned, tagged, vaccinated with Bovishield. All calves, main herd, and bulls treated with Ivomec Pour-On. Meat animals kept separate; no Ivomec for duration of fall slaughter season.

immune systems will take over on the warm, dry, early-fall pasture, well before the wet fall weather begins.

Breeding occurs on the lush fall grasses at the same time as the fall rut starts in the wild. Cattle processing occurs without having to disrupt the breeding season, but before the rainy, wet, cold late-fall weather sets in. And it's still early enough that the warble-fly treatment can be applied before the mid-November cutoff date, when the life cycle of the parasite prevents effective treatment.

By this time, the disease-carrying flies have been killed by hard frosts, so they will not be around to harass castration wounds. The dormant grass in October is still lush and provides a clean, nutritious environment on which the calves can recover from their processing trauma while the weather is still mild and relatively dry. By the time the cold, wet November weather begins, the calves are fully recovered and grazing clean, fresh, stockpiled forages with their mothers, unaffected by the pneumonia outbreaks that usually plague local feed yards at this time of year.

There is another wet period during spring, which brings further disease, mud, and stress to the cattle. For this reason, weaning and booster vaccinations for the calves, as well as the cows' vaccinations, pregnancy checking, and sorting, do not happen until the wet weather has passed. When the processing facilities are dry again, the muddy fields have dried up and the weaned calves are able to replace their mothers' milk with the first green grass shoots that emerge in spring, which makes up for any nutritional deficit left by weaning.

The cull cows and any remaining grass-finished meat animals left over from the previous year also benefit from the fresh grass growth. It allows the direct-marketing season to stretch from mid-May through the end of December, when the dormant grass has deteriorated sufficiently by way of rain leaching, humidity, and freezing that the grass-finished animals are no longer gaining sufficiently to be predictably high-quality, tender-meat animals.

The cattle year and marketing strategy on this farm meshes perfectly with nature's cycle. Calving occurs around the summer solstice at the grass peak, and breeding occurs at the fall equinox. If the calving season were shifted any later in the year, calf nutrition and yearling grass finishing would conflict with the deteriorating fall weather conditions. Breeding would shift into the muddy, wet, late fall and would no longer be in sync with the ideal photoperiod, resulting in longer acyclic periods after calving and lower conception rates. Shift the calving season any earlier and calves would have to be weaned during the wet spring weather to give the cows enough time to recuperate after calving. The cows would not have sufficient time to fatten up prior to calving without using expensive feed supplements, and their acyclic periods would be longer and their conception rates reduced. Breeding would have to occur during the summer heat, meaning photoperiod advantages of breeding during the fall rut season would be lost.

Plotting Your Cattle Production Dates

In order to match your cattle year to the ideal natural grass cycle, plot all important cattle production dates in the order listed below on a time line on which you have already indicated the grass growth cycle and nature's climatic constraints.

- Grass production curve

- Calving dates for wild grazing animals in your area

- Harsh weather constraints (e.g., mud, extreme wet, extreme cold, extreme heat, drought)

- Parasite pressure and first fly-killing frosts in fall

- Forty-two-day calving season

- Forty-two-day breeding season

- Within a month after birth, the calves' rumens become active and develop in response to the calves supplementing their mothers' milk with their first mouthfuls of grass. Thus, at one month of age calves should have access to tender, lush grass to facilitate this process.

- At three months of age the immunity from the cows' colostrum wears off and the calves' own immune systems have to take over, making this a particularly vulnerable stage in the animals' lives.

- Calf processing (e.g., branding, castrating, dehorning, tagging, vaccinations)
- Weaning, yearling vaccination booster shots
- Parasite control
- Cow pregnancy check, vaccinations, culling
- Cull cow sale dates
- Yearling commodity auction sale dates
- Grass-finished animals' slaughter dates

The Julian Calendar

The 365-day Julian calendar was introduced by Julius Caesar in 46 B.C.E. to provide a calendar that could match the length of the solar year and divide up the year into roughly equal months. Until that time, the Romans had a 355-day calendar with a complex system of leap months every other year to make up the extra days, making it difficult to schedule seasonal events. The 365.2425-day Gregorian calendar still used today is merely a Julian calendar modified to include a leap day on February 29 every four years to keep the solar year on track.

Unfortunately, the extra leap days of a Gregorian calendar somewhat complicate our livestock plan-

ning; thus, when planning your cattle production year, it is useful to utilize the consistent, 365-day Julian calendar and simply ignore the extra leap days so that every day of every month corresponds to the exact same day on the 365-day calendar rotation. By using the 365-day Julian calendar rotation, you can simply add or subtract the length of the gestation period to calculate calving and breeding dates.

The following example illustrates how to use the Julian calendar on page 299. Remember that the average gestation period of cattle is 280 days when they calve in sync with nature's seasons and is slightly longer (up to 285 days) if they calve out of sync with these seasons (see chapter 3 for a detailed discussion on gestation period length; photoperiod; and effect of summer calving on fertility, conception, and gestation). Let's assume that your cows will have a 280-day gestation period. If you want to begin calving on June 10 (day 161 on the Julian calendar), subtract the gestation period to find the breeding turnout date:

$$161 - 280 = -119$$

If the result is a negative number, add another year to determine the proper breeding turnout date:

$$-119 + 365 = \text{day 246, or September 3}$$

If you end your breeding season on October 18 (day 291), add the length of the gestation period to determine when your calving season will end the following year. If the result is greater than 365, simply subtract a year to find the correct date in the next calendar rotation:

$$291 + 280 = 571$$

If the result is greater than 365, subtract a year to determine when the calving season will end:

$$571 - 365 = \text{day 206, or July 25}$$

If you turn out your bulls on September 17 (day 260) and want to determine the calving date, add the gestation period:

$$260 + 280 = 540$$
$$540 - 365 = \text{day 175, or June 24}$$

PLANNING FOR THE TRANSITION

If you are switching your herd from a calving date required by a conventional production scenario to a summer calving date, plot a graph for each year that you are in transition between your former calving date and your target calving date in order to maximize the efficiency of those years and to make the transition as smooth as possible.

If you are switching your herd from an earlier calving season, simply postpone the bull's turnout until the date of the new breeding season. If you are switching from a later calving season (for example, if your herd is currently in a fall-calving scenario), you can breed as much as 3 weeks earlier each year without a significant effect on conception rates until you reach the target breeding date.

THE JULIAN CALENDAR

Conventional calendar dates are listed first; *Julian days are consecutive numbers from 1 to 365*

JAN		FEB		MAR		APR		MAY		JUN		JUL		AUG		SEP		OCT		NOV		DEC	
1	1	1	32	1	60	1	91	1	121	1	152	1	182	1	213	1	244	1	274	1	305	1	335
2	2	2	33	2	61	2	92	2	122	2	153	2	183	2	214	2	245	2	275	2	306	2	336
3	3	3	34	3	62	3	93	3	123	3	154	3	184	3	215	3	246	3	276	3	307	3	337
4	4	4	35	4	63	4	94	4	124	4	155	4	185	4	216	4	247	4	277	4	308	4	338
5	5	5	36	5	64	5	95	5	125	5	156	5	186	5	217	5	248	5	278	5	309	5	339
6	6	6	37	6	65	6	96	6	126	6	157	6	187	6	218	6	249	6	279	6	310	6	340
7	7	7	38	7	66	7	97	7	127	7	158	7	188	7	219	7	250	7	280	7	311	7	341
8	8	8	39	8	67	8	98	8	128	8	159	8	189	8	220	8	251	8	281	8	312	8	342
9	9	9	40	9	68	9	99	9	129	9	160	9	190	9	221	9	252	9	282	9	313	9	343
10	10	10	41	10	69	10	100	10	130	10	161	10	191	10	222	10	253	10	283	10	314	10	344
11	11	11	42	11	70	11	101	11	131	11	162	11	192	11	223	11	254	11	284	11	315	11	345
12	12	12	43	12	71	12	102	12	132	12	163	12	193	12	224	12	255	12	285	12	316	12	346
13	13	13	44	13	72	13	103	13	133	13	164	13	194	13	225	13	256	13	286	13	317	13	347
14	14	14	45	14	73	14	104	14	134	14	165	14	195	14	226	14	257	14	287	14	318	14	348
15	15	15	46	15	74	15	105	15	135	15	166	15	196	15	227	15	258	15	288	15	319	15	349
16	16	16	47	16	75	16	106	16	136	16	167	16	197	16	228	16	259	16	289	16	320	16	350
17	17	17	48	17	76	17	107	17	137	17	168	17	198	17	229	17	260	17	290	17	321	17	351
18	18	18	49	18	77	18	108	18	138	18	169	18	199	18	230	18	261	18	291	18	322	18	352
19	19	19	50	19	78	19	109	19	139	19	170	19	200	19	231	19	262	19	292	19	323	19	353
20	20	20	51	20	79	20	110	20	140	20	171	20	201	20	232	20	263	20	293	20	324	20	354
21	21	21	52	21	80	21	111	21	141	21	172	21	202	21	233	21	264	21	294	21	325	21	355
22	22	22	53	22	81	22	112	22	142	22	173	22	203	22	234	22	265	22	295	22	326	22	356
23	23	23	54	23	82	23	113	23	143	23	174	23	204	23	235	23	266	23	296	23	327	23	357
24	24	24	55	24	83	24	114	24	144	24	175	24	205	24	236	24	267	24	297	24	328	24	358
25	25	25	56	25	84	25	115	25	145	25	176	25	206	25	237	25	268	25	298	25	329	25	359
26	26	26	57	26	85	26	116	26	146	26	177	26	207	26	238	26	269	26	299	26	330	26	360
27	27	27	58	27	86	27	117	27	147	27	178	27	208	27	239	27	270	27	300	27	331	27	361
28	28	28	59	28	87	28	118	28	148	28	179	28	209	28	240	28	271	28	301	28	332	28	362
29	29			29	88	29	119	29	149	29	180	29	210	29	241	29	272	29	302	29	333	29	363
30	30			30	89	30	120	30	150	30	181	30	211	30	242	30	273	30	303	30	334	30	364
31	31			31	90			31	151			31	212	31	243			31	304			31	365

Chapter

24

Your Grazing Infrastructure

Two of the primary tools of a profitable grass-cattle enterprise are the electric fence grid and livestock water sites. These must be planned simultaneously so the grazing rotation can guarantee water to the cattle in every pasture subdivision.

Plan how each livestock water site will be supplied.

You have two fencing options: permanent electric fence subdivisions or a permanent electric fence grid that can be further subdivided with portable electric fencing. The latter allows for greater flexibility when responding to changing grass growth rates during summer and greater ability to ration the winter grazing reserve. It also costs significantly less than a permanent electric fence subdivision and is less labor-intensive to construct.

Consider how you intend to water your cattle in each pasture subdivision. Will you use "wagon-wheel" permanent fence layouts around centrally located water sites, or water alleys to reach centrally located water sites, or will you lay out a grid of polyethylene pipe under your electric fence grid so you can access water for each subdivision with a portable water tub? The latter two options allow you to use portable electric fencing to subdivide each permanent fence division. (For illustrations of each of these watering options, see pages 97 and 99.)

Planning for Your Herd's Water Requirements

Design your water system to accommodate the largest possible herd size that you may expand to so you do not experience water shortages and reduced gains as your cattle herd increases. Remember that it is advantageous to combine your cattle into a single large herd whenever possible for management ease and grazing benefits.

Determining Your Water Needs

Refer to the chart on page 86 to calculate the gallons of water consumed per day (gpd) by each of the following classes of animals in your herd (calculations should be based on your largest possible herd size to allow for future herd expansion).

Understanding your cattle's water requirements before you begin designing an electric fence grid is crucial to designing a grazing infrastructure that is capable of meeting your livestock's needs. The next step is to plan your year-round grazing infrastructure. As you map out your electric fences, include current and planned water sites so that every electric fence division and pasture subdivision will be capable of providing your livestock with adequate access to water. Then, before pounding a single nail or hanging a single wire, plan how each livestock water site will be supplied, how that water will be pumped, what kind of piping is required to transport water, what kind of trough system you will use, and

DETERMINING YOUR WATER NEEDS

Livestock Class	Number of cattle in livestock class		Gallons of water consumed per animal each day (gpd*/animal)		Total gallons of water consumed by each livestock class each day (gpd)
Dry cows and heifers		x		=	
Lactating cows		x		=	
Bulls		x		=	
Growing cattle (400 lbs)		x		=	
Growing cattle (600 lbs)		x		=	
Growing cattle (800 lbs)		x		=	
Finishing cattle (1,000 lbs)		x		=	
Finishing cattle (1,200 lbs)		x		=	
Total daily water consumption of entire herd (gpd)	Sum of water consumption of all livestock classes				(gpd)
Minimum peak flow rate to supply daily water requirements to be sustained for at least 6 hours	Refer to table on page 88				(gpm**)
Minimum on-site water storage capacity (in gallons) if water supply is unable to sustain the minimum peak flow rate for 6 hours	Double the total daily water consumption				(gallons)
Minimum working pressure required at water trough to operate water-trough fixtures (e.g., flow valve)	Check with your local supplier (note that the typical minimum working pressure for most water-trough fixtures is 15 psi***)				(psi)

*gpd = gallons per day; **gpm = gallons per minute; ***psi = pounds of pressure per square inch

how you will winterize water sites used during the frost season.

Preparing Your Farm Map

The best way to plan your electric fence and water infrastructure is to draw a map of your farm and note on it where you propose to locate each of your permanent electric fence divisions and livestock water sites. An aerial photograph of your farm will assist with your planning. Your local agricultural Extension agent or Natural Resources Conservation Service (NRCS) technician should be able to help you locate one.

Remember not to include on your map pre-existing interior fences except boundary fences unless they cannot be removed. These will restrict your planning process and future grazing management by limiting your thinking and, if they weren't designed with rotational grazing in mind, by hindering the movement of cattle. Trying to accommodate an old fence division that doesn't fit with the scheme of your new electric fence grid can add extra time and labor and miles of unnecessary fence lines. Remove any preexisting fences that do not fit into your electric fence scheme. Never electrify barbed-wire fences or combine electric and barbed wires on the same fence!

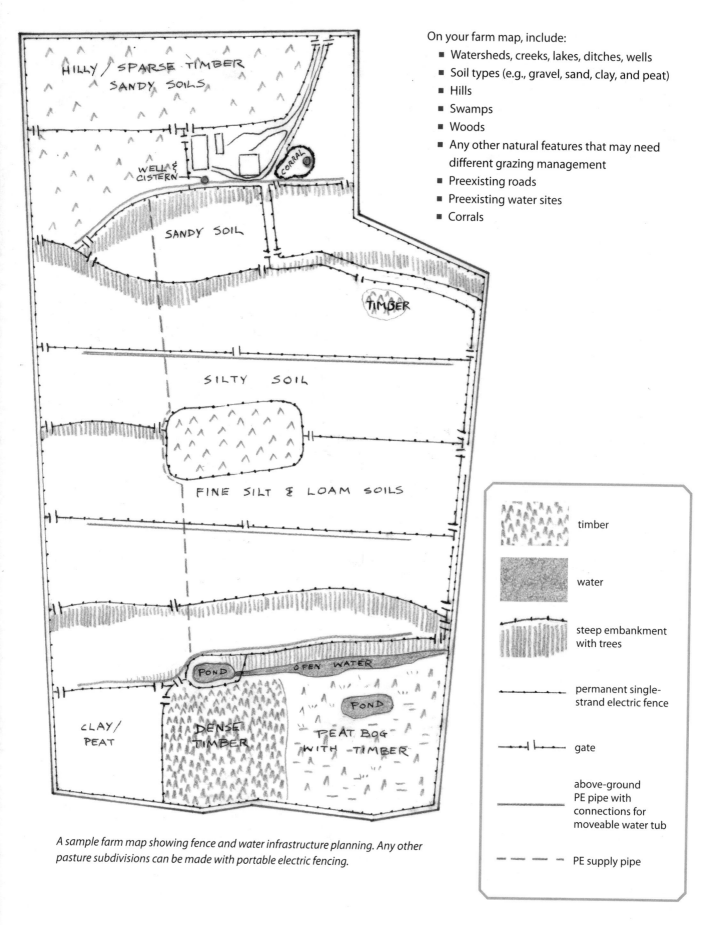

On your farm map, include:

- Watersheds, creeks, lakes, ditches, wells
- Soil types (e.g., gravel, sand, clay, and peat)
- Hills
- Swamps
- Woods
- Any other natural features that may need different grazing management
- Preexisting roads
- Preexisting water sites
- Corrals

Map labels: HILLY / SPARSE TIMBER SANDY SOILS · WELL & CISTERN · CORRAL · SANDY SOIL · TIMBER · SILTY SOIL · FINE SILT & LOAM SOILS · POND · OPEN WATER · POND · CLAY/PEAT · DENSE TIMBER · PEAT BOG WITH TIMBER

Legend:

- timber
- water
- steep embankment with trees
- permanent single-strand electric fence
- gate
- above-ground PE pipe with connections for moveable water tub
- PE supply pipe

A sample farm map showing fence and water infrastructure planning. Any other pasture subdivisions can be made with portable electric fencing.

Designing the Electric Fence Grid and Livestock Watering System

Now, draw permanent electric fence divisions on your map, including all gates, alleys, and energizer locations. Try to fence along natural boundaries so you can separately manage different ecosystems and capitalize on the advantages these natural divisions provide during the year (for example, trees for winter shelter, ridges for early-spring grazing, and sub-irrigated lowlands and swamps for grazing during the driest part of summer).

Lay out your electric fence grid with grazing in mind so your herd can rotate around the farm from one pasture to the adjacent one without having to endure long pasture moves each day. Also, plan your electric fence grid to facilitate sorting out sick and slaughter-ready animals so they can be taken back to your central corral, hospital pasture, or loading chute.

Every proposed permanent fence division and portable fencing subdivision requires access to water. Will you create water alleys to central water sites or paddocks around central water corrals, like spokes on a giant wagon wheel? Do you have an irrigation-pipe network that you can access with a portable water site or will you lay out polyethylene pipe under some of your fence lines, to which you can attach a portable trough? (See chapter 6 for an in-depth discussion of these watering options.)

The way you choose to water your livestock will determine the way in which you fence your property. As you design your pasture grid, imagine your livestock migrating through it in a smooth rotation so that wherever possible, pasture moves lead to a directly adjacent pasture and your gate and fence design facilitates easy cattle movements through the grazing system, with water access in every paddock division. (See chapter 5 for an in-depth discussion about designing and building electric fences, and chapter 9 for a discussion about how to use to your advantage cattle's perception of the world around them.)

As the grid takes shape, remember that fencing along natural boundaries such as lowlands, riparian areas, south- or north-facing slopes, hilltops, dry soils, and swamps allows you to manage the different needs of the soils and plants in them.

Also consider your seasonal needs during year-round grazing, keeping in mind that each natural division comes with different opportunities and challenges. Ridges may be drier in summer, which means they may require less grazing impact during peak summer heat to maintain a healthy grass cover and to minimize evaporation. During winter, however, these may be the areas you save for the season of deepest snow because the wind exposes their grass.

Wetlands may need protection from livestock during the wet season but may be a saving grace during the midsummer dry season. Riparian areas may need protection from manure and trampling but require quick grazes to promote grass growth for the stabilization of stream- and riverbank soil and for erosion control. Clay and organic soils may need protection from livestock trampling during the wet season, whereas gravel and sandy soils stand up to livestock during wet weather.

Densely treed or brushy areas have different plant species from grasslands and require different management. The brush may be the shady respite your cattle require during the hottest summer-grazing period. You may want to protect treed areas from livestock damage for future lumber harvests by fencing them off separately, allowing you to graze them quickly with minimal impact. Or you may want to graze the brushy areas particularly hard so the cattle will remove the undergrowth.

In fact, cattle can be used quite effectively to debrush your land, happily "logging" what you would otherwise hire a bulldozer to tackle. Historically, the northern prairie was kept free of encroaching trees by vast herds of bison, which trampled small trees, smashed bigger timber by pushing and shoving, and kept the grass growing so it choked out competing tree seedlings. With a little bit of electric fencing to direct their impact, cattle can be equally effective at clearing brush, particularly if you use mineral feeders, water sites, and supplemental feed to attract them to the areas you want cleared. When pioneers settled the treed northern prairie, cattle earned the reputation as the "poor man's bulldozer.

```
┌─────────────────────────────────────────────────────────────────────┐
│  WATER-SITE DEVELOPMENT LOG                                           │
│                                                                       │
│  Water site location:                    Piping method:              │
│  _____                  _____      │
│  _____                  _____      │
│  _____                  _____      │
│                                          _____      │
│  Water source:                                                        │
│  _____                  Water storage needed? If so, how? │
│  _____                  _____      │
│  _____                  _____      │
│                                          _____      │
│  Required gpm at water site:             _____      │
│  _____                  _____      │
│  _____                                               │
│  _____                  Frost-proof for winter grazing? Is so, how? │
│  Pumping method:                         _____      │
│  _____                  _____      │
│  _____                  _____      │
│  _____                  _____      │
│  _____                  _____      │
└─────────────────────────────────────────────────────────────────────┘
```

Plotting Your Water Access Points

On your map, draw in your livestock water access (pipes, wells, pumps, stream access points, permanent water troughs, water storage tanks) as you design each pasture subdivision. Record in your water-site development log how each pasture division will access minimum water requirements. Also address your winter water-site development plans so you can graze all corners of your land with a minimum number of winterized water sites. (See chapter 6 for detailed information about developing water sites along watercourses, dugouts, ponds, wells, water harvesters, and irrigation pipelines; a description of alternative pumping options where electricity is not available; cost-effective water-storage alternatives; and a discussion of winter water-site considerations and options.)

Piping Water

At times a water source is quite far from where it must supply your cattle. When water needs to be transported any distance through piping, friction in the pipe reduces the pressure and consequently limits the flow of water. Pumping uphill will increase pressure at the water source (at the pump) and thus decrease flow, and pumping downhill will decrease pressure at the water source (at the pump) and thus increase flow.

The Pressure Head

To design a water system that guarantees a certain pressure and flow (gallons per minute [gpm]) at the trough or storage facility, all pressure gains and losses must be converted into a *pressure head*, which is the pressure energy of the water expressed in feet equaling the height of a vertical column of water that produces the equivalent fluid pressure at its base through the force of gravity. Converting water pressure to a pressure head allows for easy calculation related to gravitational gains or losses in the system as water is pumped across varying elevations.

Imagine a giant column of water stacked above the ground. Each pressure loss or gain due to elevation, friction, or pump pressure is represented by the removal of water from or the addition of water to the column. Pressure is related to the column height by converting it into a pressure head:

1 foot = 0.433 psi, *or* 1 psi = 2.31 feet

What this means is that a 1-foot-high column of water pushing down will create 0.433 psi of pressure at the base of the column. Likewise, 1 psi of pressure pushing into a pipe can raise the water vertically by 2.31 feet before the force of gravity balances out the pressure. Converting pressure to feet provides a standard unit of measure for calculations.

Establishing the Minimum Flow Rate

The minimum flow rate required at the trough or water-storage facility determines how you will build your water system. To guarantee a particular flow, you must size the pump (if you are using one) appropriately and use pipe of proper diameter, which allows you to manipulate friction losses in the supply line and keep the pumping pressure within the safe working range for which the pipe is rated, preventing it from bursting.

Selecting the Proper Pump Pressure

To calculate the amount of pressure a pump must be able to produce to guarantee the required water volume and water pressure at a trough or reservoir, you must accommodate all the forces that increase or reduce pressure between the pump and the pipe outlet at the trough or reservoir. The calculation is completed in the following steps:

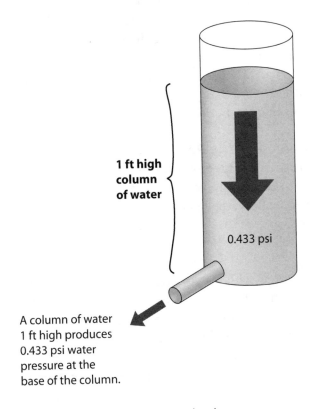

1 ft high column of water

0.433 psi

A column of water 1 ft high produces 0.433 psi water pressure at the base of the column.

Converting water pressure to a pressure head: Water pressure produced by gravity

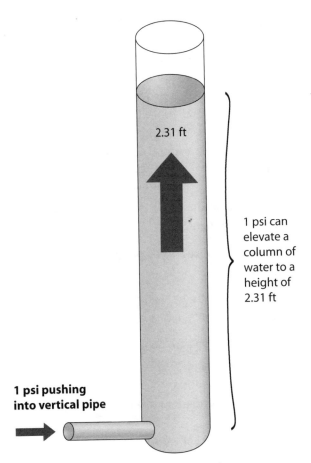

2.31 ft

1 psi can elevate a column of water to a height of 2.31 ft

1 psi pushing into vertical pipe

Converting water pressure to a pressure head: Vertical elevation gains overcome by water pressure

1. Start with the elevation gained (a positive number) or lost (a negative number) between the pump and the outlet; note that the (+) or (−) sign before the elevation indicates whether the elevation change results in a pressure gain or loss.

2. Add the vertical height the pump must lift water from the source through its suction hose to reach the pump.

3. Add the pressure lost due to friction in the pipe between the pump and the outlet (friction loss is calculated by multiplying the length of the pipe by the friction loss value found on the charts on pages 308–310 and converting this value to a pressure head).

4. Finally, add the pressure you need at the outlet of the pipe, also converted to a pressure head. (Note that the minimum water pressure required to operate most water trough fixtures is 15 psi *at the trough fixture.*)

To calculate the minimum pump pressure required to deliver 10 gpm of water at 15 psi to a trough through 1,000 feet of 1¼-inch polyethylene (PE) pipe, with a 100-foot-elevation rise from the pump to the pipe outlet and a 3-foot vertical suction from pond to pump, perform the following calculation:

1. 100-foot elevation rise from pump to pipe outlet = +100 feet

2. 3 feet of vertical suction from pond to pump = +3 feet

3. Friction loss in 1,000 feet of 1¼-inch polyethylene (PE) pipe at 10 gpm

   = 1,000 feet × 0.72 psi/100 feet
   (from chart on page 308)
   = 7.2 psi × 2.31 feet/psi
   = +16.6 feet, rounded up to +17 feet

4. Pressure required at trough, converted to pressure head

   = 15 psi × 2.31 feet/psi = +35 feet

Add:

   100 feet (1, elevation) + 3 feet (2, suction)
   + 17 feet (3, friction loss) + 35 feet
   (4, pressure at trough) = 155 ft

This pressure head is converted back to psi to determine the working pressure of the pump:

   155 feet ÷ 2.31 psi/foot = 67 psi

Thus, the pump at the water source must be pressured to 67 psi to produce the desired volume of 10 gpm through a 1¼-inch pipe while accounting for elevation gains (topography and suction) and maintaining a working pressure of 15 psi at the water trough. If this pressure is too high (i.e., higher than the maximum pressure rating on the pipe being used or than the pressure rating of pressure seals in the water system), a larger pipe size is needed. Likewise, if pumping costs become uneconomical due to high friction losses caused by high working pressure, you will need to increase pipe diameter to reduce water pressure in order to deliver the same volume of water (notice on the charts provided on pages 308–310 that values represented in parentheses or missing altogether represent pipe diameter/water flow rate combinations that are uneconomical). If the elevation had dropped, we would have subtracted instead of added the change in elevation because it would be working for us, not against us.

Selecting the Proper Pipe Diameter in a Gravity-Fed Water System

Alternatively, you can determine the pipe size to get the flow rate you require at a trough in a gravity-feed system. In this case, perform your calculations as follows:

1. Begin with the elevation drop from the water source to the water trough (which represents the pressure gain caused by gravity). Note that in this case, elevation is measured from the pipe intake in the water source to the actual pipe outlet at the trough. Topographical high points between the water source and water trough can often be overcome through siphoning, as long as air-release values are installed at the high points

to bleed out air pockets that accumulate in the pipe, which can cause air locking in a gravity-fed system. (*Air locking* is the phenomenon caused when the water suction created by water leaving the pipe outlet is insufficient to flush the trapped water pockets out of the pipe).

2. Subtract the pressure lost due to friction in the pipe between the water source and the water trough, using various pipe diameters to determine which will meet your needs (friction loss is again calculated by multiplying the length of the pipe by the friction loss value found on the charts on pages 308–310, and converting this value to a pressure head).

3. After converting this pressure head back to pressure (psi), compare with your minimum working pressure required at the trough.

If a trough requiring 12 gpm is situated 75 feet lower than and 2,000 feet away from a water source, perform the following calculations to determine the water pressure at the trough and whether you have the minimum working pressure (15 psi) required to operate the water trough fixtures (your flow valve) if you use a 1½-inch PE pipe:

1. 75-foot elevation drop from pump to pipe outlet = 75 feet

2. Friction loss in 2,000 feet of 1½-inch PE pipe at 12 gpm

   = 2,000 feet × 0.48 psi/100 feet
   (from chart on page 308)
   = 9.6 psi × 2.31 feet/psi
   = 22 feet

Add:

   75 feet (1, elevation) – 22 feet (2, friction loss)
   = 53 feet

This pressure head is converted back to psi to determine if the pressure head caused by gravity is sufficient to supply the minimum working pressure required at the trough:

   53 feet ÷ 2.31 psi/foot = 23 psi

Thus, this gravity-fed water system will produce 23 psi at the trough, which is more than enough to meet the minimum water pressure requirements of your trough fixture. If the final psi is insufficient, the pipe diameter can be increased to decrease friction losses.

PIPE FRICTION LOSS CHARTS

Use the following charts to look up friction losses in PE and PVC pipes for various pipe diameters when calculating the friction losses corresponding to specific pipe diameters and specific flow rates required at a pipe outlet. Remember that pipes are rated for the maximum pressure a pipe can withstand, but the working pressure of a pipe should always remain below 72 percent of the maximum working pressure. For example, a class or series 100 PVC pipe has a maximum pressure rating of 100 psi, but the working pressure should be no more than 72 psi to allow for the water hammer (a sudden increase in pressure) caused by pressure surges from valves, taps, and pumps when they are first turned on.

Note that while a maximum working pressure of 72 psi would be considered inadequate for most irrigation systems, it is more than adequate to operate a livestock watering system. In fact, most livestock flow valves are not rated to withstand high water pressure; they will cease to seal properly at pressure exceeding 60 to 100 psi, depending on the flow valve model you are using. Your local supplier should be able to supply you with the maximum pressure ratings of your flow valves.

To use the charts, look up the flow rate (gpm) on the left side of the table; this represents the desired flow rate *at the pipe outlet,* not at the water source. The friction losses (shown in psi/100 feet of pipe length) for various pipe sizes are listed to the right of each flow rate. Numbers in parentheses show extremely high friction losses that make pumping costs uneconomical.

Friction Losses in Polyethylene (PE) Pipe*

Flow (gpm)	PIPE DIAMETER					
	½"	¾"	1"	1¼"	1½"	2"
2	1.76	0.45	0.14	0.02		
3	3.73	0.95	0.29	0.08	0.04	0.01
4	6.35	1.62	0.50	0.13	0.06	0.02
5	(9.6)	2.44	0.76	0.20	0.09	0.03
6	(13.46)	3.43	1.06	0.28	0.13	0.04
7		4.56	1.41	0.37	0.18	0.05
8		5.84	1.80	0.47	0.22	0.07
9		(7.26)	2.24	0.59	0.28	0.08
10		(8.82)	2.73	0.72	0.34	0.10
11		(10.60)	3.27	0.86	0.41	0.12
12		(12.37)	3.82	1.01	0.48	0.14
14			(5.08)	1.34	0.63	0.19
16			(6.51)	1.71	0.81	0.24
18			(8.10)	2.13	1.01	0.30
20			(9.84)	2.59	1.22	0.36
22			(11.74)	3.09	1.46	0.43
24				(3.63)	1.72	0.51
26				(4.21)	1.99	0.59
28				(4.83)	2.28	0.68
30				(5.49)	2.59	0.77
35				(7.31)	(3.45)	1.02
40					(4.42)	1.31
45					(5.50)	1.63
50					(6.68)	1.98
55					(7.97)	(2.36)
60						(2.78)

From the *BC Sprinkler Irrigation Manual* (B.C. Ministry of Agriculture and Fisheries, Soils and Engineering Branch. Edited by Ted W. Van der Gulik. Vernon, BC: Irrigation Industry Association of British Columbia, 1989, page B-15).

*Pressure loss due to friction in psi per 100 ft of pipe (pipe diameter in inches).

Friction Losses in 125 psi PVC Pipe*

Flow (gpm)	1"	1¼"	1½"	2"
2	0.06	0.02	0.01	
3	0.13	0.04	0.02	0.01
4	0.22	0.07	0.03	0.01
5	0.33	0.10	0.05	0.02
6	0.46	0.14	0.07	0.02
7	0.62	0.19	0.09	0.03
8	0.79	0.24	0.12	0.04
9	0.98	0.30	0.15	0.05
10	1.19	0.36	0.18	0.06
11	1.42	0.43	0.22	0.07
12	1.67	0.51	0.25	0.09
14	2.22	0.67	0.34	0.11
16	2.85	0.86	0.43	0.15
18	(3.54)	1.07	0.54	0.18
20	(4.31)	1.30	0.65	0.22
22	(5.14)	1.56	0.78	0.26
24	(6.04)	1.83	0.92	0.31
26	(7.00)	2.12	1.06	0.36
28		2.43	1.22	0.41
30		(2.76)	1.39	0.47
35		(3.68)	1.84	0.62
40		(4.71)	(2.36)	0.80
45		(5.86)	(2.94)	0.99
50		(7.12)	(3.57)	1.21
55			(4.26)	1.44
60			(5.00)	1.69
65			(5.80)	(1.96)
70				(2.25)

From the *B.C. Sprinkler Irrigation Manual* (B.C. Ministry of Agriculture and Fisheries, Soils and Engineering Branch. Edited by Ted W. Van der Gulik. Vernon, BC: Irrigation Industry Association of British Columbia, 1989, page B-8).

*Pressure loss due to friction in psi per 100 ft of pipe (pipe diameter in inches).

Friction Losses in 160 psi PVC Pipe*

Flow (gpm)	1"	1¼"	1½"	2"
2	0.06	0.02	0.01	
3	0.14	0.04	0.02	
4	0.23	0.07	0.04	0.01
5	0.35	0.11	0.05	0.02
6	0.49	0.15	0.08	0.03
7	0.66	0.20	0.10	0.03
8	0.84	0.25	0.13	0.04
9	1.05	0.31	0.16	0.05
10	1.27	0.38	0.20	0.07
11	1.52	0.45	0.23	0.08
12	1.78	0.53	0.28	0.09
14	2.37	0.71	0.37	0.12
16	3.04	0.91	0.47	0.16
18	(3.78)	1.13	0.58	0.20
20	(4.59)	1.37	0.71	0.24
22	(5.48)	1.64	0.85	0.29
24	(6.44)	1.92	1.00	0.34
26		2.23	1.15	0.39
28		2.56	1.32	0.45
30		(2.91)	1.50	0.51
35		(3.87)	2.00	0.68
40		(4.95)	(2.56)	0.86
45		(6.16)	(3.19)	1.08
50			(3.88)	1.31
55			(4.62)	1.56
60			(5.43)	(1.83)
65			(6.30)	(2.12)
70				(2.44)

From the *B.C. Sprinkler Irrigation Manual* (B.C. Ministry of Agriculture and Fisheries, Soils and Engineering Branch. Edited by Ted W. Van der Gulik. Vernon, BC: Irrigation Industry Association of British Columbia, 1989, page B-10).

*Pressure loss due to friction in psi per 100 ft of pipe (pipe diameter in inches).

Friction Losses in 200 psi PVC Pipe*

Flow (gpm)	PIPE DIAMETER			
	1"	1¼"	1½"	2"
2	0.07	0.02	0.01	
3	0.14	0.04	0.02	
4	0.24	0.08	0.04	0.01
5	0.36	0.12	0.06	0.02
6	0.51	0.16	0.08	0.03
7	0.67	0.22	0.11	0.04
8	0.86	0.28	0.14	0.06
9	1.07	0.34	0.18	0.06
10	1.30	0.42	0.22	0.07
11	1.56	0.50	0.26	0.09
12	1.83	0.59	0.30	0.10
14	2.43	0.78	0.40	0.14
16	3.11	1.00	0.52	0.17
18	(3.87)	1.24	0.64	0.22
20	(4.71)	1.51	0.78	0.26
22	(5.62)	1.80	0.93	0.32
24	(6.60)	2.12	1.09	0.37
26		2.46	1.27	0.43
28		(2.82)	1.46	0.49
30		(3.20)	1.66	0.56
35		(4.26)	2.20	0.75
40		(5.45)	(2.82)	0.95
45		(6.78)	(3.51)	1.19
50			(4.26)	1.44
55			(5.09)	1.72
60			(5.97)	(2.02)
65			(6.93)	(2.35)
70				(2.69)

*Pressure loss due to friction in psi per 100 ft of pipe (pipe diameter in inches

From the *BC Sprinkler Irrigation Manual.* (B.C. Ministry of Agriculture and Fisheries, Soils and Engineering Branch. B.C. Sprinkler Irrigation Manual. Edited by Ted W. Van der Gulik. Vernon, BC: Irrigation Industry Association of British Columbia, 1989, page B-12).

Your Grazing Plan

IN THIS CHAPTER YOU WILL DEVELOP YOUR grazing plan. Note, however, that the goal of your grazing strategies during the growing season — to use your cattle as a tool to manage the grass in order to create a large, high-quality grass reserve in preparation for the dormant season — is very different from your goal during the dormant season, which is to ration your grass supply to your cattle. Consequently, you will need two separate grazing plans, one for each season, to accommodate the different grazing techniques required to achieve these different goals. (See chapter 5 and chapter 7 for detailed discussions about how to create a high-quality grass reserve during the growing season and how to effectively ration that grass reserve during the dormant season.)

Your Summer-Grazing Plan

During the growing season, you do not want your cattle to graze a pasture bare before they move on to the next pasture. Initially, this may seem counterintuitive — most of us want to force our cattle to graze each pasture completely before moving on — but this style of grazing is reserved for winter, after the grass stops growing and you are trying to ration leftover grass until it starts growing again in spring.

Your summer-grazing strategy should be designed to create the tallest possible grazing reserve in preparation for winter grazing, without allowing any of the grass to become overly mature. This must be planned. If all grass grew at the same rate during the growing season and all pastures were created equal, accomplishing this would be easy: You would simply rotate your herd through the grazing rotation fast enough to return them to each pasture before the grass in it became overly mature. The faster the grass grows, the faster your herd needs to move through the pasture rotation. To calculate how many acres must be grazed each day to avoid the grass becoming too mature, simply divide the total farm acres by the grass recovery interval (the amount of time it takes for grass to regrow after grazing).

Total farm acres ÷ grass recovery interval
= number of acres grazed per day

One of the opportunities of grazing cattle is that you can produce a salable agricultural product using marginal land unsuited for crop production. But this creates a situation where pasture yield can vary dramatically as your cattle herd rotates through your pastures (for the sake of this discussion, *pasture yield* is defined as "the amount of grass produced per acre," not the total amount of grass produced by the entire pasture). For example, consider the differences in pasture yields you can expect as a herd moves from prime, fertilized pastures, to unfertilized pastures, to marginal land, across low-yielding rangeland, through dense timber, and back onto lush grassland pastures. No two of these grazing areas produce the same amount of grass per acre, yet all of them could be included in the same grazing rotation. Consequently, you must identify a benchmark pasture yield against which all other pasture yields are compared when calculating the number of acres that you will graze each day.

In addition, the number of acres grazed by your herd each day must compensate for the changing grass growth rates that occur as the seasons turn. To do this, you will map out the changing grass recovery interval over the growing season so you can adjust the number of acres grazed each day to match this interval.

With your grass recovery intervals mapped out and a benchmark pasture yield established to which you can compare your other pastures, you will be able to calculate how many acres you must graze each day during the growing season, using the aforementioned equation, to create a high-quality grass reserve in preparation for the dormant season.

Compensating for Different Pasture Yields

In order to develop a benchmark pasture yield, you need to list all your major pasture divisions and establish how their yields compare to one another. Then, by stretching or contracting the theoretical size of each of your pastures in your calculations, you can adjust the yield of each to match the yield of your benchmark, or average, pasture. All major pasture divisions must be identified and their sizes adjusted according to the benchmark yield so that calculations can proceed as though all pastures have a uniform yield.

The worksheet below shows a number of pastures of different sizes, compares how their yields measure up against the benchmark pasture's yield, and shows how large the same pasture would be if it was stretched or contracted to match the pasture yield in the benchmark pasture.

A blank worksheet has been provided for you to use to identify your benchmark pasture, compare

PASTURE SIZES ADJUSTED TO MATCH BENCHMARK YIELD
Sample worksheet for a 633-acre farm with 11 separate major pasture divisions

Pasture name	Acres in each pasture		Yield/acre compared to benchmark pasture		Pasture size adjusted to match benchmark yield
Pasture A*	25	×	(benchmark) 1	=	25 acres
Pasture B	40	×	½	=	20 acres
Pasture C	33	×	1½	=	50 acres
Pasture D	50	×	(same as benchmark) 1	=	50 acres
Pasture E	35	×	⅔	=	23 acres
Pasture F	32	×	(same as benchmark) 1	=	32 acres
Pasture G	40	×	(same as benchmark) 1	=	40 acres
Pasture H	35	×	(same as benchmark) 1	=	35 acres
Brush #1	115	×	⅓	=	38 acres
Pasture J	188	×	¼	=	47 acres
Pasture K	40	×	(same as benchmark) 1	=	40 acres
Total number of acres in the grazing rotation after pasture sizes are adjusted to match the benchmark yield:			**Sum of pasture sizes adjusted to match benchmark yield**		400 acres

*Pasture A represents the average yield per acre and has been chosen as the benchmark pasture.

pasture yields, and adjust pasture sizes to match the benchmark pasture yield. Be sure to add up the total acreage on your farm based on the adjusted pasture sizes — you will need this total to complete your calculations of the number of acres grazed per day in the next section.

To compare your pasture yields, you will need to make your best estimate of pasture yield based on experience, or you can use the grazing reserve calculations described in the dormant-season-grazing plan on page 318 as an indication of pasture yield.

Calculating Your Grass Recovery Interval

The next step to calculating your summer-grazing schedule is to estimate how the grass recovery interval changes over the course of the seasons. (The *grass-recovery interval* is the time it takes for your pasture to regrow after grazing, from a ten- to twelve-inch grazing residual until the grass begins to go to seed.)

The worksheet below is designed to help you log grass recovery intervals throughout the grazing year. Note that during the dormant season, your grass recovery interval is the length of the entire dormant season.

PASTURE SIZES ADJUSTED TO MATCH BENCHMARK YIELD

Pasture name	Acres in each pasture		Yield/acre compared to benchmark pasture		Pasture size adjusted to match benchmark yield
		x		=	
		x		=	
		x		=	
		x		=	
		x		=	
		x		=	
		x		=	
		x		=	
		x		=	
		x		=	
		x		=	
		x		=	
		x		=	
		x		=	
		x		=	
		x		=	
		x		=	
		x		=	
		x		=	
		x		=	
Total number of acres in the grazing rotation after pasture sizes are adjusted to match the benchmark yield:			**Sum of pasture sizes adjusted to match benchmark yield**		

CALCULATING THE GRASS RECOVERY INTERVAL

Date when cattle finish grazing pasture, leaving behind a 10- to 12-inch grazing residual	Pasture location	Date when the bulk of the same pasture begins going to seed	Grass recovery interval* (count number of days between the two dates on the left)
January 1			
January 15			
February 1			
February 15			
March 1			
March 15			
April 1			
April 15			
May 1			
May 15			
June 1			
June 15			
July 1			
July 15			
August 1			
August 15			
September 1			
September 15			
October 1			
October 15			
November 1			
November 15			
December 1			
December 15			

*During the dormant season, your grass recovery interval is always recorded as the length of the entire dormant season.

Calculating the Number of Acres to Graze Each Day

Having identified the length of your grass recovery interval, you can now calculate the number of acres to graze each day during the growing season. Using the blank worksheet provided, calculate how many benchmark acres you should graze each day. A sample worksheet is provided that continues the example of the 633-acre farm.

The number of "benchmark acres to graze each day" (calculated in the previous worksheet) can be adjusted to the actual number of acres you will graze each day. Using the worksheet, convert the benchmark acres grazed per day to the actual number of acres this represents in the specific pasture that you will be grazing on that particular day. The sample worksheet on page 316 shows how to convert

CALCULATING THE GRAZING SCHEDULE:

Sample worksheet for a 633-acre farm with 11 separate major pasture divisions

Dates	Grass recovery interval	Number of benchmark acres grazed per day
May 1 to May 15	40 days	10 acres/day (400 benchmark acres ÷ 40 days)
May 16 to June 15	30 days	13.3 acres/day (400 benchmark acres ÷ 30 days)
June 16 to July 31	40 days	10 acres/day (400 benchmark acres ÷ 40 days)
Aug. 1 to Sept. 1	50 days	8 acres/day (400 benchmark acres ÷ 50 days)
Sept. 2 to Oct. 10	60 days	6.7 acres/day (400 benchmark acres ÷ 60 days)
Oct. 11 to April 30 (dormant)	201 days	2 acres/day (400 benchmark acres ÷ 201 days)

CALCULATING THE GRAZING SCHEDULE:

Dates	Grass recovery interval	Number of benchmark acres grazed per day (total benchmark acres ÷ grass recovery interval)
Jan. 1 to Jan. 15		
Jan. 16 to Jan. 31		
Feb. 1 to Feb. 15		
Feb. 16 to Feb. 28		
March 1 to March 15		
March 16 to March 31		
April 1 to April 15		
April 16 to April 30		
May 1 to May 15		
May 16 to May 31		
June 1 to June 15		
June 16 to June 30		
July 1 to July 15		
July 16 to July 31		
Aug. 1 to Aug. 15		
Aug. 16 to Aug. 31		
Sept. 1 to Sept. 15		
Sept. 16 to Sept. 30		
Oct. 1 to Oct. 15		
Oct. 16 to Oct. 31		
Nov. 1 to Nov. 15		
Nov. 16 to Nov. 30		
Dec. 1 to Dec. 15		
Dec. 16 to Dec. 31		

benchmark acres to actual acres in specific pastures on specific dates, using the 633-acre farm example.

If you are using portable electric fence subdivisions in your pastures, the actual number of acres grazed per day (calculated in the worksheet called "Actual Number of Acres Grazed Each Day") would represent the actual size of each pasture subdivision that you give to your animals each day. If you are not using daily pasture moves (though this is not rec-ommended, your pastures may not be small enough for daily pasture moves during the transition phase prior to completing your electric fencing infrastructure), the size of the pasture dictates how many days to graze each pasture. For example, if a 30-acre pasture was meant to be grazed at a rate of 10 acres per day, you would leave the herd in the pasture for 3 days (30 acres ÷ 10 actual acres grazed per day [as per calculation] = 3 days).

ACTUAL NUMBER OF ACRES TO BE GRAZED EACH DAY:
Sample worksheet for a 633-acre farm with 11 separate major pasture divisions

Date of grazing	Specific pasture identified for grazing on date shown in column on the left	Number of benchmark acres grazed per day based on grass recovery interval on that date (refer to worksheet Calculating the Grazing Schedule)		Yield/acre compared to benchmark pasture (actual pasture yield to benchmark yield, as shown in the worksheet Pasture Sizes Adjusted to Match Benchmark Yield)		Actual number of acres to be grazed each day in specific pasture identified for grazing
May 2	Pasture A	10 benchmark acres/day	÷	(benchmark) 1	=	10 acres/day
May 20	Pasture B	13.3 benchmark acres/day	÷	½	=	26.6 acres/day
June 4	Pasture C	13.3 benchmark acres/day	÷	1½	=	8.9 acres/day
June 28	Pasture D	10 benchmark acres/day	÷	(same as benchmark) 1	=	10 acres/day
July 16	Pasture E	10 benchmark acres/day	÷	⅔	=	15 acres/day
Aug. 4	Pasture F	8 benchmark acres/day	÷	(same as benchmark) 1	=	8 acres/day
Aug. 21	Pasture G	8 benchmark acres/day	÷	(same as benchmark) 1	=	8 acres/day
Sept. 12	Pasture H	6.7 benchmark acres/day	÷	(same as benchmark) 1	=	6.7 acres/day
Oct. 14	Brush #1	2 benchmark acres/day	÷	⅓	=	6 acres/day
Nov. 3	Pasture J	2 benchmark acres/day	÷	¼	=	8 acres/day
Dec. 27	Pasture K	2 benchmark acres/day	÷	(same as benchmark) 1	=	2 acres/day

ACTUAL NUMBER OF ACRES TO BE GRAZED EACH DAY:

Date of grazing	Specific pasture identified for grazing on date shown in column on the left	Number of benchmark acres grazed per day based on grass recovery interval on that date (refer to worksheet Calculating the Grazing Schedule)		Yield/acre compared to benchmark pasture (actual pasture yield to benchmark yield, as shown in the worksheet Pasture Sizes Adjusted to Match Benchmark Yield)		Actual number of acres to be grazed each day in specific pasture identified for grazing
			÷		=	
			÷		=	
			÷		=	
			÷		=	
			÷		=	
			÷		=	
			÷		=	
			÷		=	
			÷		=	
			÷		=	
			÷		=	
			÷		=	
			÷		=	
			÷		=	
			÷		=	
			÷		=	
			÷		=	
			÷		=	
			÷		=	

Your Winter-Grazing Plan: Managing the Grass Reserve

During the dormant season, the focus of your grazing management changes dramatically. Rather than trying to maximize the quality and quantity of the grass reserve by preventing it from becoming overly mature, as is done in summer, during winter the goal is to completely graze each day's ration of grass before moving on to the next slice, thereby maximizing grazing efficiency and minimizing waste. (See chapter 7 for more on winter grazing.)

Calculating the Winter Grazing Reserve

Successful winter grazing without stored feeds requires that you calculate the grazing reserve (measured in cow grazing days, or CGDs) in your pastures to allow you to adjust your herd size so that the herd's grass consumption over the winter never exceeds the total number of cow grazing days in your grazing reserve. A cow grazing day is the amount of grass required to feed a single cow/calf pair for a single day. Calculating your grass reserve in CGDs before the onset of the dormant season is as important as counting hay bales in a mechanized winter-feeding regimen. Here is one method of determining how many CGDs are left in your grass reserve.

1. Pick a general area in each pasture that represents the average growth in that pasture. If a pasture has variable growth rates (for example, dry zone, ridge zone, wet area, and brush area), assess each one separately.

2. Throw something colorful (such as a flag tied to a rock) over your shoulder so it lands in a random spot. (Selecting a random spot by eye is unreliable.)

3. In that location, decide how large an area a single cow will need in order to graze her fill in one day. Lay out some colorful posts or flagging to mark the perimeter of the CGD so you can better assess it visually. Whether you mark out the CGD with a square or a circle, the area is easy to calculate:

Area of a grazing square = length × width

Area of a grazing circle = $\pi r^2 \approx 3.14 \times radius^2$

Record the size of the area. _____

4. Repeat this random rock throwing fifteen to thirty times in every field (more if the CGD area varies greatly from one random site to another) so you can calculate an average CGD area to represent each field. (Note: A true *random sample* requires at least thirty samples to be statistically accurate.) Record how many acres each average CGD area represents (1 acre = 43,563 ft²).

5. Repeat this process for every pasture on your land.

6. Convert all livestock classes to cow/calf equivalents so that your total herd size is represented as cow/calf equivalents. (See the worksheet Converting Your Winter Herd to Cow Equivalents on page 319.)

CALIBRATING YOUR EYE

Calculating CGD areas is a subjective process that requires you to train your eye to recognize how large an area must be to feed a single cow for one day in pastures with widely varying forage yields. The first few times you use this method to calculate CGD areas, it's important to "calibrate" your eye by clipping and weighing the plant material from several CGD areas in several pastures of varying yields.

Fill a large bag with the clipped material from a single CGD area. To weigh the material, hang the bag on a weigh scale, place it on a simple bathroom scale, or stand on the bathroom scale holding the bag and subtract your weight. In each new CGD area with a visibly different forage yield, fill the empty bag with clipped material until the bag reaches the same weight as the bag from the previous CGD area. This allows you to see the exact dimensions of CGD area in different pastures with different forage yields. The consistent weight of the clipped material in the bag provides you with a measurable benchmark upon which to build your experience in estimating CGD areas.

The first chart below shows two examples of how to calculate the winter-grazing reserve (in CGDs) of individual pastures. The calculations for Pasture D show how to get from a circular CGD measurement to an average CGD per acre and finally to the number of days a herd can graze in that particular pasture by converting all the livestock classes to cow/calf pair equivalents. The calculations for Brush #1 demonstrate how to calculate a square CGD measurement.

Calculating Your Winter-Grazing Reserve

Using the next worksheet, convert all the animals in your herd to cow/calf equivalents so you have a standard animal unit for your calculations. You will use this to determine the number of CGDs in your grazing reserve, as well as to adjust your herd size to match your grazing reserve so your herd can graze through winter without stored feed or so that you can buy sufficient feed to make up for shortage.

CALCULATING THE COW-GRAZING DAYS (CGDS) OF THE WINTER-GRAZING RESERVE

Sample worksheet for herd with 258 cow/calf pairs and 176 stockers, converted to 346 cow/calf equivalents (258 + [176÷2]).

Pasture name	Pasture acreage	Measured area of 1 CGD (radius or length × width)	Calculated area of 1 CGD (length × width or πr²)	CGDs per acre (43,563 ft² ÷ average area of 1 CGD)	Total CGDs in the pasture (CGD/acre × acres/pasture)	Days entire herd can spend in pasture (total CGDs in pasture ÷ number of animals in herd, adjusted to cow/calf equivalents)
Pasture D	50 acres	15.0 ft radius	707 ft²	(43,563 ft² ÷ 708 ft²) =	62 CGDs/acre × 50 acres in pasture D =	3,100 CGD ÷ 346 cow/calf equivalents =
		14.5 ft radius	661 ft²			
		15.5 ft radius	755 ft²			
		14.0 ft radius	615 ft²			
		16.0 ft radius	804 ft²			
			Average = 708 ft²	62 CGD/acre	3,100 CGD	9 days
Brush # 1	115 acres	35.0 ft square	1,225 ft²	43,563 ft² ÷ 1,212 ft² =	36 CGD/acre × 115 acres =	4,140 CGD ÷ 346 cow/calf equivalents =
		33.5 ft square	1,122 ft²			
		34.0 ft square	1,156 ft²			
		36.0 ft square	1,296 ft²			
		35.5 ft square	1,260 ft²			
			Average = 1,212 ft²	36 CGD/acre	4,140 CGD	12 days

CONVERTING YOUR WINTER HERD TO COW EQUIVALENTS

Class of animals	Number of animals in each class	Conversion factor to cow equivalents	Number of cow equivalents
Cow/calf pairs		—	
Bulls		1:1	
Yearlings		÷ 2	
Finishing-class animals		÷ 1.3	
Total cow equivalents in your herd =			

Now, using the blank worksheet below, you can calculate the grazing reserve in each of your pastures, allowing you to adjust your herd size to match your grass reserve, find additional pasture to add to your winter-grazing rotation, or buy feed to make up for any shortages. You will need to repeat your grazing reserve calculation every year for each of your pastures before the onset of winter.

CALCULATING THE COW GRAZING DAYS (CGDS) IN YOUR WINTER-GRAZING RESERVE

Pasture name	Pasture acreage	Measured area of 1 CGD (radius or length × width)	Calculated area of 1 CGD (length × width or πr^2)	CGDs per acre (43,563 ft^2 ÷ average area of 1 CGD)	Total CGDs in the pasture (CGD/acre × acres/pasture)	Days entire herd can spend in pasture (total CGDs in pasture ÷ number of animals in herd, adjusted to cow equivalents)
				_____ CGD/acre	_____ Total CGDs in pasture	_____ days
	Average area of 1 CGD=_____ ft^2					
				_____ CGD/acre	_____ Total CGDs in pasture	_____ days
	Average area of 1 CGD=_____ ft^2					

CALCULATING THE COW GRAZING DAYS (CGDS) IN YOUR WINTER-GRAZING RESERVE continued

Pasture name	Pasture acreage	Measured area of 1 CGD (radius or length × width)	Calculated area of 1 CGD (length × width or πr^2)	CGDs per acre (43,563 ft² ÷ average area of 1 CGD)	Total CGDs in the pasture (CGD/acre × acres/pasture)	Days entire herd can spend in pasture (total CGDs in pasture ÷ number of animals in herd, adjusted to cow equivalents)
				_____ CGD/acre	_____ Total CGDs in pasture	_____ days
		Average area of 1 CGD=_____ ft²				

Calculating Shortfalls, Herd Cutbacks, and Drought Reserves

For a simple overview of your winter-grazing reserve, enter (from your previous calculations) the required information in the following worksheet, which will help you to adjust your herd size to maintain a grazing surplus as a drought reserve, or calculate how many additional pasture acres to add to your winter-grazing reserve or how much feed to purchase to make up for grazing shortfalls. This drought reserve also doubles as insurance against an unusually long winter and will be crucial to the development of your drought/disaster contingency plan in chapter 29.

Before you begin, decide how many days of drought reserve you plan to have. This number will depend on your comfort level, climate, and the predictability of your yearly grass growth. Typical drought reserves vary from as little as a few months to six months or more. In particularly arid regions, it is not unreasonable to keep as much as a year's drought reserve on hand at all times. Alternatively, you can maintain a drought reserve of stored feed, usually hay, instead of increasing your grazing reserve; this can be held over for ten years or more if it is unused. (See chapter 8 for more about developing a drought reserve.) If hay is used as a drought reserve, you should not include the drought reserve in the assessment of your grazing reserve in the following worksheet; it will be addressed separately in chapter 29.

ASSESSING YOUR WINTER-GRAZING REQUIREMENTS AND ADDRESSING SHORTFALLS

Name of item being assessed	Calculation	Result of calculation
Your actual winter-grazing reserve:		
Total number of grazing days in winter grazing reserve	Add total grazing days in all pastures (from worksheet Calculating the Cow Grazing Days in Your Winter-Grazing Reserve)	days (in grazing reserve):
Total CGDs in grazing reserve (in CGDs)	Add total CGDs in all pastures (from worksheet Calculating the Cow Grazing Days in Your Winter-Grazing Reserve)	CGD (in grazing reserve):
Your grazing requirements:		
Duration of winter season before reliable grass returns (in days)	count number of days in dormant season	days (winter):
Size of herd (in cow/calf equivalents)	herd size converted to cow/calf equivalents (from worksheet on page 319)	cow/calf equivalents:
Total CGDs required by herd to survive winter	total cow/calf equivalents × days (winter)	CGD (herd):
Your required additional drought reserve:		
Size of additional drought reserve (in days)	your decision	days (drought reserve):
Total CGDs required for drought reserve	total cow/calf equivalents × days (drought reserve)	CGD (drought reserve):
Your total required grazing reserve (including your drought reserve):		
Total size of required grazing reserve (in days)	days (winter) + days (drought reserve)	days (total required winter grazing):
Total CGDs required for herd and drought reserve	CGD (herd) + CGD (drought reserve)	CGD (total required for winter):
Assessing grazing shortfalls or excesses in your winter grazing reserve:		
Shortfalls or surpluses in grazing reserve (in days). *Note:* (+) indicates surplus, (–) indicates shortfall	days (in grazing reserve) − days (total required for winter grazing)	days (shortfall or surplus):
Shortfalls or surpluses in grazing reserve (in CGDs). *Note:* (+) indicates surplus, (–) indicates shortfall	CGDs (in grazing reserve) − CGDs (total required for winter)	CGDs (shortfall or surplus):

Name of item being assessed	Calculation	Result of calculation
Your strategy to overcome shortfalls in your grazing reserve: (note three alternative strategies listed below)		
Number of cows to sell to avoid grazing shortfalls	CGD (shortfall) ÷ days (shortfall)	cow/calf equivalents to sell:
Amount of additional pasture required to address shortfall	CGD (shortfall) ÷ CGD/acre in available additional pasture:	acres (additional):
Amount of stored feed required to address shortfall	CGD (shortfall) × feed consumption per cow per day (lbs/day) ÷ 2,000 lbs/ton:	tons feed:

The Winter-Grazing Map

To ensure that you can graze through winter without stored feed supplements, plot the winter grazing rotation on your farm map. This allows you to plan ahead for predictable weather patterns such as fall wet weather, midwinter deep snow, late-winter crusted snow, and early-spring mud.

On the map, reserve suitable sheltered pastures for extremely windy and cold days; exposed ridges for deep snow or crusted-snow days; gravel soils and south-facing slopes for early-spring and muddy days; and pastures with tall grass and trees to loosen snow on days when it is crusted. Indicate on the map which pastures you will avoid grazing to prevent damage in muddy weather and to prevent trees from being damaged by the cattle.

The winter grazing map allows you to plan a logical sequence of daily grazing slices that move outward from your winterized water sites without wasting grass to trampling. See page 111 in chapter 7 for an example of a winter grazing map.

Your Herd Nutrition Plan

AT THE SAME TIME THAT YOU DEVELOP YOUR herd's year-round grazing plan, you must also plan its nutrient and mineral supplementation program to facilitate year-round grazing. This chapter is designed to prompt you to develop your forage analyses database and develop your herd nutrition program with the help of a livestock nutritionist.

Developing Your Herd Nutrition Program

In order to plan and manage your herd's nutrition, fertility, body condition scores (BCS), and health without resorting to stored feeds during winter, you need to know the nutritional value of your grass reserve throughout the year and how it affects the body condition of your cattle. To obtain this information, you must build a database of forage analyses that reflect changes in the nutritional value of the grass that your cattle eat throughout the year.

A livestock nutritionist will use your forage analyses to calculate how the body condition scores of your individual livestock classes (e.g., cows, replacement heifers, first-calf heifers, yearlings) will diminish during the winter-grazing season and how quickly they will recover during the grass growth season. By knowing how the body condition scores of your herd change during winter, the nutritionist is able to calculate the minimum amount of protein and energy supplements your cattle will require to maintain a BCS of 5 or greater before the return of

the spring grass flush. (*Note:* The management strategy of letting your cattle's BCSs drop to 5 during the winter and allowing your cattle to supplement themselves from their own body fat stores is possible only in a summer-calving scenario, to avoid affecting your cattle's fertility and conception rates, as discussed in chapter 3.)

In a twelve-month grazing program with a summer calving season that is scheduled to give the cows time to regain a BCS of 7 in time for calving, small amounts of protein and energy supplements make up for deficiencies in the quality of your grass supply, allowing you to continue grazing throughout winter without stored feeds at a minimal supplementation cost while maintaining herd health and guaranteeing maximum conception rates in the upcoming breeding season. The required protein and energy supplements are typically so small that you can mix them into the cattle's loose mineral ration, as described in chapter 7. Without the salt in the loose mineral mix to limit the cattle's consumption of the nutrition supplements, it is very difficult to ensure that all animals consume sufficient nutritional supplements without overeating.

The forage analyses are also used by the nutritionist to calculate how your cattle's mineral supplementation program must change to compensate for the mineral and vitamin content of your grass supply. (For more on designing your supplement and mineral program, how to take forage samples, and how to avoid contaminating them, see chapter 7.)

Procuring Samples for Forage Analyses

Your forage analyses must reflect the changes in the cattle's nutritional intake from month to month. Therefore, it is important to take samples only from the pastures that the cattle graze on the day of sampling, only the specific grass varieties that the cattle select from their pasture smorgasbord, and only the parts of the grass that the cattle eat (and not the whole plant).

To build an accurate database, you need to sample your grass supply every month on the same day of the month — mark it on your calendar so that other chores, business meetings, and farm responsibilities don't interfere with this schedule. Monthly samples must be taken over at least a three-year period, after which you will have a reasonably accurate database. Then you will need to resample only every few years or whenever you add fertilizer amendments, which might affect the nutrient content of your grass. Your livestock nutritionist will tell you what nutrients your laboratory needs to test for and may even send the samples to his or her preferred laboratory for you.

What the Livestock Nutritionist Needs to Know

Before samples are tested, be sure the livestock nutritionist knows that you are winter grazing; allowing your herd's body condition to fluctuate from a BCS of 7 down to a BCS of 5 during winter (conception rates and herd health begin to be affected at a BCS of below 5), thereby forcing them to live off their body fat reserves stored during the summer grass flush; and calving during summer, *after* your cattle have regained a BCS of 7.

The livestock nutritionist will also need answers to the following questions to design a herd-specific supplement program:

- What are the frame sizes of your cattle? (used for calculating their nutrition requirements for maintenance)

- What are your calves' average birth weights? (used for calculating your cows' nutrition requirements during gestation)

- Are your cows a low-, medium-, or high-milk-producing breed? (used to calculate your cows' nutrition requirements during lactation)

- When do you wean your calves? The longer you wait to wean, the lower your cost of production because the cow can supplement the calf through winter on her milk. If, however, the BCS of your cows, replacement heifers, and first-calf heifers threaten to drop below 5 before they can begin to recover on the spring grass flush, stabilize their decreasing body condition by weaning earlier.

Locating a Livestock Nutritionist

Building a database of your grass quality through forage analyses and working with a livestock nutritionist to manage your herd's body condition fluctuations is so important that I encourage you to find a livestock nutritionist as soon as possible. Look for one who is open to working with a summer-calving herd with fluctuating body condition scores, who understands the concept of storing last summer's grass excesses as body fat to naturally supplement the grass supply during winter, and who is able to make nutrient and mineral supplement calculations to suit your natural, grass-based production philosophy. Remember, the plan you design with your livestock nutritionist for your summer-calving, year-round grazing herd is very different from the nutrition plan designed for a conventional production scenario requiring a constant BCS of 7 during the winter to ensure fertility in a herd calving out of sync with nature's grass supply. A supplement program designed for a conventional spring-calving herd that must maintain a high BCS throughout the winter to ensure the herd's fertility will not be cost-effective or appropriate for your summer-calving grass cattle enterprise. Make sure the livestock nutritionist you work with understands your herd's nutrition goals. If you have difficulty finding such a person, ask your local agriculture Extension agent, agriculture department, livestock range management agent, feed company, or forage analysis laboratory for a list of livestock nutritionists whom you can interview.

Formulating Supplements and Monitoring BCS

Using your forage analyses, your livestock nutritionist will calculate the exact daily protein, energy, and mineral supplements necessary to sustain each

KEEP THIS INFORMATION HANDY

- Name and contact information of your livestock nutritionist

- Name and contact information for a forage analysis laboratory

- Items for which your forage analysis laboratory should test based on suggestions from your nutritionist

- Sample day (the day of the month you consistently sample your pasture — a date more important than any business meeting or farm chore)

of your livestock classes based on the herd's specific grass diet. By mapping out your forage analyses over the course of the year, the livestock nutritionist can also track how the herd's BCS will change from month to month based on the changing quality of the grass supply, which will enable you to prevent the cattle's BCS from dropping below 5 during winter and ensure that they will regain a BCS of 7 before calving.

You can compare the actual BCS of the herd with the nutritionist's calculated BCS to monitor whether the cattle's body conditions are on track. If you note that the cattle's BCS begins dropping too quickly, you will be able to intervene by weaning early or increasing the cattle's supplements so you can continue grazing the grass supply instead of resorting to expensive stored feeds. This approach prevents unexpected nutritional crises and helps guarantee the herd's fertility months later during the breeding season.

When undertaking a mineral supplement program, it is also wise to have a veterinarian take random blood samples from the various age groups in your herd twice per year (once in the summer and once in midwinter) to ensure that the mineral supplements are not causing any unforeseen mineral imbalances or nutrient oversights.

Your Grass-Finishing Plan

PRODUCING HIGH-QUALITY, NUTRIENT-RICH, grass-finished beef requires planning. You must carefully consider every last detail of your grass-finishing program in your business planning process: the animals you choose, your grazing program, mineral supplementation, finishing weight, nutrition status before slaughter (are they gaining weight before slaughter to ensure tenderness, for example?), stress elimination, slaughter dates, transport, slaughter facility, and even the post-transport, settle-down period just before slaughter. If your planning is thorough, you will have a consistent product, animal after animal, year after year, satisfied customer after satisfied customer. (Before designing your grass-finishing program, reread chapter 17.)

Grass-Finishing Essentials

The following questions are designed to help you plan all the important aspects of your grass-finishing program. Write down your answers as part of your written business plan.

- What is the target finish weight of each class of your grass-finishing animals (steers, meat heifers, young slaughter cows, hamburger cows)? If you have more than one breed, what will the target finishing weights be for each class of animal in each of your breeds? (The ideal finish

weight will vary from class to class and from breed to breed.)

- Based on your target finish weights and your production conditions, what are the earliest target slaughter dates for each class of your grass-finishing animals (steers, heifers, young cows, hamburger cows)? Your forage analyses and your livestock nutritionist's calculations of fluctuating BCS and weight gains (see chapter 26) will help determine dates when your animals will be heavy enough for slaughter and how late into the fall you can continue grass-finishing and slaughtering your animals based on the quality of the grass reserve. (Remember: An animal must be gaining at the time of slaughter for its meat to be tender.) Based on your forage analyses, your livestock nutritionist will also be able to calculate how soon after the end of winter your cattle will have regained sufficient weight (and body fat) to resume slaughtering. Based on these restrictions, schedule slaughter dates to suit your marketing scenario (fall only, spring through fall, holding over a portion of your grass-finished steers and heifers

Based on your forage analyses database and your livestock nutritionist's calculations (see the discussion in chapter 26), plot out a grass-quality curve, indicating when you can grass-finish. A grass-quality curve, different from a grass-growth curve, focuses not on grass growth rate, but rather on the nutrient value of your grass supply (both summer-grazing supply and winter-grazing reserve). It illustrates how grass quality changes in response to seasonal fluctuations and environmental influences and indicates the slow degradation in quality that occurs after the end of the growing season depending on plant maturity, soil fertility, height of grass, precipitation, nutrient leaching, frost preservation, and plant species. Your grass reserve will continue to supply adequate nutrition for grass finishing long into winter, after your grass growth has ceased entirely, especially in your best-quality grass-finishing pastures.

Next, identify slaughtering seasons for each animal class based on when animals in it will reach their ideal finish weight and when you want to slaughter to meet the demands of your target market.

PLANNING YOUR SLAUGHTER SEASON

Based on your grass production year and ideal finish weights, plan your slaughter season for steers, meat heifers, young slaughter cows, and hamburger cows.

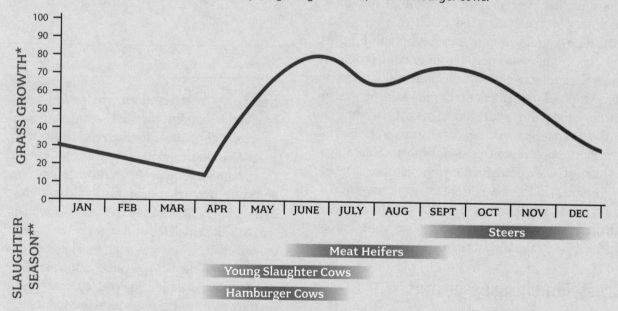

* The grass-quality curve is an indication of the nutrient value of the grass throughout the year, including the slow decline in the nutrient values of the winter-grazing reserve after the grass has been frost-killed. This allows you to plan during what seasons grass quality is sufficient to grass-finish, and therefore schedule your slaughter season.

** Slaughter season lines represent dates when each animal class will be slaughtered after grass finishing. Note that steers are 1½ years old at the time of slaughter, whereas meat heifers are 2 years old at the time of slaughter. Also note there are two slaughter seasons for hamberger cows, one in spring and another in the fall at the earliest date when the calves of the cull cows can be weaned.

through another winter to slaughter during the next growing season, slaughtering hamburger cows throughout summer, for example).

_____

_____

■ How do you intend to manage the grazing rotation of your grass-finishing animals? Will you graze them with the rest of the herd, graze in a leader–follower program, or segregate the finishing-class animals prior to slaughter? If you intend to segregate animals prior to slaughter, how soon before the slaughter date will you split them from the main herd? Write down your schedule for any changes in your grazing management prior to slaughter.

_____

_____

■ On which pastures do you intend to grass-finish? Do any of your grass-finishing pastures have any special requirements to ensure top-quality nutrition? For example, do any of them require fertilizer amendments to guarantee high-quality grass? If you plan to use your alfalfa pastures for grass-finishing in the fall, what is the approximate date of your killing frosts and the subsequent dry-down interval to minimize the bloat risk of grazing alfalfa? (See page 232 for a discussion about grazing alfalfa and preventing bloat.)

_____

_____

■ How do you plan to extend your grass-finishing season into winter? (Will you rely on special crops, alfalfa, grazing corn, or reserving specific fields for grass-finishing animals?)

_____

_____

■ What is your backup plan for drought years, when the pasture quality may not be high enough for finishing or when your grass quantity is low? Will you sell of some of your lowest-quality animals to reduce grazing pressure? Will you reserve certain pastures for grass-finishing animals or will you change your grazing management to compensate, and if so, how?

_____

_____

Adversity will come. Plan now for difficult situations so you can dependably grass-finish enough animals to ensure a stable income and avoid financial difficulties.

Monitor the volume and quality of your grazing reserve so you can introduce your backup plan before you lose the ability to grass-finish. What warning signs will you look for in your pastures and in your grazing reserve calculations to indicate that you must take measures to safeguard your grass-finishing program?

■ How will you guarantee an adequate mineral supplement program for your grass-finishing animals? Which minerals will you feed?

_____

_____

Slaughterhouse Considerations

After all the hard work that goes into producing a high-quality, tender, grass-finished animal, it is vitally important to plan how to get that animal from the pasture to the slaughterhouse, how it will be slaughtered and processed, and how the finished product will be presented to your customers. These considerations are essential to maintaining the quality of your beef product through the entire slaughter and butchering process to your customers' kitchens. The following questions are designed to help you plan this final stage of your grass-finishing operation:

- What slaughterhouse will you use to custom-slaughter your grass-finished animals for your direct-marketing program? Spend some time at the slaughterhouse to observe its procedures and handling practices. Get a tour on kill day and watch the butchers and employees as they work.

_____

_____

- Does the slaughterhouse use low-stress handling practices?

_____

_____

- Is it able to meet your meat-packaging requirements? What quality of butcher paper do you want to use? Do you require plastic wrapping in addition to butcher paper? Do you want meats vacuum-packed after they are flash-frozen? Is the packaging attractive? How will your labels be presented on the packaged products?

_____

_____

- Does the slaughterhouse have enough hanging space for your animals? Look at your slaughter schedule and calculate how many animals you expect to slaughter each week. The slaughterhouse must be able to accommodate your volume of animals on the dates required for your marketing program. Discuss these dates far ahead of time to reserve your slot early and to avoid being bumped by other customers or the flood of custom meat cutting that occurs every fall during the hunting season.

_____

_____

- Does this facility have the necessary meat inspections to suit your market (provincial/state, federal, or certified organic)?

_____

_____

- What do you need to do to prepare for transport to the slaughterhouse? Who will do the trucking?

_____

_____

- What will you do to reduce stress and noise during loading and trucking? Troubleshoot your loading corral to eliminate rattles, squeaky gates, stressful alleys, and blind spots. Can you dampen any rattling gates inside your transport vehicle? Do you have nonslip flooring in your trailer?

_____

_____

- Does the slaughterhouse allow time for your animals to settle down for at least an hour after trucking? Will it provide fresh hay, clean water, and some bedding for the animals during this period?

_____

_____

- Is the slaughterhouse able to slaughter your animals within six hours after trucking? Animals should be slaughtered absolutely no later than 24 hours after transport, although no later than six hours is ideal.

_____

_____

Your Marketing Plan

IN CHAPTER 21 YOU DEVELOPED A PROFILE of your community to help you focus your business and in chapter 22 you chose specific farm enterprises to suit your community and environment. Now it is time to develop your marketing plan to determine how you will reach your customers and sell your beef. This brief chapter is designed to provide you with a systematic framework to help you organize your marketing strategy as part of your comprehensive business plan.

You will develop a profile of the typical customer in each market niche for which you intend to advertise, design a product label, and formulate an advertising strategy. Please refer to the marketing discussion in chapter 19 as you put together your marketing plan according to the basic framework outlined here.

Developing Profiles of Your Target Customers

Profile a typical customer for each market outlet that you want to target. For example, if you are marketing beef by the side to local families, consider what characteristics, interests, concerns, passions, and needs influence most of their family shopping decisions. If you are marketing to pubs and steakhouses, characterize the typical customer who makes most of the steak and burger purchases in these establishments. Describe the typical customer who will buy small portions of meat from health food stores, farmers' markets, meat-by-the-cut outlets, and so on.

PROFILE YOUR CUSTOMER

For each market outlet you target, develop a profile of its unique customer base and how you plan to maintain it.

Market outlet:

_____

_____

_____

Customer profile:

_____

_____

_____

Typical customer:

_____

_____

_____

Hot buttons for your advertising strategy:

_____

_____

_____

Designing Your Product Label

As part of the marketing strategy outlined in your business plan, you should design your product label. Remember, the key to a successful product label is a brief, positive, eye-catching message. It should paint a beautiful picture that speaks to the needs, passions, and interests of your customers. The label should describe not what it is *not*, but what your product *is* in positive terms: its health benefits, and how natural, pure, flavorful, and tender it is. (See chapter 19 for more detailed information on creating your label.)

_____

_____

_____

_____

_____

_____

_____

_____

_____

Formulating an Advertising Strategy

When you have a clear picture of your markets and customer base, develop an effective advertising strategy for each of your chosen target markets. How will you reach your customers? Where will you advertise? How will you create buzz about your product? Whom do you need to approach and what information will you share with them?

Leave nothing to chance. Create a step-by-step plan that addresses the who, what, when, where, and how details of reaching your customer base. Answer the following questions for each specific market.

■ Where will you distribute informational bulletins and order forms so that you can reach as many potential customers as possible?

_____

_____

_____

■ Which wholesalers do you need to approach?

_____

_____

_____

■ Which managers, chefs, and order buyers will you talk to?

_____

_____

_____

- What information will you need to provide to each of your market outlets? (Preparation is key. Remember: You want to make a good first impression.)

_____

_____

_____

- What special events can you participate in to increase awareness about your product, and when do they occur?

_____

_____

_____

- What on-farm activities can you plan (an open house, farm tours, community events, cornfield labyrinths, Halloween events) to generate awareness of your product? Identify potential dates for these in your calendar so that you can start planning for them now.

_____

_____

_____

- Whom can you talk to in order to generate as much awareness about your product as possible? (Naturopaths, grocers, herbalists, and fitness trainers are all good options.)

_____

_____

_____

- To which communities, apartment complexes, gated communities, retirement communities, alternative health centers, gyms, hardware stores, bakeries, and radio stations do you plan to distribute informational bulletins?

_____

_____

_____

Chapter

Planning for Adversity

Even the best-laid business plan is vulnerable to market and weather cycles that can wreak havoc on your herd's continuity and your financial stability unless you plan in advance how you will address and overcome adversity. Despite the fact that most business planning is based on the average year, adversity is normal. Nature and markets are volatile, so while it is impossible to pinpoint exactly when you will be faced with an unexpected challenge, history shows that it will happen and that you should have a drought and disaster contingency plan on hand to rely upon to keep adversity from deteriorating into full-blown disaster.

This chapter provides a framework to help you plan for adversity so your business can remain viable, profitable, and sustainable through inevitable weather challenges and market fluctuations. This includes identifying warning signs of an impending drought, planning your drought reserve to address feed shortages, planning measures to address grass shortages, planning your grazing rotation during a drought to avoid livestock water shortages in grazing areas, and preparing a drought fund and expense record sheet to safeguard against financial catastrophes and help you manage your finances for the duration of the drought. You'll also find a worksheet that helps you calmly, rationally, and systematically evaluate your options during crises.

Drought

In chapter 25, you determined how many extra days' worth of grazing you want to keep as a drought reserve, either as an additional grass reserve or as a stored hay reserve. To create a solid drought plan that safeguards the soil, grass, herd health and fertility, and the financial viability of your farm, you need to plan precisely how you will address grazing shortfalls and ration the drought reserve until the rains return.

You also must be acutely aware of the early warning signs of drought so you don't react by doing too little too late. If you buckle down and wait, hoping the problem will take care of itself before you need to act, you'll be in trouble. (For more on planning the drought reserve, see chapter 8.)

Assessing Your Drought Reserve Needs

Begin by taking stock of your drought reserve. In your drought plan, address each of the questions listed below to identify what your needs will likely be during a drought and how you intend to address any grazing shortfalls.

To aid in this portion of the planning process, please refer to your calculations in the chapter 25 worksheet Assessing Your Winter-Grazing Requirements and Addressing Shortfalls.

■ How many extra days' worth of grazing do you intend to keep on hand as a drought reserve? How long do you expect your drought reserve to last in an emergency?

_____

_____

■ How many additional cow grazing days (CGDs = days in drought reserve × number of animals in herd expressed as cow/calf equivalents) of grass will you need as a drought reserve in addition to the CGDs your herd requires for winter grazing?

_____

_____

■ How many acres of pasture or tons of stored feed will you keep on hand at all times as your drought reserve? (The drier and more unpredictable the climate, the larger this drought reserve should be. In very arid climates, a drought reserve of an entire year is not uncommon.)

_____

_____

Warning Signs and Countermeasures to Drought

In your drought plan you should identify specific warning signs to alert you of feed shortages that could signal an impending drought, allowing you to begin implementing precautions to protect yourself long before the drought progresses into a full-scale disaster. Typically this involves monitoring your grass reserve by calculating the CGDs in your entire grass reserve several times a year using the technique in chapter 25 (see pages 318–321). If you don't have sufficient CGDs in your reserve to feed your herd until the next predictable growth season, or if you lack the additional CGDs needed for your drought reserve, begin implementing your measures.

Monitor grass growth throughout the year. This will alert you to needed interventions while the pastures still look green. The four critical grass-monitoring dates are: the midpoint of the growing season, the onset of winter, midwinter, and the onset of the growing season. They range from requiring the least severe intervention to requiring the most if your grass reserve calculations indicate an impending drought at these stages. If the rains fail or are insufficient by the fourth date — the onset of the normal growing season — you are at the doorstep of a potential full-scale drought of unknown severity, requiring your strictest disaster-prevention measures.

Decide which of these you will implement at each of the four critical monitoring dates if your grass reserve calculations show shortages. Remember, your grass reserve calculations are the most reliable indicators of an oncoming drought. Create a specific plan for each monitoring date.

For example, if the rains are insufficient or if grass growth is less than you anticipated by the date that marks the midpoint of the growing season, you should immediately begin implementing the appropriate countermeasures to safeguard your herd, grass, soil, and finances, rather than wait until your limited feed resources near depletion and then struggle to cope with a full-blown crisis. If you calculate insufficient CGDs in your grazing reserve as winter begins, adjust your herd numbers to account for the diminished grass supply or arrange for additional feed or pastures to make up for feed shortages before such shortages becomes obvious to everyone, leading to skyrocketing feed and pasture prices and a collapse of market prices as people panic and dump their excess cattle on the market.

Each of the four monitoring dates are listed below, accompanied by a set of questions to help you identify the warning signs of an impending feed shortage and describe your planned countermeasures. Refer to these questions to guide your management decisions when you discover a drought and need to begin implementing countermeasures. During the stress that inevitably accompanies a drought, these questions will help prevent impulsive, panic-driven decision making.

Midpoint in the Growing Season

■ What date marks the midpoint of your growing season?

_____

_____

■ How much grass growth must you see by this date to feel confident that you will achieve the required grass reserve by the onset of winter? Indicate the total number of CGDs contained in your pastures at this time to reinforce that grass growth is indeed adequate. This is the threshold that alerts you to insufficient grass growth during the growing season, even if the pastures still look green and there are as yet no easily recognizable signs of an impending drought. Remember, reduced grass production at this point in the growing season will lead to feed shortages many months later during the dormant season. Identifying potential feed shortages this early will allow you to avoid actual shortages with only slight reductions in your herd size.

_____

_____

_____

_____

■ Based on the shortage you discover at the midpoint of the growing season, calculate the size of the feed shortage you will face by the time you reach the end of the growing season (both in terms of days of grazing for your herd and CGDs). Indicate how many animals you will need to sell or how many additional grass-finished animals you should slaughter to guarantee adequate winter grazing and maintain your drought reserve. Alternatively, determine how many extra acres of pasture you will need to add to your grazing rotation to make up for the feed shortage.

_____

_____

_____

■ List which animal classes you will destock at this point in the growing season. Start by culling your lowest-quality and least valuable yearlings, brood cows, and heifers to help ensure that you'll have sufficient grass in the fall to be able to earn an income from your best grass-finishing animals. If more-severe culling is necessary, cull additional nonbreeding animals before further culling your quality breeding stock so you can maintain your core breeding herd through the drought. This will help to ensure a quick financial recovery when the feed shortage ends.

_____

_____

_____

_____

_____

_____

■ Describe how you can alter your grazing impact to ensure that your grass will rest sufficiently between grazes and that it is not overgrazed during each rotation. For example, you can help ensure adequate grass recovery periods by combining your herds (see chapter 9 for more on integrated herd management), by minimizing trampling and waste through daily pasture moves, or by feeding hay from your drought reserve, thereby maximizing pasture growth rates by avoiding overgrazing.

_____

_____

_____

_____

_____

_____

_____

Onset of Winter

■ What date marks the onset of winter (end of growing season, beginning of winter-grazing management)?

_____

■ What is the minimum acceptable number of CGDs in your grass reserve (winter grazing plus drought reserve) on that date? (If your grass reserve contains fewer CGDs than this threshold, implement countermeasures immediately. Do not wait until partway through winter to intervene.)

_____

_____

■ Calculate the feed shortage you must overcome to meet all your winter-grazing needs and maintain a drought reserve (determine how many days of grazing you are missing for your present herd size to survive the winter and how many CGDs this represents). Indicate how many animals you will need to sell or slaughter immediately to guarantee adequate winter grazing and maintain a precautionary drought reserve. Alternatively, determine how many extra acres of pasture you will need to add to your grazing rotation to meet all your grazing needs.

_____

_____

_____

■ List which animal classes you will destock immediately. Start by culling your lowest-quality and least valuable yearlings, brood cows, and heifers to ensure that you'll have sufficient grass in the fall to be able to earn an income from your best grass-finishing animals. Cull or slaughter all grass-finishing animals that you intended to hold through winter for spring slaughter. If more-severe culling is necessary, cull nonbreeding-class animals before culling quality breeding stock so you can maintain your core breeding herd through the drought. This will help to ensure a quick financial recovery when the drought ends.

_____

_____

_____

_____

_____

■ Describe how you can alter your grazing impact to maximize the grass you have and minimize trampling and waste. Combine your herds and ration the grass with daily pasture moves.

_____

_____

_____

_____

_____

_____

■ Indicate the minimum grass residual that will allow you to safeguard against soil and grass damage so your pastures can recover quickly at the end of the drought. Closely monitor this grazing residual, noting the minimum grazing residual in inches to avoid overgrazing.

_____

_____

_____

_____

_____

Midpoint in Winter

■ What date marks the midpoint between the onset of winter and the end of the season?

_____

■ What is the minimum number of CGDs in your grass reserve (including your drought reserve) on that date? Any dip below this will alert you to a potential feed shortage.

_____

_____

_____

■ Calculate how much less grass you have (in CGDs and days grazing) than is required to successfully complete your winter-grazing period while maintaining your drought reserve. Indicate how many animals you will need to sell to guarantee adequate winter grazing and maintain a drought reserve.

_____

_____

■ List which animal classes you will destock immediately. Start by culling any yearlings left over from last season as well as your poorest-quality breeding stock. If your calculations suggest that you will experience further shortages by winter's end, consider weaning early to reduce your herd's feed requirements. (In a dire emergency, you can wean as early as two months after calving.)

Calculate whether you'll have enough grass left over to keep the weanlings through the winter season. If necessary, sell these calves as well so you can maintain a core brood herd of highest-quality animals to help ensure a quick financial recovery when the drought ends.

_____

_____

_____

■ Calculate how much money you are willing to set aside in your budget to pay for protein and energy supplements for your herd. (Note that supplements will continue through the winter season and the full drought-reserve interval you planned.) _Do not_ spend more than this budgeted amount. Instead, destock more animals. You simply cannot afford to overextend yourself financially during a drought or feed shortage.

_____

_____

_____

_____

Return of the Growing Season

■ What date marks the beginning of the predictable growth season (when rains return and growth of lush spring grass begins)?

_____

■ What is the minimum acceptable number of CGDs left in the grass reserve for use as a drought reserve on this date in case the rains fail?

_____

_____

_____

■ If predictable rains have not arrived by this date, one of two possibilities may occur: Either grass growth will be significantly reduced if the rains arrive late or there may be no measurable grass growth if the rains fail to arrive. Either case marks the threshold between the feed shortages described in the previous three monitoring dates and the beginning of a full-scale drought of unknown severity. The more arid and unpredictable your climate, the more alarming is this signal of an impending drought. Do not wait until your drought reserve has been used up before you begin to protect yourself. When you reach this point, implement your strictest drought prevention plans. Be realistic; do not let optimism cloud your decisions. Calculate how many CGDs you have left in your drought reserve and how many days of actual grazing this corresponds to for your current herd size. Calculate how many additional animals you will sell immediately to guarantee an adequate drought reserve for the remaining animals.

_____

_____

_____

■ Indicate the minimum herd size of your best brood cows, which you will try to maintain through the drought. Begin destocking immediately to attain this minimum herd size. Don't wait until your animals begin to lose weight, grass becomes scarce for the remaining animals, or market prices collapse. Sell other classes of animals (weanlings, yearlings, grass-finishing animals) and maintain a modest breeding herd that will allow your business to recover quickly when the drought ends.

_____

_____

_____

■ Describe any additional measures that you can take to minimize grass waste and safeguard the grass and soil beyond combining your herds and rationing the grass daily.

_____

_____

_____

_____

■ Indicate the minimum height of the grazing residual that you will absolutely maintain during this

stage of the drought. If there is no growth whatsoever — that is, if the rains fail, for example — maintain a minimum 2- to 4-inch grass residual in your pastures to protect the soil from erosion and ensure that the grass can recover quickly after the rains return. If you experience only reduced growth due to insufficient rains, maintain a tall grass residual (6- to 8-inch minimum) to maximize your grass growth and minimize evaporation losses until all growth ceases. When growth ceases, switch to the short grazing residual, which you will not regraze, in the same way that you ration your grass during winter grazing.

_____

_____

_____

■ Decide in what order you will graze your pastures so you can first graze around your most unreliable livestock watering sites. Transfer your fence and water infrastructure to a blank map of your property, as discussed on page 340, and plan your drought grazing rotation to ensure that your herd has reliable water access through the drought's duration.

_____

_____

_____

■ An impending drought also indicates that you must protect yourself financially. Assess your financial drought fund (discussed in detail on page 340), monitor your expenses carefully, and reduce expenses to a bare minimum by making a list of all that you can eliminate.

_____

_____

_____

Prepare a Grazing Map of Your Planned Drought Grazing Rotation

Drought not only affects the grass supply, but also creates water shortages. Consequently, you need to safeguard yourself against unreliable livestock water sites that may fail during a drought or may be incapable of supplying adequate water for your combined herd. Plan your drought rotation on a map of your farm that includes your entire fence and water infrastructure. This rotation should occur first around the most unreliable water sites and should save for later the grass near your most dependable water sites. Keep this map with your drought plan as part of your comprehensive business plan.

Financial Preparedness

Creating a financial reserve is crucial if you are to survive a drought or market collapse. Inevitably, as a drought progresses, your livestock lose value or become worthless as more and more farmers attempt to sell their animals on the market when they run out of feed. Your equipment will undergo a similar, temporary devaluation during a drought or market collapse. Livestock and equipment provide no financial reserve during a drought or market collapse. You'll want to cut your expenses to an absolute minimum during this time, but you'll still need to pay your bills. Therefore, your financial reserve should be large enough to accommodate all necessary, unavoidable, ongoing payments throughout the course of the drought and until your farm recovers sufficiently to be able to provide you with an income again. Remember: The end of a drought doesn't necessarily signal the end of your financial woes. Your financial recovery may depend on a new crop of cattle reaching maturity and slaughter-readiness.

Planning for Your Drought Fund

Using a copy of your Yearly Expense Budget worksheet from page 282, note which expenses you will be able to forgo during a drought (such as farm improvements, drought and replacement fund payments, fuel and fertilizer costs). The expenses that remain are your unavoidable financial commitments and represent the minimum cash flow required to retain your land and cattle and meet your family's financial requirements through the duration of the drought or market collapse and the recovery period afterward, even if you are unable to earn an income during this time. The sum of these expenses is your drought fund. Also include enough money in your drought fund for a family vacation so that you will have some relief from the anxiety and stress that accompany a drought and can gain perspective on your situation.

Divide your drought fund by the number of years that you need to build up the fund. This is your drought fund allocation that you set aside annually (see chapter 22). The number of years it takes to build your drought fund will depend on your climate and drought history. If you are farming in an arid climate in which droughts are common and often severe, you may want to set aside as much as one-seventh to one-fifth or more of the drought fund each year, which reflects the repetitive nature of long-term, drought-related weather patterns in your area. In a more humid, less variable climate, however, you may spread your drought fund allocations over ten years, though no more than this. Ten years is the length of the cattle cycle, so you can expect at least one market collapse during a ten-year interval.

Use the following Drought Fund Budget worksheet to calculate your yearly drought fund allocation. Following this worksheet is a Drought Fund Expense Record, which lets you compare budgeted expenses with actual expenses so you can monitor your spending very diligently during this difficult time. Maintaining your drought fund expense budget and drought fund expense records throughout the drought forces you to cap your cash output in each expense category so that panic, stress, and desperation don't compel you to overspend. The drought fund is designed to pay for *only* the items you include in your drought fund expense budget, such as:

- Debts, lease payments, insurance, property taxes, wages, and other unavoidable cash outputs

- Family expenses (groceries, home insurance, school fees, home mortgage)

- Minimum ongoing operating expenses (mineral supplements, fuel, vehicle insurance)

- Supplemental protein and energy to allow your cattle to continue grazing a grass supply that will deteriorate in quality

- Restocking the herd when the drought ends

The drought fund is not meant to buy feed or to support other crisis management quick fixes; these typically only postpone, rather than prevent, a financial catastrophe.

DROUGHT FUND BUDGET WORKSHEET

List of expenses addressed by drought fund	Value ($)
Total drought fund expenses listed above	$
Percentage of fund to set aside each year	%
Yearly drought fund allocation	$

DROUGHT FUND EXPENSE RECORD SHEET

Budgeted expenses		Actual expenses	
Name of expense	$	Name of expense	$
Total budgeted expenses:		Total actual expenses:	
Net difference: +/– $			

Crisis Management

In chapter 20, I introduced you to an option-weighing technique that I've found very useful. It can help you work through crises and high-pressure situations when emotions might ordinarily take over, causing you to make rash decisions.

Use this worksheet to help you evaluate any important business decisions you make during a drought. It will help you evaluate individual options and follow a logical step-by-step evaluation of costs and financial gains related to each. It will also force you to identify rationally the advantages and disadvantages related to each option and decide what additional research is needed to make an informed decision about each you are considering.

OPTION-WEIGHING WORKSHEET

Option:		
Income		**Expenses**

Net gain/loss:					
Advantages	I	U	**Disadvantages**	I	U

Prep work/further research:

Planning for Change

CHANGING YOUR FARM FROM CONVENTIONAL to natural production can be a harrowing experience without a business plan to guide you and without planning the transition to natural, grass-based production methods. This chapter discusses how to make this transition as smooth as possible as you switch to summer calving, implement an intensive rotational grazing program, and extend your grazing season through the winter months, learning how to successfully graze year-round. It also discusses how to implement smoothly any additional positive changes that grow out of switching to natural, grass-based production methods, such as changing over to an organic soil fertility management program, reducing or eliminating pasture irrigation, and revitalizing old or overgrown pastures.

Your attitude, confidence, and success depend on the kind of information you take in during this big change. Stay away from the local coffee-shop talk and cancel your subscriptions to the usual doom-and-gloom agricultural publications; replace these negative influences with a support network of success stories so you can surround yourself with positive energy and enthusiasm.

Make some contacts with people who are already doing what you want to do — for example, in the local organic association and at grazing conferences or by meeting other grass-based farmers in your area. Subscribe to one of the grass-based cattle production or organic farming newspapers such as *The Stockman Grass Farmer* and *Acres USA*.

It's important to make sure everyone in your business is part of this change. You, your business partners, family members, and employees all need to embark on the same learning curve together. Everyone should move toward the same goal to ensure a successful outcome for the changes you are trying to implement.

Committing to the Change

The natural reaction to change is to take it slowly, learn as you go, make one little alteration at a time, and grow into it. I strongly recommend *against* this approach when changing your calving season and grazing strategy. A quick transition will allow you immediately to begin extending your grazing season into the winter and start learning the art of year-round grazing. A slow transition will keep you off balance for a very long time.

Living in two separate production scenarios as you change from the old to the new will complicate your life; reduce your productivity in both production systems, making both uneconomical; and prolong the turmoil of existing in uncharted territory. It is the proverbial uncommitted "one foot in and one foot out" situation that is bound to fail. In addition, when you meet with the inevitable challenges that accompany any change, your confidence will falter because you can always compare the new system with the old familiar system, still warm in your back pocket.

Plan Thoroughly

Don't change anything until after you have planned every single step between where you currently are and where you want to end up in one, three, five, and even ten years. Plan how your breeding cycle will change, what infrastructure you will alter, what your grazing strategy will be, what equipment you will free up and resell to finance changes, and in what time frame they will occur. Also figure out what effect these changes will have on your marketing and income over the entire transition period.

Only *after* you have a clear road map of the whole trip should you start making changes in your calving season, grazing management, and winter-grazing program — but once you start, try to make them happen as quickly as possible. It may be a shock to the system, but you will recover much more quickly and suffer much less financially during the transition, and your confidence can grow quickly because you will begin to see the results immediately, whereas drawing out the process would mask many of the benefits of your new system.

By committing fully to these changes, you have no choice but to make them work. They are the foundation for all other future alterations on your farm. Once the benefits of these first three changes begin to pay off, you can then gradually integrate additional ones, such as switching to organic fertilizers and phasing out your irrigation.

Moving the Calving Season

Once you have plotted your road map, switch the entire herd to its new summer-calving season as quickly as you can. You don't want to complicate your life with two separate calving seasons and manage two separate herds just when you need all your time and energy to learn the new system.

In a fall-calving scenario, you can safely move your breeding cycle forward by at least three weeks each year until you reach the ideal summer-calving season. If you have a spring- or winter-calving season, you can move the entire herd to the ideal later breeding date in one breeding season. If, however, you sold weanlings in the fall before the fiscal year's end, this inevitably means that you will defer the income from your summer calves to the following year. Plan your finances accordingly.

By switching the calving season of the herd in one fell swoop, you can offset some of the financial strain of this adjustment because your entire cow herd will no longer need as much winter feed to maintain conception rates, allowing you to extend your grazing season into the winter and learn the art of year-round grazing.

In addition to these financial savings, as soon as you implement your winter-grazing program, you can begin to reduce your forage and feed equipment to match your decreasing reliance on stored winter feed. (See chapter 7 for a discussion of how to make the transition to year-round grazing.) The savings from this excess equipment allows you to pay for infrastructure changes (fences and water sites) and provides you with some cash flow while your calf-crop income is deferred. Switching your calving season slowly will not allow you to sell off equipment until the entire herd's calving season has changed over and the entire herd has been trained to a significantly extended winter-grazing season.

Turning on Your Herd's Migration

The more pasture divisions you create, the more control you have over your grass supply, grass quality, winter grass rationing, manure and urine distribution, soil health, herd health, and so on. The faster you put your fence and water infrastructure together for improved grazing, the more viable your farm becomes. Taking time to implement change or prolonging your fence building and water-site development over several seasons will only serve to lengthen the transition and delay you from achieving your desired management benefits and financial goals. Before you begin adjusting your herd's grazing, however, plan your electric-fence and water infrastructure on a farm map to streamline operations and help develop a comprehensive and efficient grazing system, as discussed on page 301. Furthermore, it is vital that you plan your electric fence infrastructure and water infrastructure simultaneously. Planning a complete system ahead of time will save you innumerable headaches and unnecessary readjustments to your grazing infrastructure later.

Switching to Winter Grazing

This alteration can happen only after your herd has been switched to calving in sync with the grass cycle. Until your calving season is moved, your brood cows will not be able to meet all their nutritional requirements from grazing during the winter while still maintaining their fertility — unless they are given significant additional feed supplements to maintain their body condition scores.

Make the switch to winter grazing as quickly as your confidence level and experience allow. Switching back to grass in midwinter is very difficult, because the cattle are accustomed to having their food brought to them. But if you allow your cattle to continue grazing into the winter and don't give in to their pitiful bawling at the gate as soon as the first snow flies, they will soon get accustomed to digging through the snow and eating frosted grass. Save your best grass to teach them to graze through snow; they won't start digging if you give them inferior grass to graze.

The faster you train your cattle to winter-graze, the sooner you can sell your forage-making and feeding equipment to pay for infrastructure changes and to supplement your income during the transition period. Prolonging this change will only harm you financially. Yet year-round grazing is an art that must be learned (see chapter 7 for a discussion about low-cost alternatives to making and feeding stored winter feed to support your transition to winter grazing). By progressively extending the grazing season into the winter, you will learn to overcome your specific environmental and climatic challenges and gain the necessary experience to allow you to complete the transition to year-round grazing. In the meantime, protect yourself by maintaining a stored winter feed reserve until you develop the confidence and experience to consistently graze through the entire winter and complete the transition to year-round grazing.

If you are using only a small amount of feed to help you through the transition, sell your equipment and hire a custom haying or silage crew to put up feed or buy your forage until the transition is complete. The small amount of feed required does not justify keeping your capital tied up in equipment.

Making Additional Changes

While some changes should happen as quickly as possible, others are limited by nature's time frame. You can't rush a turtle. The following three changes depend on your successful implementation of an intense pasture rotation with daily pasture moves accompanied by a summer-grazing scenario and extended winter grazing that eventually lead to year-round grazing. Once these changes have been implemented, you must allow time for your pastures to adapt and benefit from these new grazing conditions. You must carefully observe your pastures to gauge when you can put these additional changes into effect without suffering losses in your pasture productivity and financial profitability.

Switching to Organic Fertilizer

Switching from synthetic fertilizers to an organic soil fertility program can be very expensive and cause a significant drop in production levels if it is attempted too quickly. Before this switch can be made, correct any soil imbalances and implement your intense grazing program in a time period that allows improvement of your soil fertility.

Nitrogen fertilizers are the most difficult and most expensive to replace with organic nitrogen sources. In addition, it takes at least three years to boost the vitality of your soil organisms after synthetic fertilization is replaced with an organic soil fertility program — hence the three-year transition period in most organic certification processes. Only a healthy grazing rotation and a comprehensive soil testing and organic soil fertility program can offset the drop in production caused by this change.

To remain financially stable and protect yourself from the increased expense of using organic fertilizers, this change should happen only at the same rate that daily pasture moves, more effective manure and urine distribution, better grazing efficiency, and other production savings (like your summer-calving scenario) take up the slack left by the absence of synthetic nitrogen and other synthetic fertilizers.

Shutting Off the Irrigation

This is another tough change to stomach, particularly after being accustomed to the predictable volume and dependable growth of an irrigated pasture. Again, it takes time for the soil organisms to adjust and for the grass species to become more drought-tolerant and deep-rooted.

Shutting off the water overnight can be catastrophic. Your irrigation requirements should be reduced slowly to reflect changes in your pastures and soils. Over time, your intensively managed pasture rotation builds organic matter in the soil and soil evaporation rates decrease, moisture infiltration increases, and the moisture-storage capacity of the soil increases.

Replanting Old Pastures and Clearing New Ones

Wait to replant or clear pastures until *after* you have perfected your herd's grazing rotation and maximized your land's potential with new grazing techniques and infrastructure. Don't get impatient. Time is an undervalued tool in the process of change. An effective grazing rotation will invigorate old pastures, reduce weed pressure, increase forage growth, and clear brush to make way for new pasture species. Give your new grazing system a chance to work for you before reaching for the tillage equipment.

SUCCESSFUL CHANGE IS a deliberate process that relies on planning every step of the way. Your comprehensive business plan, which you have designed in chapters 21 to 30, is an important part in this process. Once your planning is complete, some changes should be implemented quickly so you can begin benefiting from them as they replace previously less productive or more financially unstable production methods. Other changes require time; you, your experience, your cattle, your pastures, and your soils need to grow into them slowly while you wait for previously implemented changes to take effect.

The time frame depends on the nature of the change, which is why it is so important to fully understand and plan each new change before you launch yourself into an irreversible course of action,

decisions, and management strategies that have profound effects on your ability to achieve your goals and that significantly affect the viability and sustainability of your livestock and business enterprise as a whole. Natural, grass-based cattle production is a dynamic ecosystem in which land, animals, soils, people, and finances are all interconnected. No individual change in the system occurs in a vacuum. Each requires that you consider its far-reaching impact on the whole.

By planning each change you make with your comprehensive business plan readily available, you are well on your way to creating your own natural, grass-based cattle enterprise. With every change and step forward, you will discover new and exciting ripple effects through your entire farm system.

Epilogue

Natural, grass-based cattle farming is a tremendously exciting journey of discovery. We see so much more of life when the engines are shut off and we can walk around the pastures with our cattle and our kids.

There isn't a day that goes by without discovering a new plant, bug, or cattle behavior that adds to the richness of knowledge and enjoyment of living with the land. We find new meaning in every weed, every pounding raindrop, and every hoofprint; each tells its own story about soil conditions, grazing patterns, and its connection to all the living plants and animals surrounding it.

This method of farming reconnects us to a deeper part of ourselves that reaches back to our hunter-gatherer and nomadic shepherding history, when our survival depended more on our knowledge of the land and less on the tools we devised to conquer it, and also takes us to the forefront of our ecological awareness of plants and animals and to our most innovative technological advances. It represents an exciting blend of the old and the new, a meeting of the information age, industrial age, and our prehistoric pastoral roots.

As knowledge and good management replace machinery and diesel and child-hazardous technology and as the mind and eyes, rather than gas pedals and wrenches, become the important tools in the management of grass-based cattle, children can safely be included in the daily farm learning process. All those bugs, beetles, cows, birds, and plants, which the kids can discover and play with in their parents' trusting company, naturally suit the boundless curiosity of their young minds. Such a system fuels their ambition to be part of the farm and to bring their energy, enthusiasm, and innovation back to the land.

With the average age of North American farmers rapidly approaching the age of retirement, the need to include the young generation in farm life is not to be taken lightly. In order to sustain our industry, we must encourage the next generation of farmers to embark on agricultural careers. Yet if members of the younger generation are to choose to remain on the farm instead of following their friends to high-paying city jobs, it must be worth their while. Their farm careers should parallel the time frame of other careers, following a similar course of increasing responsibilities, a rise to management level, and eventual ownership of the business.

As soon as members of the younger generation make the decision to remain on the farm, there has to be a clearly defined, written, and scheduled transfer of responsibility, leadership, and ownership that outlines their progressive integration into the business. If they want to be part of the farm, their goals must be actively woven into the fabric of the business at the outset. Make it worth their while, for there is no more loyal and enthusiastic business partner than your own child.

As you embark on your individual journeys, I would like to close this book with an invitation to the next generation of grass-based cattle farmers to discover this exciting new agricultural trade. With your enthusiasm, you are revitalizing the agriculture industry and defining natural, grass-based cattle farming as the frontier of the modern natural/organic agriculture movement.

Metric Conversion Chart

	If you have:	Multiply by:	To find:
Flow	U.S. gallons per minute	3.78	liters per minute
	imperial gallons per minute	4.55	liters per minute
Volume	U.S. gallons	3.78	liters
	imperial gallons	4.55	liters
	U.S. pints	.47	liter
	U.S. quarts	.95	liter
	U.S. fluid ounces	29.57	milliliters
Length	inches	2.54	centimeters
	inches	25.4	millimeters
	feet	30.48	centimeters
	feet	.30	meter
	miles	1.61	kilometers
Area	acres	.405	hectare
	acres	43,560	square feet
	acres	4,046.86	square meters
Weight	pounds	.45	kilograms
	U.S. (short) tons	.91	metric tons
	ounces	28.35	grams
Energy	horsepower (HP)	.75	kilowatts

Glossary

ACYCLIC PERIOD: the cow's period of infertility after calving, prior to the resumption of postpartum cyclic ovarian function (**ESTRUS CYCLE**).

AEROBIC: refers to an environment containing oxygen or to any living organism that requires the presence of air to sustain life (e.g., aerobic microbes, aerobic bacteria).

AGROCHEMICALS: chemicals used for agricultural purposes, such as pesticides, parasiticides, herbicides, fungicides, and fertilizers.

ALLELOPATHY: the suppression of growth of one plant species by a toxin or chemical released by another plant species.

ANAEROBIC: refers to an environment that lacks oxygen or to any living organism that requires the absence of air to sustain life (e.g., anaerobic microbes, anaerobic bacteria).

ANYHYDROUS AMMONIA: one of the most common synthetic nitrogen fertilizers. It is a volatile fertilizer made by a reaction between natural gas and nitrogen, which is typically injected into the soil.

BASE SATURATION: saturation (%) of the cation exchange sites on the soil colloids (see also **CATION EXCHANGE CAPACITY** and **COLLOID**), which are occupied by basic cations such as potassium (K^{+1}), magnesium (Mg^{+2}), calcium (Ca^{+2}), and sodium (Na^{+1}).

BENTONITE CLAY: a sticky, highly absorbent clay that swells up to twelve times its dry volume when absorbing water, forming a gelatinous impermeable layer that is used to stabilize soil or line ponds and prevent water infiltration.

BOX PROGRAM: a system of preselling a regular purchase of farm products. Customers sign up and prepay for a weekly or monthly food pickup worth a predetermined value designed to suit the needs of the family (e.g., choice of $20, $40, or $60 value). The producer can then fill the order (usually by packing it into a box — hence, the term *box program*) with whatever combination of produce or meat is in season, providing variety for the customer and a predictable sales outlet for the producer. The customer picks up his/her box each week at a designated pickup location and drops off last week's empty box. Selling meat this way works particularly well if used in combination with vegetables or fruit.

BREEDER: a person who raises livestock primarily for breeding purposes (e.g., sells bulls and/or replacement heifers).

BREEDING CLASS: any livestock categorized as being among the animals responsible for a herd's reproduction and genetic continuity (e.g., brood cows, herd sires and replacement heifers), and not kept for the direct purpose of meat production.

BULL: the uncastrated adult male of domestic cattle.

CANCER EYE OR *BOVINE OCULAR NEOPLASMIA* (BON): includes a variety of benign and malignant skin tumors of the eyeball and eyelids, often characterized by a pink, fleshy growth. The most important factors affecting the susceptibility to cancerous growths in and around the eye are intense sunlight or intense ultraviolet radiation and lack of skin and/or hair pigmentation around the eyes.

CAPILLARY ACTION: the movement of water within the spaces of a porous material due to cohesive intermolecular forces between water molecules and adhesive intermolecular forces between water and other substances. In other words, water is sticky — water molecules stick to each other and to other substances such as glass, cloth, organic tissue, and soil. For example, when you dip a paper towel into water, the water will "climb" the paper towel through capillary action.

CARRYING CAPACITY: the maximum number of livestock (cattle and/or other species) that can be supported indefinitely by a particular habitat, allowing for seasonal and random changes, without degradation of the environment and without diminishing the carrying capacity in the future. This is different from the stocking rate, which merely represents the number of animals per acre in a single pasture at any given point in time. Carrying capacity is calculated by averaging the number of animals supported by a pasture throughout the year.

CATION EXCHANGE CAPACITY (CEC): the degree to which a soil can absorb and exchange cations [positively charged ions such as calcium (Ca^{+2}), magnesium (Mg^{+2}), potassium (K^{+1}), and sodium (Na^{+1})], based on the base saturation of each colloid and the total abundance of colloids in the soil (see also **BASE SATURATION** and **COLLOID**). The CEC gives an indication of the soil's potential to hold plant nutrients.

CATTLE: bovine animals, especially domesticated members of the genus *Bos*, such as *Bos indicus* (tropical cattle breeds) and *Bos taurus* (British and Continental breeds).

CATTLE CYCLE: cyclical fluctuations in cattle prices based on nationwide cattle herd inventory numbers and speculations about future cattle market prices, typically following a ten-year cyclical pattern.

CATTLE GUARD: a barrier across a road, usually a bridge (consisting of parallel metal bars) over a ditch, which allows pedestrians and vehicles to pass over the ditch, but not cattle.

CLAMP: a temporary plastic-covered silage stack made in the pasture without the permanent containment walls or permanent floor of the traditional silage bunker.

CLASS: a number of livestock grouped together into a single category because of a common purpose, common qualities, shared attributes, or shared traits, such as breeding-class vs. non-breeding-class cattle.

CLEANSING CYCLE: the estrus cycle immediately following calving, termed this because it allows the cow's reproductive tract and hormone levels to reorganize in preparation for a new pregnancy.

COCCIDIOSIS: an infection by coccidia, a parasitic protozoa, which destroys the lining of the small intestine, causing diarrhea, intestinal hemorrhage, emaciation, and sometimes fatal dysentery.

COLLOID: the smallest soil particle, made of either clay minerals or soil organic matter (particularly humus), which is larger than molecular size but small enough to be moved about by molecular forces. It carries a negative electrical charge and is therefore able to attract and hold cations (see also **BASE SATURATION, ION,** and **CATION EXCHANGE CAPACITY**).

COLOSTRUM: the thin, yellow, milky fluid produced by the cow immediately after calving. This first milk contains antibodies to infectious agents to which the cow has been exposed, which are absorbed by the newborn calf as a form of protection against disease until its own immune system takes over, at age 3 months. The level of protection is determined by the amount of colostrum ingested by the calf during the first 6 to 24 hours of life. Colostrum also contains more minerals and less fat and carbohydrate than milk.

COMMODITIES: raw or primary products that are categorized by a standardized measure of quality that are produced in large quantities by many different producers so that items from different producers are considered equivalent, allowing them to be substituted for each other without having to look at them too closely. Examples include many agricultural products (grain and cattle), fuel, minerals, and foreign currencies.

COMMODITY MARKETING SYSTEM: A standardized marketing system of trading raw or primary products (commodities), usually at designated meeting places such as auctions (e.g., livestock auctions) or stock exchanges. These products have their price determined by competitive bids and offers based on supply, demand, and speculation about future value. Purchasing decisions are made almost solely on the price and category of the product, not on differences in quality or features.

COMPENSATORY GAIN: During periods of reduced nutritional intake, cattle metabolism slows down to conserve energy and reduce nutritional demands. When these same cattle are subsequently given access to plentiful high-quality forage, their slow metabolism allows them to gain weight very rapidly for a short period of time, until their metabolism speeds up again. This rapid weight gain is called *compensatory gain,* and allows cattle that are wintered at low rates of weight gain to compensate for their winter food shortage so they will frequently weigh almost as much by midsummer as cattle wintered at a high rate of weight gain.

COW: a female of domestic cattle that has had a calf. Also commonly used as a generic reference to cattle when discussing a single member of the species of either sex and any age.

COW/CALF PAIR: the cow and her calf considered as a unit, a *cow/calf pair,* from birth until weaning.

CULL COWS: cows that you want/need to remove from the herd regardless of whether they are pregnant, such as high-maintenance cows, cows with poor genetics, sickly cows, and mean cows.

CUTTER: a term used for a cull or an open cow that is considered too old or of too poor quality to use for the prime cuts. The vast majority of the carcass is ground into hamburger by the butcher.

DEGASSING: the escape of gas from the soil into the air, such as when the soil's protective sod "skin" is broken by tillage or by poor grazing practices so that nitrogen can escape the soil as nitrogen gas and the organic carbon compounds in the soil's humus begin to be oxidized by the air and degas as carbon dioxide.

DISEASE: an impairment of health or a condition of abnormal functioning with a specific cause and symptoms resulting from infection, nutritional deficiency, toxicity, unfavorable environmental factors, and genetic or developmental errors.

DRESS PERCENTAGE: the carcass weight (hanging weight after slaughtering) as a percentage of live weight, also known as the *carcass yield.* [Dress percentage = (carcass weight ÷ live weight) × 100.] This should not be confused with the retail yield of meat from a carcass, as this depends on how the animal is butchered, how much bone is included in the meat cuts, and how many organs are included in the retail yield (see also **PERCENT CUTABILITY**). Factors affecting dress percentage are: hot vs. cold carcass at weighing, bruising losses, sex of animal, age, weight class, degree of fatness, muscularity, pregnancy status, hours without feed before slaughter, shipping distances before slaughter, feed type during finishing (grass-fed vs. grain-fed), nutritional quality of feed during finishing, and weather conditions prior to slaughter (especially temperature).

DRIFT FENCE: a fence designed as a barrier to cattle that does not form an enclosure.

DRY-DOWN: the process of grass or another crop drying up as seed formation begins due to plant growth halting and nutrients from the leaves and stem being redirected to the seed head, where they are concentrated into the plant's seeds/grain. Characterized by the browning and drying out of the vegetative parts of the plant. This term also is used to describe the drying process of a plant that has died (e.g., after a killing frost or after mowing).

DUGOUT: an artificially excavated pond used to capture and store rainwater or collect groundwater through seepage, which is used for irrigation or to supply livestock with drinking water.

EMPTY COW: any unbred cow or heifer that failed to conceive during the scheduled breeding season.

EQUINOX: literally, "equal night," the time when the sun crosses the plane of the equator, causing daylight and night to be approximately the same length around the earth. This occurs on March 21 (the vernal equinox) and September 22 (the autumnal equinox).

ESTRUS CYCLE: the series of physiological changes in the female sexual organs induced by reproductive hormones in most mammals, in which the uterus is prepared for pregnancy, ovulation occurs, and finally, in the absence of conception, the uterus lining is reabsorbed prior to the repeat of yet another estrus cycle.

EVAPORATION: the natural process by which water turns from a liquid into vapor (gas) as it is absorbed into the air.

EVAPOTRANSPIRATION: the transfer of moisture from the earth to the atmosphere by evaporation of water and transpiration from plants (see also **EVAPORATION** and **TRANSPIRATION**).

FEEDER: an animal being fattened, or suitable for fattening, usually on stored feed in a feedlot environment.

FEED TRIAL: in the conventional cattle industry, trials commonly used to evaluate the weight gain efficiency of young growing bulls as an indication of their future offspring's weight gain potential and feed efficiency. The young bulls are fed a high-energy, nutrient-rich feed ration, usually supplemented very heavily with grain and/or corn, in a controlled feeding environment such as a confinement pen (feedlot), during which time the bulls are periodically weighed to determine their daily rate of weight gain. (See chapter 3 for a detailed discussion about feed trials.)

FIGHT-OR-FLIGHT RESPONSE: the panic response to feeling crowded, to which the brain sees only two possible methods of escape: run away (flight) or attack and fend off the "attacker" (fight), particularly if fleeing does not appear to be an option.

FIRST-CALF HEIFER: a young female of domestic cattle after she gives birth to her first calf, at approximately 2 years of age, but before giving birth to her second calf, at approximately 3 years of age.

FLUSHING: a method of stimulating estrus, in which a cow receives a sudden and dramatic increase in her energy and protein intake in the week before breeding and into the breeding season.

FOOT ROT: an anaerobic bacterial infection of the feet of cattle and sheep, exacerbated by wet weather or muddy fields, often leading to the loss of the hoof if left untreated.

GEOGRIDS: a range of man-made materials woven as a grid or mesh structure, used in conjunction with gravel to stabilize soft or muddy ground. A mat is formed when the overlaying gravel mechanically interlocks with the mesh to creating a mat capable of supporting repeated cattle traffic over the underlying soft or muddy soil.

GLEY: a bluish gray sticky soil that forms as a result of water-logging and chemical reduction of the soil material due to a lack of oxygen.

GRASS TETANY: also known as hypomagnesemia, a serious livestock condition caused by low magnesium concentrations in the blood. Symptoms include low appetite, dull demeanor, staggering, stiff gait, irritability, excitability, nervousness, muscular tremors, falling over, and spasms. Grass tetany is most common in lactating cows grazing lush, early-spring pastures, because cattle require high magnesium (in addition to calcium) for lactation, and the first spring grass growth is typically deficient in magnesium because potassium is preferentially absorbed into the plants during the cold early-spring temperatures.

GRASS-FINISHING: refers to finishing cattle in a pasture-grazing system on a grass/legume diet, without grain supplements added to their diet.

GREEN MANURE: a crop of growing plants, such as ryegrass, that is incorporated/tilled into the soil before it flowers to increase soil fertility and nutrient availability.

GREENHOUSE GASES: the group of gases (both natural and man-made) in the atmosphere that trap the sun's warmth, just as glass traps warmth in a greenhouse. These gases include water vapor, carbon dioxide, methane, ozone, nitrous oxide, and chlorofluorocarbons (CFCs). Their effect is called the *greenhouse effect*, which is responsible for earth's relatively warm life-supporting temperatures.

HEIFER: a young female of domestic cattle less than two years of age, before she gives birth to her first calf.

HERD EFFECT: the positive impact on a grazing environment produced by the periodic grazing impact of livestock grazed at high stock densities and by their bunching, trampling, milling, and excited behavior.

HIDDEN HUNGER: describes soil nutrient shortages or imbalances that are not severe enough to produce easily recognizable visual signs of a nutrient deficiency in the plants growing in the soil, but nevertheless affect crop yields, cause plant health issues and pest infestations, and reduce livestock weight gains, disease resistance, and fertility.

HUMUS: the *stable* dark organic material in soils produced by the microbial decomposition of vegetable and animal matter (see also **ORGANIC MATTER**).

ICE-PREVENTER VALVE: a thermostat-controlled bypass valve used to regulate temperature in livestock watering troughs during freezing weather by allowing the warmer water from the source to bypass the water trough flow-valve to flush cold water from the trough before it begins to freeze.

INBREEDING: breeding two closely related individuals. This reduces genetic variability, thereby magnifying traits common to both and causing recessive genes common to both to be expressed. See also **LINEBREEDING**.

ION: an atom or group of atoms that has acquired a net electric charge by gaining or losing one or more electrons. A *cation* is a positively charged ion; an *anion* is a negatively charged ion.

INPUT: resources, such as fertilizer amendments, feed supplements, fuel, pesticides, herbicides, and seed, purchased from off the farm to contribute to its operation.

LEACH: to dissolve or wash out soluble components from the soil by downward-percolating ground water.

LEADER-FOLLOWER GRAZING SYSTEM: a strategy in which one class of animals, such as grass-finishing animals in preparation for slaughter, is grazed one pasture ahead of the main herd in the rotation so it can get access to better nutrition by skimming off the choice grass ahead of the main herd, which subsequently grazes the leftovers.

LIGNIFICATION: the process of depositing lignin in the cell walls of a plant, causing it to become woody and/or hard. This is a key indication of plant maturity as it typically happens during the plant's seed-formation stage. Plants also lignify in response to drying out, and cool-season grasses lignify to protect themselves during hot weather.

LINEBREEDING: breeding two closely related individuals. This reduces genetic variability, thereby magnifying traits common to both and causing recessive genes common to both to be expressed. When the outcome of this closely-related pairing produces desirable traits or magnifies good genetics, it is called linebreeding. See also **INBREEDING.**

MACRONUTRIENT: any of the chemical elements required by plants (and animals) in relatively large amounts: calcium, magnesium, nitrogen, phosphorus, potassium, sodium, and sulfur.

MAINTENANCE EFFICIENCY: an assessment of livestock nutritional efficiency, particularly as it affects our production costs.

MANURE PATCH: a small pile, left behind by a single cow, of herbivorous digestive by-product rich in minerals and symbiotic microbes from the rumen.

MICRONUTRIENT: any of the chemical elements required by plants (and animals) in relatively small concentrations, such as boron, chlorine, copper, iron, manganese, molybdenum, nickel, cobalt, zinc, selenium, iodine, vanadium, and silica.

MONOCULTURE: the cultivation of only one species in a given area, a practice common in conventional agriculture.

NODE: any thickened enlargement; the slightly enlarged portion of a stem that bears a new leaf or branch.

NONBREEDING CLASS: any livestock categorized as not being among the animals responsible for a herd's reproduction and genetic continuity (that is, steer calves, stockers, grass-finishing cattle, open and cull cows, and any other cattle being grazed for the direct purpose of meat production rather than reproduction).

OPPORTUNITY COST: the missed financial opportunity of not pursuing another, more valuable alternative.

ORGANIC MATTER: any carbon-based material of organic origin in the soil, including humus, and undecomposed or partially decomposed organic material such as roots and plant remains. See also **HUMUS.**

OXIDATION: the addition of oxygen, removal of hydrogen, or the removal of electrons from an element or compound. In the environment, organic matter is oxidized to become more stable. Oxidation is the opposite of *reduction.*

PARASITE: an organism that lives in or on the living tissue of a host plant or host animal of another species for a portion of, or the duration of, its life cycle. It obtains all its nutritional requirements from its host, generally to the disadvantage of the host's health, productivity, and well-being but without outright killing it.

PARASITICIDE: a chemical substance that destroys parasites.

PERCENT CUTABILITY: the percentage of retail yield (salable meat) from a carcass after the carcass is butchered into its respective meat cuts, and excess fat and bone not included in the cuts are removed (compare with **DRESS PERCENTAGE.**)

PEST: an unwanted or destructive insect or other small animal that harms or destroys crops or livestock.

PHOTOPERIOD: the interval in a twenty-four-hour day during which an organism is exposed to light.

PHOTOSENSITIVITY: abnormal sensitivity of the skin and/or eyes to ultraviolet light, usually following exposure to certain drugs or other sensitizing chemicals (such as certain herbs).

PHOTOSYNTHESIS: the plant process that uses sunlight and chlorophyll (the green pigment in plants) to convert carbon dioxide and water into organic materials (used as the building blocks for plant growth) and release oxygen as a by-product.

PHYTOTHERAPY: the use of plants or plant extracts for medicinal purposes, especially plants that are not part of the normal diet; herbal medicine.

PINKEYE DISEASE: or infectious bovine keratoconjunctivitis, a highly contagious infectious disease of the cattle eye, most commonly caused by the bacterium *Moraxella bovis*. Damage to the eye may lead to permanent blindness.

POLYTAPE: portable electric fencing wire in tape form, with five or more thin wires running parallel to one another through a plastic tape that is ½ inch in diameter or wider. The flat, wide polytape increases the electric fence's visibility, but it is more difficult to handle, is more prone to electrical shorts, and has more electrical resistance than **POLYWIRE.**

POLYWIRE: portable electric fencing wire formed by braiding five or more thin wires with plastic strands to form a thin, ropelike wire. Though less visible than polytape, it is more durable, less prone to tangling, easier to handle, and carries current slightly farther.

PRESSURE HEAD: the pressure energy of a fluid necessary at the distribution water system source to overcome losses and operate various equipment (e.g., irrigation or livestock watering systems) at its designed operating pressure.

PROTOZOA: any of various single-celled organisms that are larger and more complex than bacteria. They are often parasitic in nature.

RANGE SEEDER: a sowing implement specifically designed to plant seeds directly through pasture sod.

REPLACEMENT HEIFER: a heifer chosen to be added to the breeding herd to replace an old cow and a cow culled from the herd. Replacement heifers can be purchased from off farm or may be the young female offspring chosen from among the herd itself.

RIDE: cattle jumping on one another as though they are trying to breed.

ROOT ACID SOLUBILITY: refers to minerals becoming soluble in the presence of acids excreted by a plant's roots so these minerals can be absorbed by a plant in solution for its use, even if these minerals are insoluble in water alone.

RUMEN: the first stomach of a cow or other ruminant where food is softened and then regurgitated for cud chewing, and where live the cellulose-digesting enzymes that are responsible for breaking down grass's tough cellulose structure.

RUT: the annual time period when bulls, rams, and other male animals are physiologically and behaviorally capable of reproduction. It occurs during the breeding season in the same time frame as the estrus cycle, its female equivalent.

SCOURS: diarrhea in cattle, typically resulting from a contagious intestinal infection caused by a variety of viruses, bacteria, or protozoa, which can lead to potentially life-threatening dehydration and electrolyte imbalances. The most common infectious organisms that cause scours are rotavirus, coronavirus, *Cryptosporidium parvum*, salmonella bacteria, and *E. coli.* Young calves are particularly vulnerable.

SECOND-CALF HEIFER: a young female of domestic cattle after she gives birth to her second calf, at approximately three years of age, but before giving birth to her third calf, at which point she is classified as a *cow*.

SEED SCARIFICATION: hastening the sprouting of hard-covered seeds by weakening the tough outer seed coat through physical abrasion or chemical treatment to allow water and oxygen into the seed, thereby improving germination in some plant species. This happens naturally to many grass seeds as they pass through the acid environment of a ruminant's stomach.

SELL/BUY ECONOMICS: the concept of buying replacement stockers immediately after selling the previous crop of stockers, so that the difference between selling income and replacement cost generates the cash flow to pay for the year's operating expenses and guarantee a living. This requires wintering stockers rather than the traditional stocker-grazing approach of buying in the spring and selling in the fall.

SET STOCKING: a grazing technique wherein animals are placed in a single large pasture for a very long time period, instead of being rotated frequently from pasture slice to pasture slice. Although this grazing technique does not promote a sustainable, healthy grazing environment over the long term, it can be used effectively to top-graze (see **TOPGRAZING**) the slaughter-ready cattle as the final stage of a grass-finishing program, thereby maximizing weight gains and meat quality.

SHRINKAGE: the loss of livestock weight between the farm and the weigh scale at the auction, caused by moisture losses due to evaporation, sweating, dehydration, urination, passing manure and diarrhea. Shrinkage is a result of the animal's body functions and the mandatory "no feed and water period" prior to weighing at the auction, but is exacerbated by long hauling distances, long waits in the corral or auction yard, hot weather, and any form of stress including handling stress, nutritional stress and disease.

SILAGE: high-moisture forage preserved through fermentation and stored in the absence of air.

STEER: a castrated male less than two years of age, by which time it is usually slaughtered. If it lives on into adulthood, the adult castrated male is known as an *ox*.

STOCKER: a calf in the stage between weaning and the final pre-slaughter finishing phase (350 to 775 pounds), which is being grown on pasture.

SUBFERTILE: suffering from low fertility.

SUMMER SOLSTICE: the day when the planet's axis is tilted at its most toward the sun, marking the longest day of the year and the beginning of summer: June 21 in the northern hemisphere and December 22 in the southern hemisphere.

SUPERBUG: any disease-causing bacterium that has become resistant to the antibiotics usually used to control or eradicate it, also known as multi-resistant bacterium.

TAPROOT: a stout, tapering primary root that grows vertically downward with limited side branching.

TILLERING: to put forth new shoots from the roots or around the base of the original stalk.

TOPGRAZING: grazing only the highest-quality parts of plants, such as in 1) a set-stocking grazing scenario with sufficient forage to allow the cattle to be very selective about what they graze, 2) grazing a pasture only very slightly before moving on to the next pasture slice, or 3) grazing one class of animals ahead of the main herd in a leader–follower grazing scenario.

TRANSPIRATION: the process by which plants give off moisture (as vapor) to the air through the surface of their leaves.

TURBO TAPE: a high-conductivity, low-resistance variety of **POLYTAPE** used for long-distance portable electric fencing situations. Up to fifteen times more conductive than regular polytape.

TURBO WIRE: a high-conductivity, low-resistance variety of **POLYWIRE** used for long-distance portable electric fencing situations. Up to thirty times more conductive than regular polywire.

VEGETATIVE STAGE: the growth stage of a plant's life cycle, following germination and emergence and ending when the plant enters its reproductive (seed formation) stage. It is characterized by the growth of the stem and leaves.

WARBLE: a lumpy abscess under the hide of cattle, deer, and certain other animals, caused by larvae of any of several species of large parasitic hairy flies known as warble flies, also known as heel flies, botflies, gadflies, or bomb flies. These flies are found on all continents of the northern hemisphere, primarily between 25 and 60 degrees latitude.

WEANING: to cause a calf to lose the need to suckle; when its mother's milk ceases to be included in its diet.

WEANLING: a newly weaned calf.

WEED: a plant growing where it is not wanted; an undesirable plant according to our human goals for the piece of land in which it is growing or according to whatever human activity we are engaged in.

WIND POLISHING: the process in which small soil particles are blown away and the remaining soil particles are realigned by the wind or impacted by coarse soil particles blown along the soil's surface, thereby reducing soil porosity and causing a textural change (crusting) to the remaining soil surface.

YEARLING: an animal between its first and second birthday. The term applies to heifer calves, steer calves, and bull calves.

Resources

Acadian Seaplants Limited
30 Brown Avenue
Dartmouth, NS B3B 1X8
Canada
902-468-2840
www.acadianseaplants.com
Supplier of hand-harvested, solar-dried kelp meal.

Acres USA
P.O. Box 91299
Austin, TX 78709
800-355-5313
www.acresusa.com
Monthly eco-agriculture publication and very comprehensive catalog of eco-agriculture-related books.

The Low-Cost Cow/Calf Program and School for Profitable Beef Cattle Ranching
Agri-Concepts, Inc.
Dr. Dick Diven
11098 North Desert Flower Drive
Tucson, AZ 85737-7051
800-575-0864 or 520-544-0864
www.lowcostcowcalf.com
Animal nutritionist; supplement and mineral formulations. Excellent short course "Low-Cost Cow/Calf Program and School for Profitable Beef Cattle Ranching" offered in the United States and Canada.

Dr. Temple Grandin
Colorado State University
Ft. Collins, CO 80523
970-491-0243
www.grandin.com
Author and designer of low-stress livestock handling facilities.

Gallagher Animal Management Solutions
130 West 23rd Avenue
North Kansas City, MO 64116
800-531-5908 (plus the three-digit extension for your area)
www.gallagherusa.com
Call for electric-fence dealership locations.

Irrigation Industry Association of British Columbia
2330 Woodstock Drive
Abbotsford, BC V3G 2E5
Canada
604-859-8222
www.irrigationbc.com
Call to order the B.C. Sprinkler Irrigation Manual.

Kinsey's Agricultural Services, Inc.
297 County Highway 357
Charleston, MO 63834
573-683-3880
www.kinseyag.com
Soil consultant Neal Kinsey (author of Hands-On Agronomy*) provides excellent soil fertility short courses several times a year, including at the annual* Acres USA *conference.*

National Center for Appropriate Technology
Sustainable Agriculture and Rural Development
Program
3040 Continental Drive
Butte, MT 59701
406-494-4572
www.ncat.org
This program provides practical and economical solutions to everyday problems for rural America, including services and information to reduce use of pesticides, explore a diversity of crops and livestock, explore new ways to market, and improve soil fertility and water quality.

National Organic Program (NOP)
USDA-AMS-TMP-NOP
Room 4008, South Building
1400 Independence Avenue SW
Washington, DC 20250-0020
202-720-3252
www.ams.usda.gov/nop/indexIE.htm
The Web site provides a list of organic-certifying agents; organic consumer information; NOP regulations (standards); specific guidelines for producers, handlers, processors, and retailers dealing with organic products; and a wealth of other information about organic production.

Organic Materials Review Institute
P.O. Box 11558
Eugene, OR 97440
541-343-7600
www.omri.org
Lists of allowed, restricted, and prohibited products for organic use by brand name and generic name, including supplier contact information.

Ministry of Agriculture and Food
Resource Management Branch
1767 Angus Campbell Road
Abbotsford, BC V3G 2M3
Canada
604-556-3100
www.agf.gov.bc.ca/resmgmt
To order the B.C. Livestock Watering Manual.

Jo Robinson
www.eatwild.com
Directory of wild-pastured products for Canada and the United States (beef, pork, chicken, lamb, goat, bison, rabbit, dairy, and eggs); includes grass-fed nutritional information and books. Web site maintained by author Jo Robinson.

The Stockman Grass Farmer
P.O. Box 2300
Ridgeland, MS 39158
800-748-9808
www.stockmangrassfarmer.com
Monthly grass-based farming publication and a catalog of grass-based books.

Thorvin, Inc.
800-464-0417
www.thorvin.com
Producer of certified-organic kelp meal.

Bibliography and Additional Readings

Bibliography

Albrecht, William A. *The Albrecht Papers*, vol. 2: *Soil Fertility and Animal Health.* Ed. Charles Walters Jr. Kansas City, MO: Acres USA, 1975.

———. *Livestock Handling and Transport.* Wallingford, Oxfordshire, UK: CAB Publishing, 2000.

Grandin, Temple. "Australian Cattle Ranch Design." Dr. Temple Grandin's Web page. http://www.grandin.com.

Grandin, Temple, and Catherine Johnson. *Animals in Translation: Using the Mysteries of Autism to Decode Animal Behavior.* New York: Scribner, 2005.

B.C. Ministry of Agriculture and Fisheries, Soils and Engineering Branch. *B.C. Sprinkler Irrigation Manual.* Ed. Ted W. Van der Gulik. Vernon, BC: Irrigation Industry Association of British Columbia, 1989.

Diven, Dick H. *The Low-Cost Cow/Calf Program and School for Profitable Beef Cattle Ranching Course Handbook.* Tucson: Agri-Concepts, Inc., 1994.

Nation, Allan. "Allan's Observations," *The Stockman Grass Farmer,* vol. 57, no. 8, August 2000: 15–18.

Parsons, S. D. *If You Want to Be a Cowboy, Get a Job.* Harare, Zimbabwe: Franfal Publishing, 1999.

Van der Gulik, Ted; Russel Merz; Erich Schulz; and Murray Tenove. *B.C. Livestock Watering Manual.* Edited by Lance Brown. Abbotsford, BC: B.C. Ministry of Agriculture and Fisheries, Soils and Engineering Branch, 1990.

Additional Readings

Kinsey, Neal, and Charles Walters. *Neal Kinsey's Hands-On Agronomy.* Metairie, LA: Acres USA, 1999.

McCaman, Jay L. *Weeds and Why They Grow.* Sand Lake, MI: Jay L. McCaman, 1994.

Robinson, Jo. *Pasture Perfect: The Far-Reaching Benefits of Choosing Meat, Eggs and Dairy Products from Grass-fed Animals.* Vashon, WA: Vashon Island Press, 2004.

Smith, B. *Moving 'Em: A Guide to Low-Stress Animal Handling.* Kamuela, HI: The Graziers Hui, 1998. Bibliography

Related Titles from Storey Publishing

Damerow, Gail, ed. *Barnyard in Your Backyard.* North Adams, MA: Storey Publishing, 2002.

Ekarius, Carol. *How to Build Animal Housing.* North Adams, MA: Storey Publishing, 2004.

Ekarius, Carol. *Small-Scale Livestock Farming.* North Adams, MA: Storey Publishing, 1999.

Thomas, Heather Smith. *Getting Started with Beef and Dairy Cattle.* North Adams, MA: Storey Publishing, 2005.

Thomas, Heather Smith. *Storey's Guide to Raising Beef Cattle.* North Adams, MA: Storey Publishing, 1998.

Index

Page numbers in *italics* indicate an illustration;
boldface indicates a chart.

top-grazing, 358

Grazing *(continued)*
  water, distance from, 96
  during winter, 106–13, *107,* 337–38, 346

Grazing rotation. *See also* Grazing
  calving and, 82–85, *84,* 97, 147–48
  culling and, 85
  daily, importance of, 80, 82, 110, 129–30
  disease and, 83, 248
  drought and, 129–30, 339, 340
  goals of, 79–80
  grass productivity and, 58–60
  health and, 156–57, 160, 161
  herding for, 79, 81, 147–48
  herd management and, 133–34
  land, evaluating potential of, 207–8
  parasites and, 156–57, 248
  pasture rejuvenation with, 63–64
  planning for, 311–23, 345–46
  poor, *15*
  soils and, 165–66, 176, 184–85, 193–95
  in spring, 111
  stocker cattle and, 229
  in summer, 311–17
  training cattle for, 80–81, 118, 145–46, *146,* 148
  water and, 84, 97–100, 101–4, 193, 194–95
  weeds and, 184–85, 186
  in winter, 110–13, **111,** 318–23
  worksheets for, 312–17, 319–23

Greenhouse gases, 173, **173,** 236, 354
Green manure, 354
Grocery stores, 259–60
Gross profit margin analysis, 275
Grounding, 69–70, *70,* 71, *71,* 112, *112,* 120–21
Growth hormones, 236, 256
Guts, 25, 26

H

Hair, 22, 25, 27

Handling
  alleys for, 151, *151,* 230, *230*
  with all-terrain vehicles, 145
  approach for, 139–42, *140, 141*
  attitude toward, 148
  audiovisual elements of, 137–38, 140, 149, 150–51
  bruising and, 149, *149,* 152
  in corrals, 148–52, *149, 150, 151*
  with dogs, 145, *145*
  flight zone and, 138–45, *139, 140, 141,* 148, 151

hauling and, 234–35
health and, 136
herd dynamics, *141,* 142–43, *142, 143*
herding for, 79, 81, 147–48
with horses, 144
lead animal for, 142–43, *142,* 146, *147,* 149, 229, *229*
rodeo-style, 136, 137, 148, 230
shields for, 138, *138,* 151
size and, 138, 143, *144*
for slaughter, 234–35
speed and, 139, *139,* 144, 149
stocker cattle and, 229–30
stress and, 136–45, 148–52, 232, 234–35
sunlight and, 151
training cattle for, 145–46, *146,* 148
wires for, *150*

Hand-move sprinklers, 200, *200*
Hand-move stationary guns, 200, *200*
Hanging weight, 287–90
Harrowing, 156–57
Hauling, 230, 234–35, 241–42, 330

Hay
  for droughts, 130
  for feed, generally, 118–21, *120,* 210
  in organic agriculture, 249
  reserves, 226
  for stocker cattle, 229

H-braces, 74, *74*
Heads, 26

Health. *See also* Disease
  bloat and, 232–33
  of cattle, generally, 165, 177, 236
  culling to improve, 28–29, 156, 161
  of grass, 165
  grazing rotation and, 156–57, 160, 161
  handling and, 136
  herd management and, 133, 134–35
  meat quality and, 235
  nutrition and, 154, 158
  organic agriculture and, 2–3
  pharmaceuticals and, 154–55
  planning for, 324–26
  profitability and, 153
  registered-purebred breeding and, 30–31
  of soils, 60–62, **61,** 162–63, 165–81, 183–85, 188–91
  of stocker cattle, 228–29
  stress and, 136, 154, 155, 159
  temperature and, 101
  water and, 86, 89, 96, 161

Hearing, 137, 149, 150
Heart girth, 24, 26
Heated troughs, 102

Heifers
  breeding of, 23, 24–25, 36
  buying, 45
  culling, 29
  definition of, 354
  estrus cycle of, 36
  fertility of, 23, 24–25
  finishing, 135
  first-calf, 44–45, 353
  genetics of, 23, 24–25
  ideal characteristics for, 23, 24–25
  maintenance efficiency of, 23, 24–25, 44
  raising, 45
  rebreeding, 44–45
  replacement, 23, 24–25, 45, 356–57
  second-calf, 357
  separating, 135
  sexual development of, 36–39
  weight of, 36

Herbicides, 185
Herbs, 161
Herd dynamics, *141,* 142–43, *142, 143*
Herd effect, 134, *134,* 354
Herding, 79, 81, 147–48. *See also* Handling
Herd management, 132–36. *See also* Handling

Herds, *10, 28*
  combining, 130, 132–36, 300
  decomposition and, 12–14, *13*
  managing, 132–36, *134*
  migration of, 16, 80
  nutrient recycling and, 12–14, *13*
  as organisms, 9–11
  predators and, 16, 17, *17*
  remnants of, 8–9
  separating, 132–33, 135
  size of, 110, 128, 134, 300, 321–23

Herd sire, 45–47
Hidden hunger, 178, 354
Hides, 22, 25, 26

Hooves
  of bulls, 26
  of heifers, 25
  humus and, 173
  nutrient recycling and, 12–14, *13,* 61–62, *61*

Hormone-free beef, 256
Horses, 144
Horsetail, 189, *189*
Hose-reels, 202, *202*
Humidity, 11–16, *11*
Humus, 172–74, 174–75, 175–76, 190, 354
Hybrid vigor, 29
Hydraulic ram pumps, 91–92, *92*